Prespacetime Journal

Volume 2 Issue 10
October 2011

Focus Issue on Matti Pitkänen's Work:

Topological Geometrodynamics From Particle Physics Perspective

Edited by:

Editor-at-Large

Dainis Zeps, Ph.D.,

ISSN: 2153-8301

Prespacetime Journal
Published by QuantumDream, Inc.

www.prespacetime.com

Table of Contents

Editorial

Articles

Editorial

On Matti Pitkänen's Topological Geometrodynamics From Particle Physics Perspective

Dainis Zeps*

Abstract

In four articles [1,2,3,4] of current issue of Prespacetime Journal ("PSTJ"), we present Matti Pitkänen's work devoted to his new physical results after the first experimental results at Large Hadron Collider are made public. We discuss here some of Pitkanen's insights in these papers and note his excessive use of mathematical innovations as main tools in way of development of his TGD universe.

Key words: topological geometrodynamics, particle physics, mathematical physics, LHC, Higgs boson.

Introduction

Matti Pitkänen's work in the four papers of this focus issue [1,2,3,4] appears in a time when hunt after Higgs boson at LHC makes its concluding stage with expectance of not finding this boson at all. Moreover, Pitkänen's work arise as direct response to all what goes on at LHC, in the same time serving as early prediction on what we are to expect further both at LHC and physics at all.

All what go on at LHC is reflected well at PSTJ in series of articles and news issues by Philip Gibbs, e.g. [5], see [6] too. PSTJ has published already some papers concerning results of LHC hunt after Higgs boson. See e.g. Lawrence Crowell's paper [7]. Whole issue [8] is devoted to success of LHC, where in paper [9] authors give new perspectives in particle physics in the era after first fundamental results on LHC.

LHC experiments suggest higher level energy involvement in particle physics experiments, 7 TeV, that mean new phenomena in particle physics and new discoveries. And what we see does not deceive us, no Higgs particle finding up to now being only one of them. This time is not only for gathering of fruits for experimental physicists. It is time for theoreticians too, time for new theories which are to appear in direct future. In this focus issue we give word to theoretician Matti Pitkänen, whose works sufficiently often appear at PSTJ. Last year we already issued focus edition devoted to works of Matti Pitkänen "The Miracle of Existence According to Theoretical Physicist Matti Pitkänen", edited by Philip E. Gibbs, Arkadiusz Jadczyk, Dainis Zeps [10]. In editorial [11] we discussed problems we had to encounter with this work there.

On Matti Pitkänen's TGD universe

Not long time before, some half of year, before LHC started to give first results, we were used to situation that Standard Model in particle physics was main theoretical model

* Dainis Zeps, PhD, Senior Researcher, Institute of Mathematics and Computer Science, University of Latvia, Raiņa bulvāris 29, Rīga, Latvija, LV – 1459. E-Mail: dainize@mii.lu.lv

that enjoyed best experimental and theoretic-physical support. Now something has changed. Standard Model start to suffers from lack of explanation "where from masses of particles arise" and we are to look around where from new explanations could come. One of such provider of original theories is Matti Pitkänen who is ready always to give new explanations and new solutions. How he works, or how his TGD universe does work? Let us take insight in there.

Matti Pitkänen's TGD universe is mathematical theory that is developed in enormous amount of papers, part of references of which may be found in references of here presented papers [1,2,3,4]. His only published book devoted to this subject is [12], some mathematical background may be found in [13]. Here we present papers [1,2,3,4] that are devoted to newest results concerning his particle physics predictions just after getting first results from LHC experiments.
Let us take insight there.

Matti Pitkänen's TGD universe and new predictions of mass calculations and new physics emergence

Matti Pitkänen bases his mass calculations mainly on three points. Firstly, conformal symmetries are most essential there. As mass squared operator essentially infinitesimal conformal scaling operator is applied. Mass spectrum before thermalization is obtained from conformal symmetry. Besides mass states and extremely heavy states as in string models are suggested and maintained there. Secondly, mass squared is p-adic thermal expectation of conformal weight in p-adic thermodynamics. Thirdly, TGD universe supports p-adic length scale hypothesis stating that p-adic primes near powers of two are of specific physical importance. Mersenne primes are applied there too.

Mass predictions main criterion according Pitkänen sounds like this. Masses may be predicted correctly within accuracy of percent in the case that they are known. Otherwise mass prediction might be highly non-trivial since mass scales are exponentially sensitive to the integer in by defining p-adic length scale. No fitting is possible.

Besides, in all his TGD universe suggestions, volens nolens new physics emergence is heavily stressed. Among all, new space-time concept is suggested, where traditional Standard Model quantum numbers application is replaced with the use of imbedding space geometry. Generalized Feynman graphs with new space-time topology and geometry are introduced and developed. Setting all this to work in TGD universe, one assumingly reaches desirable goal in theoretical physics: space-time is no more fixed arena of physics but dynamical object in all scales.

Among the problems cured like this are hot topics in LHC verifiable physics e.g. issues of SUSY and Higgs: TGD is at a point to predict light sneutrinos and 250 GeV selectron. TGD universe suggests Higgs field replacement by use of gauge boson where photon develops small mass too with effective compensations of all masses in total. According this model no whatever Higgs-like boson should be fixed by LHC, that supposedly should be confirmed already within this year.

Further, well know problem of proton stability solves in TGD universe as not being at all. Proton is stable. The problem becomes solved applying separate conservation rules for quarks and leptons, i.e., not using grand unified gauge symmetry to both as by Standard Model, according which quark and lepton belong to the same multiplet of grand unifying gauge group.

Using partonic 2-surfaces of different genera mass scaling in wide difference may be effectively covered by appropriate gauge symmetry relating different families of particles, thus solving mass scaling differences for electrons, muons, tau particles, neutrinos and quarks. New view on Cabibbo-Kobayashi-Maskawa mixing of quarks by reduction to mixing of topologies of partonic 2-surfaces is suggested.

New view on color quantum numbers suggested there too. According TGD universe, color angular momentum should be treated like quantum number in degrees of freedom rather than spin-like as in QCD.

As it is predicted by author of TGD universe, successful application of fractality to effect to get scaled copies of hadron physics may be expected, i.e., elementary particles , in particular quarks, could appear in several p-adic length scales. Here M. Pitkänen appeals to p-adic length scale hypothesis and fractally scaled variants of hadron physics. He asks – "Will LHC find scaled variant of hadron physics in TeV range? Is 145 GeV bump there or not?" This might be good indications for M. Pitkänen's new M_{89} hadron physics.

TGD universe as mathematical world

Behind all outer innovations stands M. Pitkänen's view on how new physics should work. The main tools are appropriate choices of mathematical instruments and conditions that serve as constraints for mathematical model that should stand as good description for physical world as precisely as possible. What distinguishes Pitkänen from other theoretical physicist is the range of innovations how these tools should be found and applied. If scientific community mainly tend to keep to more or less conventional traditions how physical phenomena should be treated, or which are these physical entities that should be kept as unchanged, say, space, time, or already settled as fixed to some convenient model, say, Standard Model, or at least stay to the point that all at once can't be changed in a moment at such a serious thing as physics, Pitkänen takes another position: if physical entities are so changing and unsafe for what may serve as ground notions for good description for experimental results, let us choose something that may stand for all this.

By view of Pitkänen this place should be taken by mathematics. The difference what makes Pitkänen's approach new is that he uses prevalence of mathematics without exception everywhere to everything. All is made by mathematics and before all stands mathematics. In order to reach the goals in such settlement mathematical tools should be chosen as powerful as possible. Because of that so excessive use of mathematical tools like conformal invariance and symmetries, prime numeric theoretical concepts, fractals and whatever else what we find in contemporary mathematics. Besides of this no constraints of how to apply all this in new physical innovations, the only concern of physicist is to build mathematically conditioned world of only one option for world that were modeled by theory. Mathematical definiteness should be as dense as to give place only to the world detected by experimental physics. In this respect M. Pitkänen uses to say: -- Number theory is extremely powerful constraint. Because of this so many instruments in TGD universe are taken from number theory, say, p-adic applications, infinite primes, Mersenne primes and so on.

Of course, this is used and applied by Pitkänen with some confinement on his own resources too, and this is his developed TGD universe. What he is not ready to change as easy as physical entities is his ground assumptions that are laid in the ground of TGD universe as a mathematical model. In other words Pitkänen is ready to replace particles with partonic surfaces or whatever mathematical construction with more easy than to change somewhat in his TGD universe that he had laid there some twenty or thirty years before. It may have some explanation. If someone works in solitude he has to economize his recourses after all. One more thing everyone who visits TGD universe would like to see could be presence of more proofs that all mathematics involved there works properly.

References

1. Matti Pitkänen, Overall View About TGD from Particle Physics Perspective. Prespacetime Journal, V2(10): pp. 1457-1528.
2. Matti Pitkänen, New Particle Physics Predicted by TGD: Part I. Prespacetime Journal, V2(10): pp. 1529-1604.
3. Matti Pitkänen, New Particle Physics Predicted by TGD: Part II. Prespacetime Journal, V2(10): pp. 1605-1647.
4. Matti Pitkänen, Particle Massivation in TGD Universe. Prespacetime Journal, V2(10): pp. 1648-1723.
5. Philip E. Gibbs, Exciting New Era of Particle Physics, Vol 2, No 7 (2011), pp. 940-945.
6. Philip E. Gibbs, http://blog.vixra.org/
7. *Lawrence B. Crowell,* Physics in a Higgsless World, Prespacetime Journal, July 2011, Vol. 2, Issue 7, pp. 953-955
8. Huping Hu, Philip E. Gibbs (editors), Great Success of LHC & Tevatron & a Brave New World of Particle Physics, Prespacetime Journal | July 2011, Vol 2, No 7 (2011), pp. 940-1183
9. *Huping Hu, Maoxin Wu,* The Dawn of a Brave New World in Particle Physics, Vol 2, No 7 (2011), pp. 946-952
10. Matti Pitkänen, The Miracle of Existence According to Theoretical Physicist Matti Pitkänen, ed. by Philip E. Gibbs, Arkadiusz Jadczyk, Dainis Zeps, PSTJ, 2010, Vol 1,No 4, 519-722
11. Philip E. Gibbs, Arkadiusz Jadczyk, Dainis Zeps "The Miracle of Existence According to Theoretical Physicist Matti Pitkänen" Editorial Comment, *PSTJ, 2010, Vol 1,No 4, 516-518*
12. Matti Pitkänen, Topological Geometrodynamics, Luniver Press, 2006, 824 pp.
13. M. Pitkänen, The Geometry of CP2 and its Relationship to Standard Mode, PSTJ, 2010, Vol 1, No 4, pp. 713-722

Article

Overall View About TGD from Particle Physics Perspective

Matti Pitkänen [1]

Abstract

Topological Geometrodynamics is able to make rather precise and often testable predictions. In this and two other articles I want to describe the recent over all view about the aspects of quantum TGD relevant for particle physics.

In the first article I will concentrate the heuristic picture about TGD with emphasis on particle physics.

- First I will represent briefly the basic ontology: the motivations for TGD and the notion of many-sheeted space-time, the concept of zero energy ontology, the identification of dark matter in terms of hierarchy of Planck constant which now seems to follow as a prediction of quantum TGD, the motivations for p-adic physics and its basic implications, and the identification of space-time surfaces as generalized Feynman diagrams and the basic implications of this identification.
- Symmetries of quantum TGD are discussed. Besides the basic symmetries of the imbedding space geometry allowing to geometrize standard model quantum numbers and classical fields there are many other symmetries. General Coordinate Invariance is especially powerful in TGD framework allowing to realize quantum classical correspondence and implies effective 2-dimensionality realizing strong form of holography. Super-conformal symmetries of super string models generalize to conformal symmetries of 3-D light-like 3-surfaces and one can understand the generalization of Equivalence Principle in terms of coset representations for the two super Virasoro algebras associated with lightlike boundaries of so called causal diamonds defined as intersections of future and past directed lightcones (CDs) and with light-like 3-surfaces. Super-conformal symmetries imply generalization of the space-time supersymmetry in TGD framework consistent with the supersymmetries of minimal supersymmetric variant of the standard model. Twistorial approach to gauge theories has gradually become part of quantum TGD and the natural generalization of the Yangian symmetry identified originally as symmetry of $\mathcal{N}=4$ SYMs is postulated as basic symmetry of quantum TGD.
- The so called weak form of electric-magnetic duality has turned out to have extremely far reaching consequences and is responsible for the recent progress in the understanding of the physics predicted by TGD. The duality leads to a detailed identification of elementary particles as composite objects of massless particles and predicts new electro-weak physics at LHC. Together with a simple postulate about the properties of preferred extremals of Kähler action the duality allows also to realized quantum TGD as almost topological quantum field theory giving excellent hopes about integrability of quantum TGD.
- There are two basic visions about the construction of quantum TGD. Physics as infinite-dimensional Kähler geometry of world of classical worlds (WCW) endowed with spinor structure and physics as generalized number theory. These visions are briefly summarized as also the practical constructing involving the concept of Dirac operator. As a matter fact, the construction of TGD involves three Dirac operators. The Kähler Dirac equation holds true in the interior of space-time surface and its solutions havea natural interpretation in terms of description of matter, in particular condensed matter. Chern-Simons Dirac action is associated with the light-like 3-surfaces and space-like 3-surfaces at ends of space-time surface at light-like boundaries of CD. One can assign to it a generalized eigenvalue equation and the matrix valued eigenvalues correspond to the the action of Dirac operator on momentum eigenstates. Momenta are however not usual momenta but pseudo-momenta very much analogous to region momenta of twistor approach. The third Dirac operator is associated with super Virasoro generators and super Virasoro conditions define Dirac equation in WCW. These conditions characterize zero energy states as modes of WCW spinor fields and code for the generalization of S-matrix to a collection of what I call M-matrices defining the rows of unitary U-matrix defining unitary process.

[1] Correspondence: E-mail:matpitka@luukku.com

- Twistor approach has inspired several ideas in quantum TGD during the last years and it seems that the Yangian symmetry and the construction of scattering amplitudes in terms of Grassmannian integrals generalizes to TGD framework. This is due to ZEO allowing to assume that all particles have massless fermions as basic building blocks. ZEO inspires the hypothesis that incoming and outgoing particles are bound states of fundamental fermions associated with wormhole throats. Virtual particles would also consist of on mass shell massless particles but without bound state constraint. This implies very powerful constraints on loop diagrams and there are excellent hopes about their finiteness. Twistor approach also inspires the conjecture that quantum TGD allows also formulation in terms of 6-dimensional holomorphic surfaces in the product $CP_3 \times CP_3$ of two twistor spaces and general arguments allow to identify the partial different equations satisfied by these surfaces.

Contents

1 Introduction

Topological Geometrodynamics is able to make rather precise and often testable predictions. In this and two other articles I want to describe the recent over all view about the aspects of quantum TGD relevant for particle physics.

During these 32 years TGD has become quite an extensive theory involving also applications to quantum biology and quantum consciousness theory. Therefore it is difficult to decide in which order to proceed. Should one represent first the purely mathematical theory as done in the articles in Prespacetime Journal [45, 46, 49, 50, 47, 44, 48, 51]? Or should one start from the TGD inspired heuristic view about space-time and particle physics and represent the vision about construction of quantum TGD briefly after that? In this and other two articlesI have chosen the latter approach since the emphasis is on the applications on particle physics.

Second problem is to decide about how much material one should cover. If the representation is too brief no-one understands and if it is too detailed no-one bothers to read. I do not know whether the outcome was a success or whether there is any way to success but in any case I have been sweating a lot in trying to decide what would be the optimum dose of details.

In the first article I concentrate the heuristic picture about TGD with emphasis on particle physics.

- First I represent briefly the basic ontology: the motivations for TGD and the notion of many-sheeted space-time, the concept of zero energy ontology, the identification of dark matter in terms of hierarchy of Planck constant which now seems to follow as a prediction of quantum TGD, the motivations for p-adic physics and its basic implications, and the identification of space-time surfaces as generalized Feynman diagrams and the basic implications of this identification.

- Symmetries of quantum TGD are discussed. Besides the basic symmetries of the imbedding space geometry allowing to geometrize standard model quantum numbers and classical fields there are many other symmetries. General Coordinate Invariance is especially powerful in TGD framework allowing to realize quantum classical correspondence and implies effective 2-dimensionality realizing strong form of holography. Super-conformal symmetries of super string models generalize to conformal symmetries of 3-D light-like 3-surfaces and one can understand the generalization of Equivalence Principle in terms of coset representations for the two super Virasoro algebras associated with lightlike boundaries of so called causal diamonds defined as intersections of future and past directed lightcones (CDs) and with light-like 3-surfaces. Super-conformal symmetries imply generalization of the space-time supersymmetry in TGD framework consistent with the supersymmetries of minimal supersymmetric variant of the standard model. Twistorial approach to gauge theories has gradually become part of quantum TGD and the natural generalization of the Yangian symmetry identified originally as symmetry of $\mathcal{N} = 4$ SYMs is postulated as basic symmetry of quantum TGD.

- The so called weak form of electric-magnetic duality has turned out to have extremely far reaching consequences and is responsible for the recent progress in the understanding of the physics predicted by TGD. The duality leads to a detailed identification of elementary particles as composite objects of massless particles and predicts new electro-weak physics at LHC. Together with a simple postulate about the properties of preferred extremals of Kähler action the duality allows also to realized quantum TGD as almost topological quantum field theory giving excellent hopes about integrability of quantum TGD.

- There are two basic visions about the construction of quantum TGD. Physics as infinite-dimensional Kähler geometry of world of classical worlds (WCW) endowed with spinor structure and physics as generalized number theory. These visions are briefly summarized as also the practical constructing involving the concept of Dirac operator. As a matter fact, the construction of TGD involves three Dirac operators. The Kähler Dirac equation holds true in the interior of space-time surface and its solutions have a natural interpretation in terms of description of matter, in particular condensed matter. Chern-Simons Dirac action is associated with the light-like 3-surfaces and space-like 3-surfaces at ends of space-time surface at light-like boundaries of CD. One can assign to it a generalized eigenvalue equation and the matrix valued eigenvalues correspond to the the action of Dirac operator on momentum eigenstates. Momenta are however not usual momenta but pseudo-momenta very much analogous to region momenta of twistor approach. The third Dirac operator is associated with super Virasoro generators and super Virasoro conditions define Dirac equation in WCW. These conditions characterize zero energy states as modes of WCW spinor fields and code for the generalization of S-matrix to a collection of what I call M-matrices defining the rows of unitary U-matrix defining unitary process.

- Twistor approach has inspired several ideas in quantum TGD during the last years and it seems that the Yangian symmetry and the construction of scattering amplitudes in terms of Grassmannian integrals generalizes to TGD framework. This is due to ZEO allowing to assume that all particles have massless fermions as basic building blocks. ZEO inspires the hypothesis that incoming and outgoing particles are bound states of fundamental fermions associated with wormhole throats. Virtual particles would also consist of on mass shell massless particles but without bound state constraint. This implies very powerful constraints on loop diagrams and there are excellent hopes about their finiteness. Twistor approach also inspires the conjecture that quantum TGD allows also formulation in terms of 6-dimensional holomorphic surfaces in the product $CP_3 \times CP_3$ of two twistor spaces and general arguments allow to identify the partial different equations satisfied by these surfaces.

The discussion of this articleis rather sketchy and the reader interesting in details can consult the books about TGD [40, 29, 23, 19, 30, 34, 38].

2 Some aspects of quantum TGD

In the following I summarize very briefly those basic notions of TGD which are especially relevant for the applications to particle physics. The representation will be practically formula free. The article series published in Prespacetime Journal [45, 46, 49, 50, 47, 44, 48, 51] describes the mathematical theory behind TGD. The seven books about TGD [40, 29, 23, 30, 33] provide a detailed summary about the recent state of TGD.

2.1 New space-time concept

The physical motivation for TGD was what I have christened the energy problem of General Relativity. The notion of energy is ill-defined because the basic symmetries of empty space-time are lost in the presence of gravity. The way out is based on assumption that space-times are imbeddable as 4-surfaces to certain 8-dimensional space by replacing the points of 4-D empty Minkowski space with 4-D very small internal space. This space -call it S- is unique from the requirement that the theory has the symmetries of standard model: $S = CP_2$, where CP_2 is complex projective space with 4 real dimensions [51] , is the unique choice.

The replacement of the abstract manifold geometry of general relativity with the geometry of surfaces brings the shape of surface as seen from the perspective of 8-D space-time and this means additional degrees of freedom giving excellent hopes of realizng the dream of Einstein about geometrization of fundamental interactions.

The work with the generic solutions of the field equations assignable to almost any general coordinate invariant variational principle led soon to the realization that the space-time in this framework is much more richer than in general relativity.

1. Space-time decomposes into space-time sheets with finite size: this lead to the identification of physical objects that we perceive around us as space-time sheets. For instance, the outer boundary of the table is where that particular space-time sheet ends. Besides sheets also string like objects and elementary particle like objects appear so that TGD can be regarded also as a generalization of string models obtained by replacing strings with 3-D surfaces.

2. Elementary particles are identified as topological inhomogenities glued to these space-time sheets. In this conceptual framework material structures and shapes are not due to some mysterious substance in slightly curved space-time but reduce to space-time topology just as energy- momentum currents reduce to space-time curvature in general relativity.

3. Also the view about classical fields changes. One can assign to each material system a field identity since electromagnetic and other fields decompose to topological field quanta. Examples are magnetic and electric flux tubes and flux sheets and topological light rays representing light propagating along tube like structure without dispersion and dissipation making em ideal tool for communications [24]. One can speak about field body or magnetic body of the system.

Field body indeed becomes the key notion distinguishing TGD inspired model of quantum biology from competitors but having applications also in particle physics since also leptons and quarks possess field bodies. The is evidence for the Lamb shift anomaly of muonic hydrogen [92] and the color magnetic body of u quark whose size is somethat larger than the Bohr radius could explain the anomaly [18].

2.2 Zero energy ontology

In standard ontology of quantum physics physical states are assumed to have positive energy. In zero energy ontology physical states decompose to pairs of positive and negative energy states such that all net values of the conserved quantum numbers vanish. The interpretation of these states in ordinary ontology would be as transitions between initial and final states, physical events. By quantum classical correspondences zero energy states must have space-time and imbedding space correlates.

1. Positive and negative energy parts reside at future and past light-like boundaries of causal diamond (CD) defined as intersection of future and past directed light-cones and visualizable as double cone. The analog of CD in cosmology is big bang followed by big crunch. CDs for a fractal hierarchy containing CDs within CDs. Disjoint CDs are possible and CDs can also intersect.

2. p-Adic length scale hypothesis [20] motivates the hypothesis that the temporal distances between the tips of the intersecting light-cones come as octaves $T = 2^n T_0$ of a fundamental time scale T_0 defined by CP_2 size R as $T_0 = R/c$. One prediction is that in the case of electron this time scale is .1 seconds defining the fundamental biorhythm. Also in the case u and d quarks the time scales correspond to biologically important time scales given by 10 ms for u quark and by and 2.5 ms for d quark [2]. This means a direct coupling between microscopic and macroscopic scales.

Zero energy ontology conforms with the crossing symmetry of quantum field theories meaning that the final states of the quantum scattering event are effectively negative energy states. As long as one can restrict the consideration to either positive or negative energy part of the state ZEO is consistent with positive energy ontology. This is the case when the observer characterized by a particular CD studies the physics in the time scale of much larger CD containing observer's CD as a sub-CD. When the time scale sub-CD of the studied system is much shorter that the time scale of sub-CD characterizing the observer, the interpretation of states associated with sub-CD is in terms of quantum fluctuations.

ZEO solves the problem which results in any theory assuming symmetries giving rise to to conservation laws. The problem is that the theory itself is not able to characterize the values of conserved quantum numbers of the initial state. In ZEO this problem disappears since in principle any zero energy state is obtained from any other state by a sequence of quantum jumps without breaking of conservation laws. The fact that energy is not conserved in general relativity based cosmologies can be also understood since each CD is characterized by its own conserved quantities. As a matter fact, one must be speak about average values of conserved quantities since one can have a quantum superposition of zero energy states with the quantum numbers of the positive energy part varying over some range.

For thermodynamical states this is indeed the case and this leads to the idea that quantum theory in ZEO can be regarded as a "complex square root" of thermodynamics obtained as a product of positive diagonal square root of density matrix and unitary S-matrix. M-matrix defines time-like entanglement coefficients between positive and negative energy parts of the zero energy state and replaces S-matrix as the fundamental observable. In standard quantum measurement theory this time-like entanglement would be reduced in quantum measurement and regenerated in the next quantum jump if one accepts Negentropy Maximization Principle (NMP) [17] as the fundamental variational principle. Various M-matrices define the rows of the unitary U matrix characterizing the unitary process part of quantum jump. From the point of view of consciousness theory the importance of ZEO is that conservation laws in principle pose no restrictions for the new realities created in quantum jumps: free will is maximal.

2.3 The hierarchy of Planck constants

The motivations for the hierarchy of Planck constants come from both astrophysics and biology [28, 8]. In astrophysics the observation of Nottale [101] that planetary orbits in solar system seem to correspond to Bohr orbits with a gigantic gravitational Planck constant motivated the proposal that Planck constant might not be constant after all [31, 25].

This led to the introduction of the quantization of Planck constant as an independent postulate. It has however turned that quantized Planck constant in effective sense could emerge from the basic structure of TGD alone. Canonical momentum densities and time derivatives of the imbedding space coordinates are the field theory analogs of momenta and velocities in classical mechanics. The extreme non-linearity and vacuum degeneracy of Kähler action imply that the correspondence between canonical momentum densities and time derivatives of the imbedding space coordinates is 1-to-many: for vacuum extremals themselves 1-to-infinite.

A convenient technical manner to treat the situation is to replace imbedding space with its n-fold singular covering. Canonical momentum densities to which conserved quantities are proportional would be same at the sheets corresponding to different values of the time derivatives. At each sheet of the covering Planck constant is effectively $\hbar = n\hbar_0$. This splitting to multisheeted structure can be seen as a phase transition reducing the densities of various charges by factor $1/n$ and making it possible to have perturbative phase at each sheet (gauge coupling strengths are proportional to $1/\hbar$ and scaled down by $1/n$). The connection with fractional quantum Hall effect [95] is almost obvious. At the more detailed level one finds that the spectrum of Planck constants would be given by $\hbar = n_a n_b \hbar_0$ [10].

This has many profound implications, which are wellcome from the point of view of quantum biology but the implications would be profound also from particle physics perspective and one could say that living matter represents zoome up version of quantum world at elementary particle length scales.

1. Quantum coherence and quantum superposition become possible in arbitrary long length scales. One can speak about zoomed up variants of elementary particles and zoomed up sizes make it possible to satisfy the overlap condition for quantum length parameters used as a criterion for the presence of macroscopic quantum phases. In the case of quantum gravitation the length scale involved are astrophysical. This would conform with Penrose's intuition that quantum gravity is fundamental for the understanding of consciousness and also with the idea that consciousness cannot be localized to brain.

2. Photons with given frequency can in principle have arbitrarily high energies by $E = hf$ formula, and this would explain the strange anomalies associated with the interaction of ELF em fields with living matter [104]. Quite generally the cyclotron frequencies which correspond to energies much below the thermal energy for ordinary value of Planck constant could correspond to energies above thermal threshold.

3. The value of Planck constant is a natural characterizer of the evolutionary level and biological evolution would mean a gradual increase of the largest Planck constant in the hierarchy characterizing given quantum system. Evolutionary leaps would have interpretation as phase transitions increasing the maximal value of Planck constant for evolving species. The space-time correlate would be the increase of both the number and the size of the sheets of the covering associated with the system so that its complexity would increase.

4. The phase transitions changing Planck constant change also the length of the magnetic flux tubes. The natural conjecture is that biomolecules form a kind of Indra's net connected by the flux tubes and $\hbar$ changing phase transitions are at the core of the quantum bio-dynamics. The contraction of the magnetic flux tube connecting distant biomolecules would force them near to each other making possible for the bio-catalysis to proceed. This mechanism could be central for DNA replication and other basic biological processes. Magnetic Indra's net could be also responsible for the coherence of gel phase and the phase transitions affecting flux tube lengths could induce the contractions and expansions of the intracellular gel phase. The reconnection of flux tubes would allow the restructing of the signal pathways between biomolecules and other subsystems and would be also involved with ADP-ATP transformation inducing a transfer of negentropic entanglement [13]. The braiding of the magnetic flux tubes could make possible topological quantum computation like processes and analog of computer memory realized in terms of braiding patterns [9].

5. p-Adic length scale hypothesis and hierarchy of Planck constants suggest entire hierarchy of zoomed up copies of standard model physics with range of weak interactions and color forces scaling like $\hbar$. This is not conflict with the known physics for the simple reason that we know very little about dark matter (partly because we might be making misleading assumptions about its nature). One implication is that it might be someday to study zoomed up variants particle physics at low energies using dark matter.

   Dark matter would make possible the large parity breaking effects manifested as chiral selection of bio-molecules [93]. What is required is that classical Z^0 and W fields responsible for parity breaking effects are present in cellular length scale. If the value of Planck constant is so large that weak scale is some biological length scale, weak fields are effectively massless below this scale and large parity breaking effects become possible.

   For the solutions of field equations which are almost vacuum extremals Z^0 field is non-vanishing and proportional to electromagnetic field. The hypothesis that cell membrane corresponds to a space-time sheet near a vacuum extremal (this corresponds to criticality very natural if the cell membrane is to serve as an ideal sensory receptor) leads to a rather successful model for cell membrane as sensory receptor with lipids representing the pixels of sensory qualia chart. The surprising prediction is that bio-photons [103] and bundles of EEG photons can be identified as different decay products of dark photons with energies of visible photons. Also the peak frequencies of sensitivity for photoreceptors are predicted correctly [28].

2.4 p-Adic physics and number theoretic universality

p-Adic physics [19, 37] has become gradually a key piece of TGD inspired biophysics. Basic quantitative predictions relate to p-adic length scale hypothesis and to the notion of number theoretic entropy. Basic ontological ideas are that life resides in the intersection of real and p-adic worlds and that p-adic space-time sheets serve as correlates for cognition and intentionality. Number theoretical universality requires the fusion of real physics and various p-adic physics to single coherent whole. On implication is the generalization of the notion of number obtained by fusing real and p-adic numbers to a larger structure.

2.4.1 p-Adic number fields

p-Adic number fields Q_p [69] -one for each prime p- are analogous to reals in the sense that one can speak about p-adic continuum and that also p-adic numbers are obtained as completions of the field of rational numbers. One can say that rational numbers belong to the intersection of real and p-adic numbers. p-Adic number field Q_p allows also an infinite number of its algebraic extensions. Also transcendental extensions are possible. For reals the only extension is complex numbers.

p-Adic topology defining the notions of nearness and continuity differs dramatically from the real topology. An integer which is infinite as a real number can be completely well defined and finite as a p-adic number. In particular, powers p^n of prime p have p-adic norm (magnitude) equal to p^{-n} in Q_p so that at the limit of very large n real magnitude becomes infinite and p-adic magnitude vanishes.

p-Adic topology is rough since p-adic distance $d(x,y) = d(x-y)$ depends on the lowest pinary digit of $x-y$ only and is analogous to the distance between real points when approximated by taking into account only the lowest digit in the decimal expansion of $x-y$. A possible interpretation is in terms of a finite measurement resolution and resolution of sensory perception. p-Adic topology looks somewhat strange. For instance, p-adic spherical surface is not infinitely thin but has a finite thickness and p-adic surfaces possess no boundary in the topological sense. Ultrametricity is the technical term characterizing the basic properties of p-adic topology and is coded by the inequality $d(x-y) \leq Min\{d(x), d(y)\}$. p-Adic topology brings in mind the decomposition of perceptive field to objects.

2.4.2 Motivations for p-adic number fields

The physical motivations for p-adic physics came from the observation that p-adic thermodynamics - not for energy but infinitesimal scaling generator of so called super-conformal algebra [64] acting as symmetries of quantum TGD [29] - predicts elementary particle mass scales and also masses correctly under very general assumptions [19]. The calculations are discussed in more detail in the second article of the series. In particular, the ratio of proton mass to Planck mass, the basic mystery number of physics, is predicted correctly. The basic assumption is that the preferred primes characterizing the p-adic number fields involved are near powers of two: $p \simeq 2^k$, k positive integer. Those nearest to power of two correspond to Mersenne primes $M_n = 2^n - 1$. One can also consider complex primes known as Gaussian primes, in particular Gaussian Mersennes $M_{G,n} = (1+i)^n - 1$.

It turns out that Mersennes and Gaussian Mersennes are in a preferred position physically in TGD based world order. What is especially interesting that the length scale range 10 nm-2.5 μ assignable to DNA contains as many as 4 Gaussian Mersennes corresponding to $n = 151, 157, 163, 167$ [28]. This number theoretical miracle supports the view that p-adic physics is especially important for the understanding of living matter.

The philosophical for p-adic numbers fields come from the question about the possible physical correlates of cognition and intention [22]. Cognition forms representations of the external world which have finite cognitive resolution and the decomposition of the perceptive field to objects is an essential element of these representations. Therefore p-adic space-time sheets could be seen as candidates of thought bubbles, the mind stuff of Descartes. One can also consider p-adic space-time sheets as correlates of intentions. The quantum jump in which p-adic space-time sheet is replaced with a real one could serve as a quantum correlate of intentional action. This process is forbidden by conservation laws in standard ontology: one cannot even compare real and p-adic variants of the conserved quantities like energy in the general case. In zero energy ontology the net values of conserved quantities for zero energy states vanish so that conservation laws allow these transitions.

Rational numbers belong to the intersection of real and p-adic continua. An obvious generalization of this statement applies to real manifolds and their p-adic variants. When extensions of p-adic numbers are allowed, also some algebraic numbers can belong to the intersection of p-adic and real worlds. The notion of intersection of real and p-adic worlds has actually two meanings.

1. The intersection could consist of the rational and possibly some algebraic points in the intersection of real and p-adic partonic 2-surfaces at the ends of CD. This set is in general discrete. The interpretation could be as discrete cognitive representations.

2. The intersection could also have a more abstract meaning. For instance, the surfaces defined by rational functions with rational coefficients have a well-defined meaning in both real and p-adic context and could be interpreted as belonging to this intersection. There is strong temptation to assume that intentions are transformed to actions only in this intersection. One could say that life resides in the intersection of real and p-adic worlds in this abstract sense.

Additional support for the idea comes from the observation that Shannon entropy $S = -\sum p_n log(p_n)$ allows a p-adic generalization if the probabilities are rational numbers by replacing $log(p_n)$ with $-log(|p_n|_p)$, where $|x|_p$ is p-adic norm. Also algebraic numbers in some extension of p-adic numbers can be allowed. The unexpected property of the number theoretic Shannon entropy is that it can be negative and its unique minimum value as a function of the p-adic prime p it is always negative. Entropy transforms to information!

In the case of number theoretic entanglement entropy there is a natural interpretation for this. Number theoretic entanglement entropy would measure the information carried by the entanglement whereas ordinary entanglement entropy would characterize the uncertainty about the state of either entangled system. For instance, for p maximally entangled states both ordinary entanglement entropy and number theoretic entanglement negentropy are maximal with respect to R_p norm. Entanglement carries maximal information. The information would be about the relationship between the systems, a rule. Schrödinger cat would be dead enough to know that it is better to not open the bottle completely.

Negentropy Maximization Principle [17] coding the basic rules of quantum measurement theory implies that negentropic entanglement can be stable against the effects of quantum jumps unlike entropic entanglement. Therefore living matter could be distinguished from inanimate matter also by negentropic entanglement possible in the intersection of real and p-adic worlds. In consciousness theory negentropic entanglement could be seen as a correlate for the experience of understanding or any other positively colored experience, say love.

Negentropically entangled states are stable but binding energy and effective loss of relative translational degrees of freedom is not responsible for the stability. Therefore bound states are not in question. The distinction between negentropic and bound state entanglement could be compared to the difference between unhappy and happy marriage. The first one is a social jail but in the latter case both parties are free to leave but do not want to. The special characterics of negentropic entanglement raise the question whether the problematic notion of high energy phosphate bond [102] central for metabolism could be understood in terms of negentropic entanglement. This would also allow an information theoretic interpretation of metabolism since the transfer of metabolic energy would mean a transfer of negentropy [13].

3 Symmetries of quantum TGD

Symmetry principles play key role in the construction of WCW geometry have become and deserve a separate explicit treament even at the risk of repetitions.

3.1 General Coordinate Invariance

General coordinate invariance is certainly of the most important guidelines and is much more powerful in TGD framework thanin GRT context.

1. General coordinate transformations as a gauge symmetry so that the diffeomorphic slices of space-time surface equivalent physically. 3-D light-like 3-surfaces defined by wormhole throats define preferred slices and allows to fix the gauge partially apart from the remaining 3-D variant of general coordinate invariance and possible gauge degeneracy related to the choice of the light-like 3-surface due to the Kac-Moody invariance. This would mean that the random light-likeness represents gauge degree of freedome except at the ends of the light-like 3-surfaces.

2. GCI can be strengthed so that the pairs of space-like ends of space-like 3-surfaces at CDs are equivalent with light-like 3-surfaces connecting them. The outcome is effective 2-dimensionality because their intersections at the boundaries of CDs must carry the physically relevant information.

3.2 Generalized conformal symmetries

One can assign Kac-Moody type conformal symmetries to light-like 3-surfaces as isometries of H localized with respect to light-like 3-surfaces. Kac Moody algebra essentially the Lie algebra of gauge group with central extension meaning that projective representation in which representation matrices are defined only modulo a phase factor. Kac-Moody symmetry is not quite a pure gauge symmetry.

One can assign a generalization of Kac-Moody symmetries to the boundaries of CD by replacing Lie-group of Kac Moody algebra with the group of symplectic (contact-) tranformations [70, 67, 66] of H_+ provided with a degenerate Kähler structure made possible by the effective 2-dimensionality of δM^4_+. The light-like radial coordinate of δM^4_+ plays the role of the complex coordinate of conformal transformations or their hyper-complex analogs. These symmetries are also localized with respect to the internal coordinates of the partonic 2-surface so that rather huge symmetry group is in question. The basic hypothesis is that these transformations with possible some restrictions on the depedence on the coordinates of X^2 define the isometries of WCW.

A further physically well-motivated hypothesis inspired by holography and extended GCI is that these symmetries extend so that they apply at the entire space-time sheet. This requires the slicing of space-time surface by partonic 2- surfaces and by stringy world sheets such that each point of stringy world sheet defines a partonic 2-surface and vice versa. This slicing has deep physical motivations since it realizes geometrically standard facts about gauge invariance (partonic 2-surface defines the space of physical polarizations and stringy space-time sheet corresponds to non-physical polarizations) and its existence is a hypothesis about the properties of the preferred extremals of Kähler action. There is a similar decomposition also at the level of CD and so called Hamilton-Jacobi coordinates for M^4_+ [3] define this kind of slicings. This slicing can induced the slicing of the space-time sheet. The number theoretic vision gives a further justification for this hypothesis and also strengthens it by postulating the presence of the preferred time direction having interpretation in terms of real unit of octonions. In ZEO this time direction corresponds to the time-like vector connecting the tips of CD.

Conformal symmetries would provide the realization of WCW as a union of symmetric spaces. Symmetric spaces are coset spaces of form G/H. The natural identification of G and H is as groups of X^2-local symplectic transformations and local Kac-Moody group of X^2-local H isometries. Quantum fluctuating (metrically non-trivial) degrees of freedom would correspond to symplectic transformations of H_+ and induced Kähler form at X^2 would define a local representation for zero modes: not necessarily all of them.

3.3 Equivalence Principle and super-conformal symmetries

Equivalence Principle (EP) is a second corner stone of General Relativity and together with GCI leads to Einstein's equations. What EP states is that inertial and gravitational masses are identical. In this form it is not well-defined even in GRT since the definition of gravitational and inertial four-momenta is highly problematic because Noether theorem is not avaible. The realization is in terms of local equations identifying energy momentum tensor with Einstein tensor.

Whether EP is realized in TGD has been a longstanding open question [39]. The problem has been that at the classical level EP in its GRT form can hold true only in long enough length scales and it took long to time to realize that only the stringy form of this principle is required. The first question is how to identify the gravitational and inertial four-momenta. This is indeed possible. One can assign to the two types super-conformal symmetries assigned with light-like 3-surfaces and space-like 3-surfaces four-momenta to both. EP states that these four momenta are identical and is equivalent with the generalization of GCI and effective 2-dimensionality. The condition generalizes so that it applies to the generators of super-conformal algebras associated with the two super-conformal symmetries. This leads to a generalization of a standard mathematical construction of super-conformal theories known as coset representation [76]. What the construction states is that the differences of super-conformal generators defined by super-symmetric algebra and Kac-Moody algebra annihilate physical states.

3.4 Extension to super-conformal symmetries

The original idea behind the extension of conformal symmetries to super-conformal symmetries was the observation that isometry currents defining infinitesimal isometries of WCW have natural super-counterparts obtained by contracting the Killing vector fields with the complexified gamma matrices of the imbedding space.

This vision has generalized considerably as the construction of WCW spinor structure in terms of modified Dirac action has developed. The basic philosophy behind this idea is that configuration space spinor structure must relate directly to the fermionic sector of quantum physics. In particular, modified gamma matrices should be expressible in terms of the fermionic oscillator operators associated with the second quantized induced spinor fields. The explicit realization of this program leads to an identification of rich spectrum of super-conformal symmetries and generalization of the ordinary notion of space-time supersymmetry. What happens that all fermionic oscillator operator generate broken super-symmetries whereas in SUSYs there is only finite number of them. One can however identify sub-algebra of super-conformal symmetries associated with right handed neutrino and this gives $\mathcal{N}=1$ super-symmetry [82] of SUSYs [12].

3.5 Space-time supersymmetry in TGD Universe

It has been clear from the beginning that the notion of super-conformal symmetry crucial for the successes of super-string models generalizes in TGD framework. The answer to the question whether space-time SUSY makes sense in TGD framework has not been obvious at all but it seems now that the answer is affirmative. The evolution of the ideas relevant for the formulation of SUSY in TGD framework is summarized in the chapters of [30]. The chapters are devoted to the notion of bosonic emergence [26], to the SUSY QFT limit of TGD [12] , to twistor approach to TGD [41], and to the generalization of Yangian symmetry of $\mathcal{N}=4$ SYM manifest in the Grassmannian twistor approach [84] to a multi-local variant of super-conformal symmetries [42] represent a gradual development of the ideas about how super-symmetric M-matrix could be constructed in TGD framework. A warning to the reader is in order. In their recent form these chapters do not represent the final outcome but just an evolution of ideas proceeding by trial and error. There are however good reasons to believe that the chapter about Yangian symmetry is nearest to the correct physical interpretation and mathematical formulation.

Contrary to the original expectations, TGD seems to allow a generalization of the space-time super-symmetry. This became clear with the increased understanding of the modified Dirac action [4, 11, 7]. The appearance of the momentum and color quantum numbers in the measurement interaction part of the modified Dirac action associated with the light-like wormhole throats [11] couples space-time degrees of freedom to quantum numbers and allows also to define SUSY algebra at fundamental level as anti-commutation relations of fermionic oscillator operators. Depending on the situation $\mathcal{N} = 2N$ SUSY algebra (an inherent cutoff on the number of fermionic modes at light-like wormhole throat) or fermionic part of super-conformal algebra with infinite number of oscillator operators results. The addition of fermion in particular mode would define particular super-symmetry. This super-symmetry is badly broken due to the dynamics of the modified Dirac operator which also mixes M^4 chiralities inducing massivation. Since right-handed neutrino has no electro-weak couplings the breaking of the corresponding super-symmetry should be weakest.

Zero energy ontology combined with the analog of the twistor approach to $\mathcal{N} = 4$ SYMs and weak form of electric-magnetic duality has actually led to this kind of formulation [42]. What is new that also virtual particles have massless fermions as their building blocks. This implies manifest finiteness of loop integrals so that the situation simplifies dramatically. What is also new element that physical particles and also string like objects correspond to bound states of massless fermions.

The question is whether this SUSY has a realization as a SUSY algebra at space-time level and whether the QFT limit of TGD could be formulated as a generalization of SUSY QFT. There are several problems involved.

1. In TGD framework super-symmetry means addition of fermion to the state and since the number of spinor modes is larger states with large spin and fermion numbers are obtained. This picture does not fit to the standard view about super-symmetry. In particular, the identification of theta parameters as Majorana spinors and super-charges as Hermitian operators is not possible.

2. The belief that Majorana spinors are somehow an intrinsic aspect of super-symmetry is however only a belief. Weyl spinors meaning complex theta parameters are also possible. Theta parameters can also carry fermion number meaning only the supercharges carry fermion number and are non-hermitian. The general classification of super-symmetric theories indeed demonstrates that for $D = 8$ Weyl spinors and complex and non-hermitian super-charges are possible. The original motivation for Majorana spinors might come from MSSM assuming that right handed neutrino does not exist. This belief might have also led to string theories in $D = 10$ and $D = 11$ as the only possible candidates for TOE after it turned out that chiral anomalies cancel. It indeed turns out that TGD view about space-time SUSY is internally consistent. Even more, the separate conservation of quark and lepton number is essential for the internal consistency of this view [12].

3. The massivation of particles is the basic problem of both SUSYs and twistor approach. I have discussed several solutions to this problem [41, 42]. The simplest and most convincing solution of the problem is following and inspired by twistor Grassmannian approach to $\mathcal{N} = 4$ SYM and the generalization of the Yangian symmetry of this theory. In zero energy ontology one can construct physical particles as bound states of massless particles associated with the opposite wormhole throats. If the particles have opposite 3-momenta the resulting state is automatically massive. In fact, this forces massivation of also spin one bosons since the fermion and antifermion must move in opposite directions for their spins to be parallel so that the net mass is non-vanishing: note that this means that even photon, gluons, and graviton have small mass. This mechanism makes topologically condensed fermions massive and padic thermodynamics allows to describe the massivation in terms of zero energy states and M-matrix. Bosons receive to their mass besides the small mass coming from thermodynamics also a contribution which is counterpart of the contribution coming from Higgs vacuum expectation value and Higgs gives rise to longitudinal polarizations. No Higgs potential is however needed. The cancellation of infrared divergences necessary for exact Yangian symmetry and the observation that even photon receives small mass suggest that scalar Higgs would disappear completely from the spectrum.

3.5.1 Basic data bits

Let us first summarize the data bits about possible relevance of super-symmetry for TGD before the addition of the 3-D measurement interaction term to the modified Dirac action [4, 11].

1. Right-handed covariantly constant neutrino spinor ν_R defines a super-symmetry in CP_2 degrees of freedom in the sense that Dirac equation is satisfied by covariant constancy and there is no need

for the usual ansatz $\Psi = D\Psi_0$ giving $D^2\Psi = 0$. This super-symmetry allows to construct solutions of Dirac equation in CP_2 [78, 74, 77, 73].

2. In $M^4 \times CP_2$ this means the existence of massless modes $\Psi = \not{p}\Psi_0$, where Ψ_0 is the tensor product of M^4 and CP_2 spinors. For these solutions M^4 chiralities are not mixed unlike for all other modes which are massive and carry color quantum numbers depending on the CP_2 chirality and charge. As matter fact, covariantly constant right-handed neutrino spinor mode is the only color singlet. The mechanism leading to non-colored states for fermions is based on super-conformal representations for which the color is neutralized [16, 21]. The negative conformal weight of the vacuum also cancels the enormous contribution to mass squared coming from mass in CP_2 degrees of freedom.

3. Right-handed covariantly constant neutrino allows to construct the gamma matrices of the world of classical worlds (WCW) as fermionic counterparts of Hamiltonians of WCW. This gives rise super-symplectic symmetry algebra having interpretation also as a conformal algebra. Also more general super-conformal symmetries exist.

4. Space-time (in the sense of Minkowski space M^4) super-symmetry in the conventional sense of the word is impossible in TGD framework since it would require require Majorana spinors. In 8-D space-time with Minkowski signature of metric Majorana spinors are definitely ruled out by the standard argument leading to super string model. Majorana spinors would also break separate conservation of lepton and baryon numbers in TGD framework.

3.5.2 Could one generalize super-symmetry?

Could one then consider a more general space-time super-symmetry with "space-time" identified as space-time surface rather than Minkowski space?

1. The TGD variant of the super-symmetry could correspond quite concretely to the addition to fermion and boson states right-handed neutrinos. Since right-handed neutrinos do not have electro-weak interactions, the addition might not appreciably affect the mass formula although it could affect the p-adic prime defining the mass scale.

2. The problem is to understand what this addition of the right-handed neutrino means. To begin with, notice that in TGD Universe fermions reside at light-like 3-surfaces at which the signature of induced metric changes. Bosons correspond to pairs of light-like wormhole throats with wormhole contact having Euclidian signature of the induced metric. The long standing problem has been that for bosons with parallel light-like four-momenta with same sign of energy the spins of fermion and antifermion are opposite so that one would obtain only scalar bosons!

   I have considered several solutions to the problem but the final solution came from the basic problem of twistor approach to $\mathcal{N} = 4$ SUSY. This theory is believed to be UV finite but has IR divergences spoiling the Yangian SUSY. These infinities cancel if the physical particles are bound states of pairs of wormhole throats with light-like momenta. Just the requirement that spin is equal to one forces massivation. This is true for all spin 1 particles, also those regarded as massless. Massivation of the photon is not a problem if the mass corresponds to the IR cutoff determined by the largest causal diamond (CD) defining the measurement resolution. For electron the size of CD corresponds to the size scale of Earth. The basic prediction is that Higgs disappears completely from the spectrum so that this mechanism is testable at LHC.

   The first proposal to the solution of problem was that either fermion or antifermion in the boson state carries what might be called un-physical polarization in the standard conceptual framework. This means that it has negative energy but three-momentum parallel to that of the second wormhole throat. The assumption that the bosonic wormhole throats correspond to positive and negative energy space-time sheets realizes this constraint in the framework of zero energy ontology. It however turned out that for light-like momenta these states have more natural interpretation in terms of virtual bosons able to have space-like momenta. This means that one can realize virtual particles as pairs of on mass shell wormhole throats with either sign of energy and 3-momentum so that the basic condition of twistorial approach is satisfied. The conservation of 4-momentum at vertices gives extremely powerful kinematical constraints so that there are excellent hopes about cancellation of UV divergences of loop integrals.

3. The super-symmetry as an addition to the fermion state a second wormhole throats carrying right handed neutrino quantum numbers does not make sense since the resulting state cannot be distinguished from gauge boson or Higgs type particle. The light-like 3-surfaces can however carry fermion numbers up to the number of modes of the induced spinor field, which is expected to be infinite inside string like objects having wormhole throats at ends and finite when one has space time sheets containing the throats [11]. In very general sense one could say that each mode defines a very large broken N-super-symmetry with the value of N depending on state and light-like 3-surface. The breaking of this super-symmetry would come from electro-weak - , color - , and gravitational interactions. Right-handed neutrino would by its electro-weak and color inertness define a minimally broken super-symmetry.

4. What this addition of the right handed neutrinos or more general fermion modes could precisely mean? One cannot assign fermionic oscillator operators to right handed neutrinos which are covariantly constant in both M^4 and CP_2 degrees of freedom since the modes with vanishing energy (frequency) cannot correspond to fermionic oscillator operator creating a physical state since one would have $a = a^\dagger$. The intuitive view is that all the spinor modes move in an exactly collinear manner -somewhat like quarks inside hadron do approximately.

3.5.3 Modified Dirac equation briefly

The answer to the question what "collinear motion" means mathematically emerged from the recent progress in the understanding of the modified Dirac equation.

1. The modified Dirac action involves two terms. Besides the original 4-D modified Dirac action there is measurement interaction which can be localized to wormhole throat or to any light-like 3-surfaces "parallel" to it in the slicing of space-time sheet by light-like 3-surfaces. This term correlates space-time geometry with quantum numbers assignable to super-conformal representations and is also necessary to obtain almost-stringy propagator.

2. The modified Dirac equation with measurement action added reads as

$$\begin{aligned} D_K\Psi &= 0 \ , \\ D_3\Psi &= (D_{C-S} + Q \times O)\Psi = 0 \ , \\ [D_3, D_K]\,\Psi &= 0 \ . \end{aligned} \tag{3.1}$$

   (a) D_K corresponds formally to 4-D massless Dirac equation in X^4. D_3 realizes measurement interaction. D_{C-S} is the 3-D modified Dirac action defined by Chern-Simons action.

   (b) Q is linear in Cartan algebra generators of the isometry algebra of imbedding space (color isospin and hypercharge plus four-momentum or two components of four momentum and spin and boost in direction of 3-momentum). Q is expressible as

$$Q = Q_A \partial_\alpha h^k g^{AB} j_{Bk} \hat{\Gamma}^\alpha_{CS} \ . \tag{3.2}$$

   Here Q_A is Cartan algebra generator acting on physical states. Physical states must be eigen states of Q_A since otherwise the equations do not make sense. g^{AB} is the inverse of the matrix defined by the imbedding space inner product of Killing vector fields j^k_A and j^l_B: its existence allows only Cartan algebra charges. $\hat{\Gamma}^\alpha_{CS}$ is the modified gamma matrix associated with the Chern-Simons action.

(c) In general case the modified gamma matrices are defined in terms of action density L as

$$\hat{\Gamma}^{\alpha} = \frac{\partial L}{\partial_{\alpha} h^k}\gamma^k \ . \tag{3.3}$$

γ^k denotes imbedding space gamma matrices.

(d) The operator O characterizes the conserved fermionic current to which Cartan algebra generators of isometries couple. The simplest conserved currents correspond to quark or lepton currents and corresponding vectorial isospin- and spin currents [11]. Besides this there is an infinite hierarchy of conserved currents relating to quantum criticality and in one-one correspondence with vanishing second variations of Kähler action for preferred extremal. These couplings allow to represent measurement interaction for any observable.

3. The equation $D_3\nu_R = 0$ would reduce for vanishing color charges and covariantly constant spinor to the analog of algebraic fermionic on mass shell condition $p_A\gamma^A\nu_R = 0$ since Q is obtained by projecting the total four-momentum of the parton state interpreted as a vector-field of H to the space-time surface and by replacing ordinary gamma matrices with the modified ones. This equation cannot be exact since Q depends on the point of the light-like 3-surface so that covariant constancy fails and D_{C-S} cannot annihilate the state. This is the space-time correlate for the breaking of super-symmetry. The action of the Cartan algebra generators is purely algebraic and on the state of super-conformal representations rather than that of a differential operator on spinor field. The modified equation implies that all spinor modes represent fermions moving collinearly in the sense an equation with the same total four-momentum and total color quantum numbers is satisfied by all of them. Note that p_A represents the total four-momentum of the state rather than individual four-momenta of fermions.

3.5.4 TGD counterpart of space-time super-symmetry

This picture allows to define more precisely what one means with the approximate super-symmetries in TGD framework.

1. One can in principle construct many-fermion states containing both fermions and anti-fermions at given light-like 3-surface. The four-momenta of states related by super-symmetry need not be same. Super-symmetry breaking is present and has as the space-time correlate the deviation of the modified gamma matrices from the ordinary M^4 gamma matrices. In particular, the fact that $\hat{\Gamma}^{\alpha}$ possesses CP_2 part in general means that different M^4 chiralities are mixed: a space-time correlate for the massivation of the elementary particles.

2. For right-handed neutrino super-symmetry breaking is expected to be smallest but also in the case of the right-handed neutrino mode mixing of M^4 chiralities takes place and breaks the TGD counterpart of super-symmetry.

3. The fact that all helicities in the state are physical for a given light-like 3-surface has important implications. For instance, the addition of a right-handed antineutrino to right-handed (left-handed) electron state gives scalar (spin 1) state. Also states with fermion number two are obtained from fermions. For instance, for e_R one obtains the states $\{e_R, e_R\nu_R\overline{\nu}_R, e_R\overline{\nu}_R, e_R\nu_R\}$ with lepton numbers $(1, 1, 0, 2)$ and spins $(1/2, 1/2, 0, 1)$. For e_L one obtains the states $\{e_L, e_L\nu_R\overline{\nu}_R, e_L\overline{\nu}_R, e_L\nu_R\}$ with lepton numbers $(1, 1, 0, 2)$ and spins $(1/2, 1/2, 1, 0)$. In the case of gauge boson and Higgs type particles -allowed by TGD but not required by p-adic mass calculations- gauge boson has 15 super partners with fermion numbers $[2, 1, 0, -1, -2]$.

The cautious conclusion is that the recent view about quantum TGD allows the analog of super-symmetry which is necessary broken and for which the multiplets are much more general than for the ordinary super-symmetry. Right-handed neutrinos might however define something resembling ordinary super-symmetry to a high extent. The question is how strong prediction one can deduce using quantum TGD and proposed super-symmetry.

1. For a minimal breaking of super-symmetry only the p-adic length scale characterizing the super-partner differs from that for partner but the mass of the state is same. This would allow only a discrete set of masses for various super-partners coming as half octaves of the mass of the particle in question. A highly predictive model results.

2. The quantum field theoretic description should be based on QFT limit of TGD formulated in terms of bosonic emergence [26]. This formulation should allow to calculate the propagators of the super-partners in terms of fermionic loops.

3. This TGD variant of space-time super-symmetry resembles ordinary super-symmetry in the sense that selection rules due to the right-handed neutrino number conservation and analogous to the conservation of R-parity hold true. The states inside super-multiplets have identical electro-weak and color quantum numbers but their p-adic mass scales can be different. It should be possible to estimate reaction reaction rates using rules very similar to those of super-symmetric gauge theories.

4. It might be even possible to find some simple generalization of standard super-symmetric gauge theory to get rough estimates for the reaction rates. There are however problems. The fact that spins $J = 0, 1, 2, 3/2, 2$ are possible for super-partners of gauge bosons forces to ask whether these additional states define an analog of non-stringy strong gravitation. Note that graviton in TGD framework corresponds to a pair of wormhole throats connected by flux tube (counterpart of string) and for gravitons one obtains 2^8-fold degeneracy.

3.6 Twistorial approach, Yangian symmetry, and generalized Feynman diagrams

There has been impressive steps in the understanding of $\mathcal{N} = 4$ maximally sypersymmetric YM theory possessing 4-D super-conformal symmetry. This theory is related by AdS/CFT duality to certain string theory in $AdS_5 \times S^5$ background. Second stringy representation was discovered by Witten and is based on 6-D Calabi-Yau manifold defined by twistors. The unifying proposal is that so called Yangian symmetry is behind the mathematical miracles involved.

The notion of Yangian symmetry would have a generalization in TGD framework obtained by replacing conformal algebra with appropriate super-conformal algebras. Also a possible realization of twistor approach and the construction of scattering amplitudes in terms of Yangian invariants defined by Grassmannian integrals is considered in TGD framework and based on the idea that in zero energy ontology one can represent massive states as bound states of massless particles. There is also a proposal for a physical interpretation of the Cartan algebra of Yangian algebra allowing to understand at the fundamental level how the mass spectrum of n-particle bound states could be understood in terms of the n-local charges of the Yangian algebra.

Twistors were originally introduced by Penrose to characterize the solutions of Maxwell's equations. Kähler action is Maxwell action for the induced Kähler form of CP_2. The preferred extremals allow a very concrete interpretation in terms of modes of massless non-linear field. Both conformally compactified Minkowski space identifiable as so called causal diamond and CP_2 allow a description in terms of twistors. These observations inspire the proposal that a generalization of Witten's twistor string theory relying on the identification of twistor string world sheets with certain holomorphic surfaces assigned with Feynman diagrams could allow a formulation of quantum TGD in terms of 3-dimensional holomorphic surfaces of $CP_3 \times CP_3$ mapped to 6-surfaces dual $CP_3 \times CP_3$, which are sphere bundles so that they are projected in a natural manner to 4-D space-time surfaces. Very general physical and mathematical arguments lead to a highly unique proposal for the holomorphic differential equations defining the complex 3-surfaces conjectured to correspond to the preferred extremals of Kähler action.

3.6.1 Background

I am outsider as far as concrete calculations in $\mathcal{N} = 4$ SUSY are considered and the following discussion of the background probably makes this obvious. My hope is that the reader had patience to not care about this and try to see the big pattern.

The developments began from the observation of Parke and Taylor [89] that n-gluon tree amplitudes with less than two negative helicities vanish and those with two negative helicities have unexpectedly simple form when expressed in terms of spinor variables used to represent light-like momentum. In fact, in the formalism based on Grassmanian integrals the reduced tree amplitude for two negative helicities is just "1" and defines Yangian invariant. The article *Perturbative Gauge Theory As a String Theory In Twistor Space* [91] by Witten led to so called Britto-Cachazo-Feng-Witten (BCFW) recursion relations for tree level amplitudes [90, 85, 90] allowing to construct tree amplitudes using the analogs of Feynman rules in which vertices correspond to maximally helicity violating tree amplitudes (2 negative helicity gluons) and propagator is massless Feynman propagator for boson. The progress inspired the idea that the theory might be completely integrable meaning the existence of infinite-dimensional un-usual symmetry. This symmetry would be so called Yangian symmetry [42] assigned to the super counterpart of the conformal group of 4-D Minkowski space.

Drumond, Henn, and Plefka represent in the article *Yangian symmetry of scattering amplitudes in $\mathcal{N} = 4$ super Yang-Mills theory* [86] an argument suggesting that the Yangian invariance of the scattering amplitudes ins an intrinsic property of planar $\mathcal{N} = 4$ super Yang Mills at least at tree level.

The latest step in the progress was taken by Arkani-Hamed, Bourjaily, Cachazo, Carot-Huot, and Trnka and represented in the article *Yangian symmetry of scattering amplitudes in $\mathcal{N} = 4$ super Yang-Mills theory* [84]. At the same day there was also the article of Rutger Boels entitled *On BCFW shifts of integrands and integrals* [83] in the archive. Arkani-Hamed *et al* argue that a full Yangian symmetry of the theory allows to generalize the BCFW recursion relation for tree amplitudes to all loop orders at planar limit (planar means that Feynman diagram allows imbedding to plane without intersecting lines). On mass shell scattering amplitudes are in question.

3.6.2 Yangian symmetry

The notion equivalent to that of Yangian was originally introduced by Faddeev and his group in the study of integrable systems. Yangians are Hopf algebras which can be assigned with Lie algebras as the deformations of their universal enveloping algebras. The elegant but rather cryptic looking definition is in terms of the modification of the relations for generating elements [42]. Besides ordinary product in the enveloping algebra there is co-product Δ which maps the elements of the enveloping algebra to its tensor product with itself. One can visualize product and co-product is in terms of particle reactions. Particle annihilation is analogous to annihilation of two particle so single one and co-product is analogous to the decay of particle to two. Δ allows to construct higher generators of the algebra.

Lie-algebra can mean here ordinary finite-dimensional simple Lie algebra, Kac-Moody algebra or Virasoro algebra. In the case of SUSY it means conformal algebra of M^4- or rather its super counterpart. Witten, Nappi and Dolan have described the notion of Yangian for super-conformal algebra in very elegant and and concrete manner in the article *Yangian Symmetry in D=4 superconformal Yang-Mills theory* [87]. Also Yangians for gauge groups are discussed.

In the general case Yangian resembles Kac-Moody algebra with discrete index n replaced with a continuous one. Discrete index poses conditions on the Lie group and its representation (adjoint representation in the case of $\mathcal{N} = 4$ SUSY). One of the conditions conditions is that the tensor product $R \otimes R^*$ for representations involved contains adjoint representation only once. This condition is non-trivial. For $SU(n)$ these conditions are satisfied for any representation. In the case of $SU(2)$ the basic branching rule for the tensor product of representations implies that the condition is satisfied for the product of any representations.

Yangian algebra with a discrete basis is in many respects analogous to Kac-Moody algebra. Now however the generators are labelled by non-negative integers labeling the light-like incoming and outgoing momenta of scattering amplitude whereas in in the case of Kac-Moody algebra also negative values are allowed. Note that only the generators with non-negative conformal weight appear in the construction of states of Kac-Moody and Virasoro representations so that the extension to Yangian makes sense.

The generating elements are labelled by the generators of ordinary conformal transformations acting in M^4 and their duals acting in momentum space. These two sets of elements can be labelled by conformal weights $n = 0$ and $n = 1$ and and their mutual commutation relations are same as for Kac-Moody algebra. The commutators of $n = 1$ generators with themselves are however something different for a non-vanishing deformation parameter h. Serre's relations characterize the difference and involve the deformation parameter h. Under repeated commutations the generating elements generate infinite-dimensional symmetric algebra, the Yangian. For $h = 0$ one obtains just one half of the Virasoro algebra or Kac-Moody algebra. The generators with $n > 0$ are $n + 1$-local in the sense that they involve $n + 1$-forms of local generators assignable to the ordered set of incoming particles of the scattering amplitude. This non-locality generalizes the notion of local symmetry and is claimed to be powerful enough to fix the scattering amplitudes completely.

3.6.3 How to generalize Yangian symmetry in TGD framework?

As far as concrete calculations are considered, I have nothing to say. I am just perplexed. It is however possible to keep discussion at general level and still say something interesting (as I hope!). The key question is whether it could be possible to generalize the proposed Yangian symmetry and geometric picture behind it to TGD framework.

1. The first thing to notice is that the Yangian symmetry of $\mathcal{N} = 4$ SUSY in question is quite too limited since it allows only single representation of the gauge group and requires massless particles. One must allow all representations and massive particles so that the representation of symmetry algebra must involve states with different masses, in principle arbitrary spin and arbitrary internal quantum numbers. The candidates are obvious: Kac-Moody algebras [57] and Virasoro algebras [64] and their super counterparts. Yangians indeed exist for arbitrary super Lie algebras. In TGD framework conformal algebra of Minkowski space reduces to Poincare algebra and its extension to Kac-Moody allows to have also massive states.

2. The formal generalization looks surprisingly straightforward at the formal level. In zero energy ontology one replaces point like particles with partonic two-surfaces appearing at the ends of light-like orbits of wormhole throats located to the future and past light-like boundaries of causal diamond ($CD \times CP_2$ or briefly CD). Here CD is defined as the intersection of future and past directed light-cones. The polygon with light-like momenta is naturally replaced with a polygon with more general momenta in zero energy ontology and having partonic surfaces as its vertices. Non-point-likeness forces to replace the finite-dimensional super Lie-algebra with infinite-dimensional Kac-Moody algebras and corresponding super-Virasoro algebras assignable to partonic 2-surfaces.

3. This description replaces disjoint holomorphic surfaces in twistor space with partonic 2-surfaces at the boundaries of $CD \times CP_2$ so that there seems to be a close analogy with Cachazo-Svrcek-Witten picture. These surfaces are connected by either light-like orbits of partonic 2-surface or space-like 3-surfaces at the ends of CD so that one indeed obtains the analog of polygon.

What does this then mean concretely (if this word can be used in this kind of context;-)?

1. At least it means that ordinary Super Kac-Moody and Super Virasoro algebras associated with isometries of $M^4 \times CP_2$ annihilating the scattering amplitudes must be extended to a co-algebras with a non-trivial deformation parameter. Kac-Moody group is thus the product of Poincare and color groups. This algebra acts as deformations of the light-like 3-surfaces representing the light-like orbits of particles which are extremals of Chern-Simon action with the constraint that weak form of electric-magnetic duality holds true. I know so little about the mathematical side that I cannot tell whether the condition that the product of the representations of Super-Kac-Moody and Super-Virasoro algedbras ontains adjoint representation only once, holds true in this case. In any case, it would allow all representations of finite-dimensional Lie group in vertices whereas $\mathcal{N} = 4$ SUSY would allow only the adjoint.

2. Besides this ordinary kind of Kac-Moody algebra there is the analog of Super-Kac-Moody algebra associated with the light-cone boundary which is metrically 3-dimensional. The finite-dimensional Lie group is in this case replaced with infinite-dimensional group of symplectomorphisms of $\delta M^4_{+/-}$ made local with respect to the internal coordinates of partonic 2-surface. A coset construction is applied to these two Virasoro algebras so that the differences of the corresponding Super-Virasoro

generators and Kac-Moody generators annihilate physical states. This implies that the corresponding four-momenta are same: this expresses the equivalence of gravitational and inertial masses. A generalization of the Equivalence Principle is in question. This picture also justifies p-adic thermodynamics applied to either symplectic or isometry Super-Virasoro and giving thermal contribution to the vacuum conformal and thus to mass squared.

3. The construction of TGD leads also to other super-conformal algebras and the natural guess is that the Yangians of all these algebras annihilate the scattering amplitudes.

4. Obviously, already the starting point symmetries look formidable but they still act on single partonic surface only. The discrete Yangian associated with this algebra associated with the closed polygon defined by the incoming momenta and the negatives of the outgoing momenta acts in multi-local manner on scattering amplitudes. It might make sense to speak about polygons defined also by other conserved quantum numbers so that one would have generalized light-like curves in the sense that state are massless in 8-D sense.

3.6.4 Is there any hope about description in terms of Grassmannians?

At technical level the successes of the twistor approach rely on the observation that the amplitudes can be expressed in terms of very simple integrals over sub-manifolds of the space consisting of k-dimensional planes of n-dimensional space defined by delta function appearing in the integrand. These integrals define super-conformal Yangian invariants appearing in twistorial amplitudes and the belief is that by a proper choice of the surfaces of the twistor space one can construct all invariants. One can construct also the counterparts of loop corrections by starting from tree diagrams and annihilating pair of particles by connecting the lines and quantum entangling the states at the ends in the manner dictated by the integration over loop momentum. These operations can be defined as operations for Grassmannian integrals in general changing the values of n and k. This description looks extremely powerful and elegant and nosta importantly involves only the external momenta.

The obvious question is whether one could use similar invariants in TGD framework to construct the momentum dependence of amplitudes.

1. The first thing to notice is that the super algebras in question act on infinite-dimensional representations and basically in the world of classical worlds assigned to the partonic 2-surfaces correlated by the fact that they are associated with the same space-time surface. This does not promise anything very practical. On the other hand, one can hope that everything related to other than M^4 degrees of freedom could be treated like color degrees of freedom in $\mathcal{N}=4$ SYM and would boil down to indices labeling the quantum states. The Yangian conditions coming from isometry quantum numbers, color quantum numbers, and electroweak quantum numbers are of course expected to be highly non-trivial and could fix the coefficients of various singlets resulting in the tensor product of incoming and outgoing states.

2. The fact that incoming particles can be also massive seems to exclude the use of the twistor space. The following observation however raises hopes. The Dirac propagator for wormhole throat is massless propagator but for what I call pseudo momentum. It is still unclear how this momentum relates to the actual four-momentum. Could it be actually equal to it? The recent view about pseudo-momentum does not support this view but it is better to keep mind open. In any case this finding suggests that twistorial approach could work in in more or less standard form. What would be needed is a representation for massive incoming particles as bound states of massless partons. In particular, the massive states of super-conformal representations should allow this kind of description.

Could zero energy ontology allow to achieve this dream?

1. As far as divergence cancellation is considered, zero energy ontology suggests a totally new approach producing the basic nice aspects of QFT approach, in particular unitarity and coupling constant evolution. The big idea related to zero energy ontology is that all virtual particle particles correspond

to wormhole throats, which are pairs of on mass shell particles. If their momentum directions are different, one obtains time-like continuum of virtual momenta and if the signs of energy are opposite one obtains also space-like virtual momenta. The on mass shell property for virtual partons (massive in general) implies extremely strong constraints on loops and one expect that only very few loops remain and that they are finite since loop integration reduces to integration over much lower-dimensional space than in the QFT approach. There are also excellent hopes about Cutkoski rules.

2. Could zero energy ontology make also possible to construct massive incoming particles from massless ones? Could one construct the representations of the super conformal algebras using only massless states so that at the fundamental level incoming particles would be massless and one could apply twistor formalism and build the momentum dependence of amplitudes using Grassmannian integrals.

   One could indeed construct on mass shell massive states from massless states with momenta along the same line but with three-momenta at opposite directions. Mass squared is given by $M^2 = 4E^2$ in the coordinate frame, where the momenta are opposite and of same magnitude. One could also argue that partonic 2-surfaces carrying quantum numbers of fermions and their superpartners serve as the analogs of point like massless particles and that topologically condensed fermions and gauge bosons plus their superpartners correspond to pairs of wormhole throats. Stringy objects would correspond to pairs of wormhole throats at the same space-time sheet in accordance with the fact that space-time sheet allows a slicing by string worlds sheets with ends at different wormhole throats and definining time like braiding.

The weak form of electric magnetic duality indeed supports this picture. To understand how, one must explain a little bit what the weak form of electric magnetic duality means.

1. Elementary particles correspond to light-like orbits of partonic 2-surfaces identified as 3-D surfaces at which the signature of the induced metric of space-time surface changes from Euclidian to Minkowskian and 4-D metric is therefore degenerate. The analogy with black hole horizon is obvious but only partial. Weak form of electric-magnetic duality states that the Kähler electric field at the wormhole throat and also at space-like 3-surfaces defining the ends of the space-time surface at the upper and lower light-like boundaries of the causal diamond is proportonial to Kähler magnetic field so that Kähler electric flux is proportional Kähler magnetic flux. This implies classical quantization of Kähler electric charge and fixes the value of the proportionality constant.

2. There are also much more profound implications. The vision about TGD as almost topological QFT suggests that Kähler function defining the Kähler geometry of the "world of classical worlds" (WCW) and identified as Kähler action for its preferred extremal reduces to the 3-D Chern-Simons action evaluted at wormhole throats and possible boundary components. Chern-Simons action would be subject to constraints. Wormhole throats and space-like 3-surfaces would represent extremals of Chern-Simons action restricted by the constraint force stating electric-magnetic duality (and realized in terms of Lagrange multipliers as usual).

   If one assumes that Kähler current and other conserved currents are proportional to current defining Beltrami flow whose flow lines by definition define coordinate curves of a globally defined coordinate, the Coulombic term of Kähler action vanishes and it reduces to Chern-Simons action if the weak form of electric-magnetic duality holds true. One obtains almost topological QFT. The absolutely essential attribute "almost" comes from the fact that Chern-Simons action is subject to constraints. As a consequence, one obtains non-vanishing four-momenta and WCW geometry is non-trivial in M^4 degrees of freedom. Otherwise one would have only topological QFT not terribly interesting physically.

Consider now the question how one could understand stringy objects as bound states of massless particles.

1. The observed elementary particles are not Kähler monopoles and there much exist a mechanism neutralizing the monopole charge. The only possibility seems to be that there is opposite Kähler magnetic charge at second wormhole throat. The assumption is that in the case of color neutral

particles this throat is at a distance of order intermediate gauge boson Compton length. This throat would carry weak isospin neutralizing that of the fermion and only electromagnetic charge would be visible at longer length scales. One could speak of electro-weak confinement. Also color confinement could be realized in analogous manner by requiring the cancellation of monopole charge for many-parton states only. What comes out are string like objects defined by Kähler magnetic fluxes and having magnetic monopoles at ends. Also more general objects with three strings branching from the vertex appear in the case of baryons. The natural guess is that the partons at the ends of strings and more general objects are massless for incoming particles but that the 3-momenta are in opposite directions so that stringy mass spectrum and representations of relevant super-conformal algebras are obtained. This description brings in mind the description of hadrons in terms of partons moving in parallel apart from transversal momentum about which only momentum squared is taken as observable.

2. Quite generally, one expects for the preferred extremals of Kähler action the slicing of space-time surface with string world sheets with stringy curves connecting wormhole throats. The ends of the stringy curves can be identified as light-like braid strands. Note that the strings themselves define a space-like braiding and the two braidings are in some sense dual. This has a concrete application in TGD inspired quantum biology, where time-like braiding defines topological quantum computer programs and the space-like braidings induced by it its storage into memory. Stringlike objects defining representations of super-conformal algebras must correspond to states involving at least two wormhole throats. Magnetic flux tubes connecting the ends of magnetically charged throats provide a particular realization of stringy on mass shell states. This would give rise to massless propagation at the parton level. The stringy quantization condition for mass squared would read as $4E^2 = n$ in suitable units for the representations of super-conformal algebra associated with the isometries. For pairs of throats of the same wormhole contact stringy spectrum does not seem plausible since the wormhole contact is in the direction of CP_2. One can however expect generation of small mass as deviation of vacuum conformal weight from half integer in the case of gauge bosons.

If this picture is correct, one might be able to determine the momentum dependence of the scattering amplitudes by replacing free fermions with pairs of monopoles at the ends of string and topologically condensed fermions gauge bosons with pairs of this kind of objects with wormhole throat replaced by a pair of wormhole throats. This would mean suitable number of doublings of the Grassmannian integrations with additional constraints on the incoming momenta posed by the mass shell conditions for massive states.

3.6.5 Could zero energy ontology make possible full Yangian symmetry?

The partons in the loops are on mass shell particles have a discrete mass spectrum but both signs of energy are possible for opposite wormhole throats. This implies that in the rules for constructing loop amplitudes from tree amplitudes, propagator entanglement is restricted to that corresponding to pairs of partonic on mass shell states with both signs of energy. As emphasized in [84] , it is the Grassmannian integrands and leading order singularities of $\mathcal{N} = 4$ SYM, which possess the full Yangian symmetry. The full integral over the loop momenta breaks the Yangian symmetry and brings in IR singularities. Zero energy ontologist finds it natural to ask whether QFT approach shows its inadequacy both via the UV divergences and via the loss of full Yangian symmetry. The restriction of virtual partons to discrete mass shells with positive or negative sign of energy imposes extremely powerful restrictions on loop integrals and resembles the restriction to leading order singularities. Could this restriction guarantee full Yangian symmetry and remove also IR singularities?

3.6.6 Could Yangian symmetry provide a new view about conserved quantum numbers?

The Yangian algebra has some properties which suggest a new kind of description for bound states. The Cartan algebra generators of $n = 0$ and $n = 1$ levels of Yangian algebra commute. Since the co-product Δ maps $n = 0$ generators to $n = 1$ generators and these in turn to generators with high value of n, it seems that they commute also with $n \geq 1$ generators. This applies to four-momentum, color isospin and color hyper charge, and also to the Virasoro generator L_0 acting on Kac-Moody algebra of isometries and defining mass squared operator.

Could one identify total four momentum and Cartan algebra quantum numbers as sum of contributions from various levels? If so, the four momentum and mass squared would involve besides the local term assignable to wormhole throats also n-local contributions. The interpretation in terms of n-parton bound states would be extremely attractive. n-local contribution would involve interaction energy. For instance, string like object would correspond to $n = 1$ level and give $n = 2$-local contribution to the momentum. For baryonic valence quarks one would have 3-local contribution corresponding to $n = 2$ level. The Yangian view about quantum numbers could give a rigorous formulation for the idea that massive particles are bound states of massless particles.

4 Weak form electric-magnetic duality and color and weak forces

The notion of electric-magnetic duality [81] was proposed first by Olive and Montonen and is central in $\mathcal{N} = 4$ supersymmetric gauge theories. It states that magnetic monopoles and ordinary particles are two different phases of theory and that the description in terms of monopoles can be applied at the limit when the running gauge coupling constant becomes very large and perturbation theory fails to converge. The notion of electric-magnetic self-duality is more natural since for CP_2 geometry Kähler form is self-dual and Kähler magnetic monopoles are also Kähler electric monopoles and Kähler coupling strength is by quantum criticality renormalization group invariant rather than running coupling constant. The notion of electric-magnetic (self-)duality emerged already two decades ago in the attempts to formulate the Kähler geometric of world of classical worlds. Quite recently a considerable step of progress took place in the understanding of this notion [5]. What seems to be essential is that one adopts a weaker form of the self-duality applying at partonic 2-surfaces. What this means will be discussed in the sequel.

Every new idea must be of course taken with a grain of salt but the good sign is that this concept leads to precise predictions. The point is that elementary particles do not generate monopole fields in macroscopic length scales: at least when one considers visible matter. The first question is whether elementary particles could have vanishing magnetic charges: this turns out to be impossible. The next question is how the screening of the magnetic charges could take place and leads to an identification of the physical particles as string like objects identified as pairs magnetic charged wormhole throats connected by magnetic flux tubes.

1. The first implication is a new view about electro-weak massivation reducing it to weak confinement in TGD framework. The second end of the string contains particle having electroweak isospin neutralizing that of elementary fermion and the size scale of the string is electro-weak scale would be in question. Hence the screening of electro-weak force takes place via weak confinement realized in terms of magnetic confinement.

2. This picture generalizes to the case of color confinement. Also quarks correspond to pairs of magnetic monopoles but the charges need not vanish now. Rather, valence quarks would be connected by flux tubes of length of order hadron size such that magnetic charges sum up to zero. For instance, for baryonic valence quarks these charges could be $(2, -1, -1)$ and could be proportional to color hyper charge.

3. The highly non-trivial prediction making more precise the earlier stringy vision is that elementary particles are string like objects in electro-weak scale: this should become manifest at LHC energies.

4. The weak form electric-magnetic duality together with Beltrami flow property of Kähler leads to the reduction of Kähler action to Chern-Simons action so that TGD reduces to almost topological QFT and that Kähler function is explicitly calculable. This has enormous impact concerning practical calculability of the theory.

5. One ends up also to a general solution ansatz for field equations from the condition that the theory reduces to almost topological QFT. The solution ansatz is inspired by the idea that all isometry currents are proportional to Kähler current which is integrable in the sense that the flow parameter

associated with its flow lines defines a global coordinate. The proposed solution ansatz would describe a hydrodynamical flow with the property that isometry charges are conserved along the flow lines (Beltrami flow). A general ansatz satisfying the integrability conditions is found. The solution ansatz applies also to the extremals of Chern-Simons action and and to the conserved currents associated with the modified Dirac equation defined as contractions of the modified gamma matrices between the solutions of the modified Dirac equation. The strongest form of the solution ansatz states that various classical and quantum currents flow along flow lines of the Beltrami flow defined by Kähler current (Kähler magnetic field associated with Chern-Simons action). Intuitively this picture is attractive. A more general ansatz would allow several Beltrami flows meaning multi-hydrodynamics. The integrability conditions boil down to two scalar functions: the first one satisfies massless d'Alembert equation in the induced metric and the the gradients of the scalar functions are orthogonal. The interpretation in terms of momentum and polarization directions is natural.

6. The general solution ansatz works for induced Kähler Dirac equation and Chern-Simons Dirac equation and reduces them to ordinary differential equations along flow lines. The induced spinor fields are simply constant along flow lines of indued spinor field for Dirac equation in suitable gauge. Also the generalized eigen modes of the modified Chern-Simons Dirac operator can be deduced explicitly if the throats and the ends of space-time surface at the boundaries of CD are extremals of Chern-Simons action. Chern-Simons Dirac equation reduces to ordinary differential equations along flow lines and one can deduce the general form of the spectrum and the explicit representation of the Dirac determinant in terms of geometric quantities characterizing the 3-surface (eigenvalues are inversely proportional to the lengths of strands of the flow lines in the effective metric defined by the modified gamma matrices).

4.1 Could a weak form of electric-magnetic duality hold true?

Holography means that the initial data at the partonic 2-surfaces should fix the configuration space metric. A weak form of this condition allows only the partonic 2-surfaces defined by the wormhole throats at which the signature of the induced metric changes. A stronger condition allows all partonic 2-surfaces in the slicing of space-time sheet to partonic 2-surfaces and string world sheets. Number theoretical vision suggests that hyper-quaternionicity *resp.* co-hyperquaternionicity constraint could be enough to fix the initial values of time derivatives of the imbedding space coordinates in the space-time regions with Minkowskian *resp.* Euclidian signature of the induced metric. This is a condition on modified gamma matrices and hyper-quaternionicity states that they span a hyper-quaternionic sub-space.

4.1.1 Definition of the weak form of electric-magnetic duality

One can also consider alternative conditions possibly equivalent with this condition. The argument goes as follows.

1. The expression of the matrix elements of the metric and Kähler form of WCW in terms of the Kähler fluxes weighted by Hamiltonians of $\delta M^4_\pm$ at the partonic 2-surface X^2 looks very attractive. These expressions however carry no information about the 4-D tangent space of the partonic 2-surfaces so that the theory would reduce to a genuinely 2-dimensional theory, which cannot hold true. One would like to code to the WCW metric also information about the electric part of the induced Kähler form assignable to the complement of the tangent space of $X^2 \subset X^4$.

2. Electric-magnetic duality of the theory looks a highly attractive symmetry. The trivial manner to get electric magnetic duality at the level of the full theory would be via the identification of the flux Hamiltonians as sums of of the magnetic and electric fluxes. The presence of the induced metric is however troublesome since the presence of the induced metric means that the simple transformation properties of flux Hamiltonians under symplectic transformations -in particular color rotations- are lost.

3. A less trivial formulation of electric-magnetic duality would be as an initial condition which eliminates the induced metric from the electric flux. In the Euclidian version of 4-D YM theory this

duality allows to solve field equations exactly in terms of instantons. This approach involves also quaternions. These arguments suggest that the duality in some form might work. The full electric magnetic duality is certainly too strong and implies that space-time surface at the partonic 2-surface corresponds to piece of CP_2 type vacuum extremal and can hold only in the deep interior of the region with Euclidian signature. In the region surrounding wormhole throat at both sides the condition must be replaced with a weaker condition.

4. To formulate a weaker form of the condition let us introduce coordinates (x^0, x^3, x^1, x^2) such (x^1, x^2) define coordinates for the partonic 2-surface and (x^0, x^3) define coordinates labeling partonic 2-surfaces in the slicing of the space-time surface by partonic 2-surfaces and string world sheets making sense in the regions of space-time sheet with Minkowskian signature. The assumption about the slicing allows to preserve general coordinate invariance. The weakest condition is that the generalized Kähler electric fluxes are apart from constant proportional to Kähler magnetic fluxes. This requires the condition

$$J^{03}\sqrt{g_4} = KJ_{12} . \tag{4.1}$$

A more general form of this duality is suggested by the considerations of [14] reducing the hierarchy of Planck constants to basic quantum TGD and also reducing Kähler function for preferred extremals to Chern-Simons terms [79] at the boundaries of CD and at light-like wormhole throats. This form is following

$$J^{n\beta}\sqrt{g_4} = K\epsilon \times \epsilon^{n\beta\gamma\delta}J_{\gamma\delta}\sqrt{g_4} . \tag{4.2}$$

Here the index n refers to a normal coordinate for the space-like 3-surface at either boundary of CD or for light-like wormhole throat. ϵ is a sign factor which is opposite for the two ends of CD. It could be also opposite of opposite at the opposite sides of the wormhole throat. Note that the dependence on induced metric disappears at the right hand side and this condition eliminates the potentials singularity due to the reduction of the rank of the induced metric at wormhole throat.

5. Information about the tangent space of the space-time surface can be coded to the configuration space metric with loosing the nice transformation properties of the magnetic flux Hamiltonians if Kähler electric fluxes or sum of magnetic flux and electric flux satisfying this condition are used and K is symplectic invariant. Using the sum

$$J_e + J_m = (1+K)J , \tag{4.3}$$

where J can denotes the Kähler magnetic flux, makes it possible to have a non-trivial configuration space metric even for $K = 0$, which could correspond to the ends of a cosmic string like solution carrying only Kähler magnetic fields. This condition suggests that it can depend only on Kähler magnetic flux and other symplectic invariants. Whether local symplectic coordinate invariants are possible at all is far from obvious, If the slicing itself is symplectic invariant then K could be a non-constant function of X^2 depending on string world sheet coordinates. The light-like radial coordinate of the light-cone boundary indeed defines a symplectically invariant slicing and this slicing could be shifted along the time axis defined by the tips of CD.

4.1.2 Electric-magnetic duality physically

What could the weak duality condition mean physically? For instance, what constraints are obtained if one assumes that the quantization of electro-weak charges reduces to this condition at classical level?

1. The first thing to notice is that the flux of J over the partonic 2-surface is analogous to magnetic flux

$$Q_m = \frac{e}{\hbar}\oint BdS = n \ .$$

n is non-vanishing only if the surface is homologically non-trivial and gives the homology charge of the partonic 2-surface.

2. The expressions of classical electromagnetic and Z^0 fields in terms of Kähler form [43] ,[43] read as

$$\begin{aligned} \gamma &= \frac{eF_{em}}{\hbar} = 3J - sin^2(\theta_W)R_{03} \ , \\ Z^0 &= \frac{g_Z F_Z}{\hbar} = 2R_{03} \ . \end{aligned} \quad (4.4)$$

Here R_{03} is one of the components of the curvature tensor in vielbein representation and F_{em} and F_Z correspond to the standard field tensors. From this expression one can deduce

$$J = \frac{e}{3\hbar}F_{em} + sin^2(\theta_W)\frac{g_Z}{6\hbar}F_Z \ . \quad (4.5)$$

3. The weak duality condition when integrated over X^2 implies

$$\begin{aligned} \frac{e^2}{3\hbar}Q_{em} + \frac{g_Z^2 p}{6}Q_{Z,V} &= K\oint J = Kn \ , \\ Q_{Z,V} &= \frac{I_V^3}{2} - Q_{em} \ , \ p = sin^2(\theta_W) \ . \end{aligned} \quad (4.6)$$

Here the vectorial part of the Z^0 charge rather than as full Z^0 charge $Q_Z = I_L^3 + sin^2(\theta_W)Q_{em}$ appears. The reason is that only the vectorial isospin is same for left and right handed components of fermion which are in general mixed for the massive states.

The coefficients are dimensionless and expressible in terms of the gauge coupling strengths and using $\hbar = r\hbar_0$ one can write

$$\begin{aligned} \alpha_{em}Q_{em} &+ p\frac{\alpha_Z}{2}Q_{Z,V} = \frac{3}{4\pi} \times rnK \ , \\ \alpha_{em} &= \frac{e^2}{4\pi\hbar_0} \ , \ \alpha_Z = \frac{g_Z^2}{4\pi\hbar_0} = \frac{\alpha_{em}}{p(1-p)} \ . \end{aligned} \quad (4.7)$$

4. There is a great temptation to assume that the values of Q_{em} and Q_Z correspond to their quantized values and therefore depend on the quantum state assigned to the partonic 2-surface. The linear coupling of the modified Dirac operator to conserved charges implies correlation between the geometry of space-time sheet and quantum numbers assigned to the partonic 2-surface. The assumption of standard quantized values for Q_{em} and Q_Z would be also seen as the identification of the fine structure constants α_{em} and α_Z. This however requires weak isospin invariance.

4.1.3 The value of K from classical quantization of Kähler electric charge

The value of K can be deduced by requiring classical quantization of Kähler electric charge.

1. The condition that the flux of $F^{03} = (\hbar/g_K)J^{03}$ defining the counterpart of Kähler electric field equals to the Kähler charge g_K would give the condition $K = g_K^2/\hbar$, where g_K is Kähler coupling constant which should invariant under coupling constant evolution by quantum criticality. Within experimental uncertainties one has $\alpha_K = g_K^2/4\pi\hbar_0 = \alpha_{em} \simeq 1/137$, where α_{em} is finite structure constant in electron length scale and $\hbar_0$ is the standard value of Planck constant.

2. The quantization of Planck constants makes the condition highly non-trivial. The most general quantization of r is as rationals but there are good arguments favoring the quantization as integers corresponding to the allowance of only singular coverings of CD andn CP_2. The point is that in this case a given value of Planck constant corresponds to a finite number pages of the "Big Book". The quantization of the Planck constant implies a further quantization of K and would suggest that K scales as $1/r$ unless the spectrum of values of Q_{em} and Q_Z allowed by the quantization condition scales as r. This is quite possible and the interpretation would be that each of the r sheets of the covering carries (possibly same) elementary charge. Kind of discrete variant of a full Fermi sphere would be in question. The interpretation in terms of anyonic phases [27] supports this interpretation.

3. The identification of J as a counterpart of $eB/\hbar$ means that Kähler action and thus also Kähler function is proportional to $1/\alpha_K$ and therefore to $\hbar$. This implies that for large values of $\hbar$ Kähler coupling strength $g_K^2/4\pi$ becomes very small and large fluctuations are suppressed in the functional integral. The basic motivation for introducing the hierarchy of Planck constants was indeed that the scaling $\alpha \to \alpha/r$ allows to achieve the convergence of perturbation theory: Nature itself would solve the problems of the theoretician. This of course does not mean that the physical states would remain as such and the replacement of single particles with anyonic states in order to satisfy the condition for K would realize this concretely.

4. The condition $K = g_K^2/\hbar$ implies that the Kähler magnetic charge is always accompanied by Kähler electric charge. A more general condition would read as

$$K = n \times \frac{g_K^2}{\hbar}, n \in Z \ . \qquad (4.8)$$

This would apply in the case of cosmic strings and would allow vanishing Kähler charge possible when the partonic 2-surface has opposite fermion and antifermion numbers (for both leptons and quarks) so that Kähler electric charge should vanish. For instance, for neutrinos the vanishing of electric charge strongly suggests $n = 0$ besides the condition that abelian Z^0 flux contributing to em charge vanishes.

It took a year to realize that this value of K is natural at the Minkowskian side of the wormhole throat. At the Euclidian side much more natural condition is

$$K = \frac{1}{hbar} \ . \qquad (4.9)$$

In fact, the self-duality of CP_2 Kähler form favours this boundary condition at the Euclidian side of the wormhole throat. Also the fact that one cannot distinguish between electric and magnetic charges in Euclidian region since all charges are magnetic can be used to argue in favor of this form. The same constraint arises from the condition that the action for CP_2 type vacuum extremal has the value required by the argument leading to a prediction for gravitational constant in terms of the square of CP_2 radius and α_K the effective replacement $g_K^2 \to 1$ would spoil the argument.

The boundary condition $J_E = J_B$ for the electric and magnetic parts of Kählwer form at the Euclidian side of the wormhole throat inspires the question whether all Euclidian regions could be self-dual so that the density of Kähler action would be just the instanton density. Self-duality follows if the deformation of the metric induced by the deformation of the canonically imbedded CP_2 is such that in CP_2 coordinates for the Euclidian region the tensor $(g^{\alpha\beta}g^{\mu\nu} - g^{\alpha\nu}g^{\mu\beta})/\sqrt{g}$ remains invariant. This is certainly the case for CP_2 type vacuum extremals since by the light-likeness of M^4 projection the metric remains invariant. Also conformal scalings of the induced metric would satisfy this condition. Conformal scaling is not consistent with the degeneracy of the 4-metric at the wormhole throat. Full self-duality is indeed an un-necessarily strong condition.

4.1.4 Reduction of the quantization of Kähler electric charge to that of electromagnetic charge

The best manner to learn more is to challenge the form of the weak electric-magnetic duality based on the induced Kähler form.

1. Physically it would seem more sensible to pose the duality on electromagnetic charge rather than Kähler charge. This would replace induced Kähler form with electromagnetic field, which is a linear combination of induced K"ahler field and classical Z^0 field

$$\begin{aligned} \gamma &= 3J - sin^2\theta_W R_{03} \ , \\ Z^0 &= 2R_{03} \ . \end{aligned} \tag{4.10}$$

   Here $Z_0 = 2R_{03}$ is the appropriate component of CP_2 curvature form [43]. For a vanishing Weinberg angle the condition reduces to that for Kähler form.

2. For the Euclidian space-time regions having interpretation as lines of generalized Feynman diagrams Weinberg angle should be non-vanishing. In Minkowskian regions Weinberg angle could however vanish. If so, the condition guaranteing that electromagnetic charge of the partonic 2-surfaces equals to the above condition stating that the em charge assignable to the fermion content of the partonic 2-surfaces reduces to the classical Kähler electric flux at the Minkowskian side of the wormhole throat. One can argue that Weinberg angle must increase smoothly from a vanishing value at both sides of wormhole throat to its value in the deep interior of the Euclidian region.

3. The vanishing of the Weinberg angle in Minkowskian regions conforms with the physical intuition. Above elementary particle length scales one sees only the classical electric field reducing to the induced Kähler form and classical Z^0 fields and color gauge fields are effectively absent. Only in phases with a large value of Planck constant classical Z^0 field and other classical weak fields and color gauge field could make themselves visible. Cell membrane could be one such system [28]. This conforms with the general picture about color confinement and weak massivation.

The GRT limit of TGD suggests a further reason for why Weinberg angle should vanish in Minkowskian regions.

1. The value of the Kähler coupling strength mut be very near to the value of the fine structure constant in electron length scale and these constants can be assumed to be equal.

2. GRT limit of TGD with space-time surfaces replaced with abstract 4-geometries would naturally correspond to Einstein-Maxwell theory with cosmological constant which is non-vanishing only in Euclidian regions of space-time so that both Reissner-Nordström metric and CP_2 are allowed as simplest possible solutions of field equations [39]. The extremely small value of the observed cosmological constant needed in GRT type cosmology could be equal to the large cosmological constant associated with CP_2 metric multiplied with the 3-volume fraction of Euclidian regions.

3. Also at GRT limit quantum theory would reduce to almost topological QFT since Einstein-Maxwell action reduces to 3-D term by field equations implying the vanishing of the Maxwell current and of the curvature scalar in Minkowskian regions and curvature scalar + cosmological constant term in Euclidian regions. The weak form of electric-magnetic duality would guarantee also now the preferred extremal property and prevent the reduction to a mere topological QFT.

4. GRT limit would make sense only for a vanishing Weinberg angle in Minkowskian regions. A non-vanishing Weinberg angle would make sense in the deep interior of the Euclidian regions where the approximation as a small deformation of CP_2 makes sense.

The weak form of electric-magnetic duality has surprisingly strong implications for the basic view about quantum TGD as following considerations show.

4.2 Magnetic confinement, the short range of weak forces, and color confinement

The weak form of electric-magnetic duality has surprisingly strong implications if one combines it with some very general empirical facts such as the non-existence of magnetic monopole fields in macroscopic length scales.

4.2.1 How can one avoid macroscopic magnetic monopole fields?

Monopole fields are experimentally absent in length scales above order weak boson length scale and one should have a mechanism neutralizing the monopole charge. How electroweak interactions become short ranged in TGD framework is still a poorly understood problem. What suggests itself is the neutralization of the weak isospin above the intermediate gauge boson Compton length by neutral Higgs bosons. Could the two neutralization mechanisms be combined to single one?

1. In the case of fermions and their super partners the opposite magnetic monopole would be a wormhole throat. If the magnetically charged wormhole contact is electromagnetically neutral but has vectorial weak isospin neutralizing the weak vectorial isospin of the fermion only the electromagnetic charge of the fermion is visible on longer length scales. The distance of this wormhole throat from the fermionic one should be of the order weak boson Compton length. An interpretation as a bound state of fermion and a wormhole throat state with the quantum numbers of a neutral Higgs boson would therefore make sense. The neutralizing throat would have quantum numbers of $X_{-1/2} = \nu_L \overline{\nu}_R$ or $X_{1/2} = \overline{\nu}_L \nu_R$. $\nu_L \overline{\nu}_R$ would not be neutral Higgs boson (which should correspond to a wormhole contact) but a super-partner of left-handed neutrino obtained by adding a right handed neutrino. This mechanism would apply separately to the fermionic and anti-fermionic throats of the gauge bosons and corresponding space-time sheets and leave only electromagnetic interaction as a long ranged interaction.

2. One can of course wonder what is the situation situation for the bosonic wormhole throats feeding gauge fluxes between space-time sheets. It would seem that these wormhole throats must always appear as pairs such that for the second member of the pair monopole charges and I_V^3 cancel each other at both space-time sheets involved so that one obtains at both space-time sheets magnetic dipoles of size of weak boson Compton length. The proposed magnetic character of fundamental particles should become visible at TeV energies so that LHC might have surprises in store!

4.2.2 Magnetic confinement and color confinement

Magnetic confinement generalizes also to the case of color interactions. One can consider also the situation in which the magnetic charges of quarks (more generally, of color excited leptons and quarks) do not vanish and they form color and magnetic singles in the hadronic length scale. This would mean that magnetic charges of the state $q_{\pm 1/2} - X_{\mp 1/2}$ representing the physical quark would not vanish and magnetic confinement would accompany also color confinement. This would explain why free quarks are not observed. To how degree then quark confinement corresponds to magnetic confinement is an interesting question.

For quark and antiquark of meson the magnetic charges of quark and antiquark would be opposite and meson would correspond to a Kähler magnetic flux so that a stringy view about meson emerges. For valence quarks of baryon the vanishing of the net magnetic charge takes place provided that the magnetic net charges are $(\pm 2, \mp 1, \mp 1)$. This brings in mind the spectrum of color hyper charges coming as $(\pm 2, \mp 1, \mp 1)/3$ and one can indeed ask whether color hyper-charge correlates with the Kähler magnetic charge. The geometric picture would be three strings connected to single vertex. Amusingly, the idea that color hypercharge could be proportional to color hyper charge popped up during the first year of TGD when I had not yet discovered CP_2 and believed on $M^4 \times S^2$.

p-Adic length scale hypothesis and hierarchy of Planck constants defining a hierarchy of dark variants of particles suggest the existence of scaled up copies of QCD type physics and weak physics. For p-adically scaled up variants the mass scales would be scaled by a power of $\sqrt{2}$ in the most general case. The dark variants of the particle would have the same mass as the original one. In particular, Mersenne primes $M_k = 2^k - 1$ and Gaussian Mersennes $M_{G,k} = (1+i)^k - 1$ has been proposed to define zoomed copies of these physics. At the level of magnetic confinement this would mean hierarchy of length scales for the magnetic confinement.

One particular proposal is that the Mersenne prime M_{89} should define a scaled up variant of the ordinary hadron physics with mass scaled up roughly by a factor $2^{(107-89)/2} = 512$. The size scale of color confinement for this physics would be same as the weal length scale. It would look more natural that the weak confinement for the quarks of M_{89} physics takes place in some shorter scale and M_{61} is the first Mersenne prime to be considered. The mass scale of M_{61} weak bosons would be by a factor $2^{(89-61)/2} = 2^{14}$ higher and about 1.6×10^4 TeV. M_{89} quarks would have virtually no weak interactions but would possess color interactions with weak confinement length scale reflecting themselves as new kind of jets at collisions above TeV energies.

In the biologically especially important length scale range 10 nm -2500 nm there are as many as four Gaussian Mersennes corresponding to $M_{G,k}$, $k = 151, 157, 163, 167$. This would suggest that the existence of scaled up scales of magnetic-, weak- and color confinement. An especially interesting possibly testable prediction is the existence of magnetic monopole pairs with the size scale in this range. There are recent claims about experimental evidence for magnetic monopole pairs [98].

4.2.3 Magnetic confinement and stringy picture in TGD sense

The connection between magnetic confinement and weak confinement is rather natural if one recalls that electric-magnetic duality in super-symmetric quantum field theories means that the descriptions in terms of particles and monopoles are in some sense dual descriptions. Fermions would be replaced by string like objects defined by the magnetic flux tubes and bosons as pairs of wormhole contacts would correspond to pairs of the flux tubes. Therefore the sharp distinction between gravitons and physical particles would disappear.

The reason why gravitons are necessarily stringy objects formed by a pair of wormhole contacts is that one cannot construct spin two objects using only single fermion states at wormhole throats. Of course, also super partners of these states with higher spin obtained by adding fermions and anti-fermions at the wormhole throat but these do not give rise to graviton like states [12]. The upper and lower wormhole throat pairs would be quantum superpositions of fermion anti-fermion pairs with sum over all fermions. The reason is that otherwise one cannot realize graviton emission in terms of joining of the ends of light-like 3-surfaces together. Also now magnetic monopole charges are necessary but now there is no need to assign the entities $X_\pm$ with gravitons.

Graviton string is characterized by some p-adic length scale and one can argue that below this length scale the charges of the fermions become visible. Mersenne hypothesis suggests that some Mersenne prime is in question. One proposal is that gravitonic size scale is given by electronic Mersenne prime M_{127}. It is however difficult to test whether graviton has a structure visible below this length scale.

What happens to the generalized Feynman diagrams is an interesting question. It is not at all clear how closely they relate to ordinary Feynman diagrams. All depends on what one is ready to assume about what happens in the vertices. One could of course hope that zero energy ontology could allow some very simple description allowing perhaps to get rid of the problematic aspects of Feynman diagrams.

1. Consider first the recent view about generalized Feynman diagrams which relies zero energy ontology. A highly attractive assumption is that the particles appearing at wormhole throats are on mass shell particles. For incoming and outgoing elementary bosons and their super partners they would be positive it resp. negative energy states with parallel on mass shell momenta. For virtual bosons they the wormhole throats would have opposite sign of energy and the sum of on mass shell states would give virtual net momenta. This would make possible twistor description of virtual particles

allowing only massless particles (in 4-D sense usually and in 8-D sense in TGD framework). The notion of virtual fermion makes sense only if one assumes in the interaction region a topological condensation creating another wormhole throat having no fermionic quantum numbers.

2. The addition of the particles $X^{\pm}$ replaces generalized Feynman diagrams with the analogs of stringy diagrams with lines replaced by pairs of lines corresponding to fermion and $X_{\pm 1/2}$. The members of these pairs would correspond to 3-D light-like surfaces glued together at the vertices of generalized Feynman diagrams. The analog of 3-vertex would not be splitting of the string to form shorter strings but the replication of the entire string to form two strings with same length or fusion of two strings to single string along all their points rather than along ends to form a longer string. It is not clear whether the duality symmetry of stringy diagrams can hold true for the TGD variants of stringy diagrams.

3. How should one describe the bound state formed by the fermion and $X^{\pm}$? Should one describe the state as superposition of non-parallel on mass shell states so that the composite state would be automatically massive? The description as superposition of on mass shell states does not conform with the idea that bound state formation requires binding energy. In TGD framework the notion of negentropic entanglement has been suggested to make possible the analogs of bound states consisting of on mass shell states so that the binding energy is zero [17]. If this kind of states are in question the description of virtual states in terms of on mass shell states is not lost. Of course, one cannot exclude the possibility that there is infinite number of this kind of states serving as analogs for the excitations of string like object.

4. What happens to the states formed by fermions and $X_{\pm 1/2}$ in the internal lines of the Feynman diagram? Twistor philosophy suggests that only the higher on mass shell excitations are possible. If this picture is correct, the situation would not change in an essential manner from the earlier one.

The highly non-trivial prediction of the magnetic confinement is that elementary particles should have stringy character in electro-weak length scales and could behaving to become manifest at LHC energies. This adds one further item to the list of non-trivial predictions of TGD about physics at LHC energies [18].

5 Quantum TGD very briefly

There are two basic approaches to the construction of quantum TGD. The first approach relies on the vision of quantum physics as infinite-dimensional Kähler geometry [58] for the "world of classical worlds" (WCW) identified as the space of 3-surfaces in in certain 8-dimensional space. Essentially a generalization of the Einstein's geometrization of physics program is in question. The second vision is the identification of physics as a generalized number theory involving p-adic number fields and the fusion of real numbers and p-adic numbers to a larger structure, classical number fields, and the notion of infinite prime.

With a better resolution one can distinguish also other visions crucial for quantum TGD. Indeed, the notion of finite measurement resolution realized in terms of hyper-finite factors, TGD as almost topological quantum field theory, twistor approach, zero energy ontology, and weak form of electric-magnetic duality play a decisive role in the actual construction and interpretation of the theory. One can however argue that these visions are not so fundamental for the formulation of the theory than the first two.

5.1 Physics as infinite-dimensional geometry

It is good to start with an attempt to give overall view about what the dream about physics as infinite-dimensional geometry is. The basic vision is generalization of the Einstein's program for the geometrization of classical physics so that entire quantum physics would be geometrized. Finite-dimensional geometry is certainly not enough for this purposed but physics as infinite-dimensional geometry of what might be called world of classical worlds (WCW) -or more neutrally configuration space of 3-surfaces of some higher-dimensional imbeddign space- might make sense. The requirement that the Hermitian conjugation of quantum theories has a geometric realization forces Kähler geometry for WCW. WCW defines the fixed arena of quantum physics and physical states are identified as spinor fields in WCW. These spinor fields are classical and no second quantization is needed at this level. The justification comes from the observation that infinite-dimensional Clifford algebra [54] generated by gamma matrices allows a natural identification as fermionic oscillator algebra.

The basic challenges are following.

1. Identify WCW.
2. Provide WCW with Kähler metric and spinor structure
3. Define what spinors and spinor fields in WCW are.

There is huge variety of finite-dimensional geometries and one might think that in infinite-dimensional case one might be drowned with the multitude of possibilities. The situation is however exactly opposite. The loop spaces associated with groups have a unique Kähler geometry due to the simple condition that Riemann connection exists mathematically [72]. This condition requires that the metric possesses maximal symmetries. Thus raises the vision that infinite-dimensional Kähler geometric existence is unique once one poses the additional condition that the resulting geometry satisfies some basic constraints forced by physical considerations.

The observation about the uniqueness of loop geometries leads also to a concrete vision about what this geometry could be. Perhaps WCW could be refarded as a union of symmetric spaces [65] for which every point is equivalent with any other. This would simplify the construction of the geometry immensely and would mean a generalization of cosmological principle to infinite-D context [14] ,[46].

This still requires an answer to the question why $M^4 \times CP_2$ is so unique. Something in the structure of this space must distinguish it in a unique manner from any other candidate. The uniqueness of M^4 factor can be understood from the miraculous conformal symmetries of the light-cone boundary but in the case of CP_2 there is no obvious mathematical argument of this kind although physically CP_2 is unique [51]. The observation that $M^4 \times CP_2$ has dimension 8, the space-time surfaces have dimension 4, and partonic 2-surfaces, which are the fundamental objects by holography have dimension 2, suggests that classical number fields [60, 55, 62] are involved and one can indeed end up to the choice $M^4 \times CP_2$ from physics as generalized number theory vision by simple arguments [37] ,[47]. In particular, the choices M^8 -a subspace of complexified octonions (for octonions see [60]), which I have used to call hyper-octonions- and $M^4 \times CP_2$ can be regarded as physically equivalent: this "number theoretical compactification" is analogous to spontaneous compactification in M-theory. No dynamical compactification takes place so that $M^8 - H$ duality is a more appropriate term.

5.2 Physics as generalized number theory

Physics as a generalized number theory (for an overview about number theory see [59]) program consists of three separate threads: various p-adic physics and their fusion together with real number based physics to a larger structure [36] ,[50] , the attempt to understand basic physics in terms of classical number fields [37] ,[47] (in particular, identifying associativity condition as the basic dynamical principle), and infinite primes [35] ,[44] , whose construction is formally analogous to a repeated second quantization of an arithmetic quantum field theory. In this article a summary of the philosophical ideas behind this dream and a summary of the technical challenges and proposed means to meet them are discussed.

The construction of p-adic physics and real physics poses formidable looking technical challenges: p-adic physics should make sense both at the level of the imbedding space, the "world of classical worlds" (WCW), and space-time and these physics should allow a fusion to a larger coherent whole. This forces to generalize the notion of number by fusing reals and p-adics along rationals and common algebraic numbers. The basic problem that one encounters is definition of the definite integrals and harmonic analysis [56] in the p-adic context [20]. It turns out that the representability of WCW as a union of symmetric spaces [65] provides a universal group theoretic solution not only to the construction of the Kähler geometry of WCW but also to this problem. The p-adic counterpart of a symmetric space is obtained from its discrete invariant by replacing discrete points with p-adic variants of the continuous symmetric space. Fourier analysis [56] reduces integration to summation. If one wants to define also integrals at space-time level, one must pose additional strong constraints which effectively reduce the partonic 2-surfaces and perhaps even space-time surfaces to finite geometries and allow assign to a given partonic 2-surface a unique power of a unique p-adic prime characterizing the measurement resolution in angle variables. These integrals might make sense in the intersection of real and p-adic worlds defined by algebraic surfaces.

The dimensions of partonic 2-surface, space-time surface, and imbedding space suggest that classical number fields might be highly relevant for quantum TGD. The recent view about the connection is based on hyper-octonionic representation of the imbedding space gamma matrices, and the notions of associative and co-associative space-time regions defined as regions for which the modified gamma matrices span quaternionic or co-quaternionic plane at each point of the region. A further condition is that the tangent space at each point of space-time surface contains a preferred hyper-complex (and thus commutative) plane identifiable as the plane of non-physical polarizations so that gauge invariance has a purely number theoretic interpretation. WCW can be regarded as the space of sub-algebras of the local octonionic Clifford algebra [54] of the imbedding space defined by space-time surfaces with the property that the local sub-Clifford algebra spanned by Clifford algebra valued functions restricted at them is associative or co-associative in a given region.

The recipe for constructing infinite primes is structurally equivalent with a repeated second quantization of an arithmetic super-symmetric quantum field theory. At the lowest level one has fermionic and bosonic states labeled by finite primes and infinite primes correspond to many particle states of this theory. Also infinite primes analogous to bound states are predicted. This hierarchy of quantizations can be continued indefinitely by taking the many particle states of the previous level as elementary particles at the next level. Construction could make sense also for hyper-quaternionic and hyper-octonionic primes although non-commutativity and non-associativity pose technical challenges. One can also construct infinite number of real units as ratios of infinite integers with a precise number theoretic anatomy. The fascinating finding is that the quantum states labeled by standard model quantum numbers allow a representation as wave fuctions in the discrete space of these units. Space-time point becomes infinitely richly structured in the sense that one can associate to it a wave function in the space of real (or octonionic) units allowing to represent the WCW spinor fields. One can speak about algebraic holography or number theoretic Brahman=Atman identity and one can also say that the points of imbedding space and space-time surface are subject to a number theoretic evolution.

5.3 Questions

The experience has shown repeatedly that a correct question and identification of some weakness of existing vision is what can only lead to a genuine progress. In the following I discuss the basic questions, which have stimulated progress in the challenge of constructing WCW geometry.

5.3.1 What is WCW?

Concerning the identification of WCW I have made several guesses and the progress has been basically due to the gradual realization of various physical constraints and the fact that standard physics ontology is not enough in TGD framework.

1. The first guess was that WCW corresponds to all possible space-like 3-surfaces in $H = M^4 \times CP_2$, where M^4 denotes Minkowski space and CP_2 denotes complex projective space of two complex dimensions having also representation as coset space $SU(3)/U(2)$ (see the separate article summarizing the basic facts about CP_2 and how it codes for standard model symmetries [43] ,[48, 43]). What led to the this particular choice H was the observation that the geometry of H codes for standard model quantum numbers and that the generalization of particle from point like particle to 3-surface allows to understand also remaining quantum numbers having no obvious explanation in standard model (family replication phenomenon). What is important to notice is that Poincare symmetries act as exact symmetries of M^4 rather than space-time surface itself: this realizes the basic vision about Poincare invariant theory of gravitation. This lifting of symmetries to the level of imbedding space and the new dynamical degrees of freedom brought by the sub-manifold geometry of space-time surface are absolutely essential for entire quantum TGD and distinguish it from

general relativity and string models. There is however a problem: it is not obvious how to get cosmology.

2. The second guess was that WCW consists of space-like 3-surfaces in $H_+ = M^4_+ \times CP2$, where M^4_+ future light-cone having interpretation as Big Bang cosmology at the limit of vanishing mass density with light-cone property time identified as the cosmic time. One obtains cosmology but loses exact Poincare invariance in cosmological scales since translations lead out of future light-cone. This as such has no practical significance but due to the metric 2-dimensionality of light-cone boundary δM^4_+ the conformal symmetries of string model assignable to finite-dimensional Lie group generalize to conformal symmetries assignable to an infinite-dimensional symplectic group of $S^2 \times CP_2$ and also localized with respect to the coordinates of 3-surface. These symmetries are simply too beautiful to be important only at the moment of Big Bang and must be present also in elementary particle length scales. Note that these symmetries are present only for 4-D Minkowski space so that a partial resolution of the old conundrum about why space-time dimension is just four emerges.

3. The third guess was that the light-like 3-surfaces in H or H_+ are more attractive than space-like 3-surfaces. The reason is that the infinite-D conformal symmetries characterize also light-like 3-surfaces because they are metrically 2-dimensional. This leads to a generalization of Kac-Moody symmetries [57] of super string models with finite-dimensional Lie group replaced with the group of isometries of H. The natural identification of light-like 3-surfaces is as 3-D surfaces defining the regions at which the signature of the induced metric changes from Minkowskian $(1, -1, -1, -1)$ to Euclidian $(-1-1-1-1)$- I will refer these surfaces as throats or wormhole throats in the sequel. Light-like 3-surfaces are analogous to blackhole horizons and are static because strong gravity makes them light-like. Therefore also the dimension 4 for the space-time surface is unique.

   This identification leads also to a rather unexpected physical interpretation. Single lightlike wormhole throat carriers elementary particle quantum numbers. Fermions and their superpartners are obtained by glueing Euclidian regions (deformations of so called CP_2 type vacuum extremals of Kähhler action) to the background with Minkowskian signature. Bosons are identified as wormhole contacts with two throats carrying fermion *resp.* antifermionic quantum numbers. These can be identified as deformations of CP_2 vacuum extremals between between two parallel Minkowskian space-time sheets. One can say that bosons and their superpartners emerge. This has dramatic implications for quantum TGD [6] and QFT limit of TGD [26].

   The question is whether one obtains also a generalization of Feynman diagrams. The answer is affirmative. Light-like 3-surfaces or corresponding Euclidian regions of space-time are analogous to the lines of Feynman diagram and vertices are replaced by 2-D surface at which these surfaces glued together. One can speak about Feynman diagrams with lines thicknened to light-like 3-surfaces and vertices to 2-surfaces. The generalized Feynman diagrams are singular as 3-manifolds but the vertices are non-singular as 2-manifolds. Same applies to the corresponding space-time surfaces and space-like 3-surfaces. Therefore one can say that WCW consists of generalized Feynman diagrams-something rather different from the original identification as space-like 3-surfaces and one can wonder whether these identification could be equvalent.

4. The fourth guess was a generalization of the WCW combining the nice aspects of the identifications $H = M^4 \times CP_2$ (exact Poincare invariance) and $H = M^4_+ \times CP_2$ (Big Bang cosmology). The idea was to generalize WCW to a union of basic building bricks -causal diamonds (CDs) - which themselves are analogous to Big Bang-Big Crunch cosmologies breaking Poincare invariance, which is however regained by the allowance of union of Poincare trnsforms of the causal diamonds.

   The starting point is General Coordinate Invariance (GCI). It does not matter, which 3-D slice of the space-time surface one choose to represent physical data as long as slices are related by a diffeomorphism of the space-time surface. This condition implies holography in the sense that 3-D slices define holograms about 4-D reality.

   The question is whether one could generalize GCI in the sense that the descriptions using space-like and light-like 3-surfaces would be equivalent physically. This requires that finite-sized space-like

3-surfaces are somehow equivalent with light-like 3-surfaces. This suggests that the light-like 3-surfaces must have ends. Same must be true for the space-time surfaces and must define preferred space-like 3-surfaces just like wormhole throats do. This makes sense only if the 2-D intersections of these two kinds of 3-surfaces -call them partonic 2-surfaces- and their 4-D tangent spaces carry the information about quantum physics. A strenghening of holopraphy principle would be the outcome. The challenge is to understand, where the intersections defining the partonic 2-surfaces are located.

Zero energy ontology (ZEO) allows to meet this challenge.

(a) Assume that WCW is union of sub-WCWs identified as the space of light-like 3-surfaces assignable to $CD \times CP_2$ with given CD defined as an intersection of future and past directed lightcones of M^4. The tips of CDs have localization in M^4 and one can perform for CD both translations and Lorentz boost for CDs. Space-time surfaces inside CD define the basic building brick of WCW. Also unions of CDs allowed and the CDs belonging to the union can intersect. One can of course consider the possibility of intersections and analogy with the set theoretic realization of topology.

(b) ZEO property means that the light-like boundaries of these objects carry positive and negative energy states, whose quantum numbers are opposite. Everything can be created from vacuum and can be regarded as quantum fluctuations in the standard vocabulary of quantum field theories.

(c) Space-time surfaces inside CDs begin from the lower boundary and end to the upper boundary and in ZEO it is natural to identify space-like 3-surfaces as pairs of space-like 3-surfaces at these boundaries. Light-like 3-surfaces connect these boundaries.

(d) The generalization of GCI states that the descriptions based on space-like 3-surfaces must be equivalent with that based on light-like 3-surfaces. Therefore only the 2-D intersections of light-like and space-like 3-surfaces - partonic 2-surfaces- and their 4-D tangent spaces (4-surface is there!) matter. Effective 2-dimensionality means a strengthened form of holography but does not imply exact 2-dimensionality, which would reduce the theory to a mere string model like theory. Once these data are given, the 4-D space-time surface is fixed and is analogous to a generalization of Bohr orbit to infinite-D context. This is the first guess. The situation is actually more delicate due to the non-determinism of Kähler action motivating the interaction of the hierarchy of CDs within CDs.

In this framework one obtains cosmology: CDs represent a fractal hierarchy of big bang-big crunch cosmologies. One obtains also Poincare invariance. One can also interpret the non-conservation of gravitational energy in cosmology which is an empirical fact but in conflict with exact Poincare invariance as it is realized in positive energy ontology [39, 32]. The reason is that energy and four-momentum in zero energy ontology correspond to those assignable to the positive energy part of the zero energy state of a particular CD. The density of energy as cosmologist defines it is the statistical average for given CD: this includes the contibutions of sub-CDs. This average density is expected to depend on the size scale of CD density is should therefore change as quantum dispersion in the moduli space of CDs takes place and leads to large time scale for any fixed sub-CD.

Even more, one obtains actually quantum cosmology! There is large variety of CDs since they have position in M^4 and Lorentz transformations change their shape. The first question is whether the M^4 positions of both tips of CD can be free so that one could assign to both tips of CD momentum eigenstates with opposite signs of four-momentum. The proposal, which might look somewhat strange, is that this not the case and that the proper time distance between the tips is quantized in octaves of a fundamental time scale $T = R/c$ defined by CP_2 size R. This would explains p-adic length scale hypothesis which is behind most quantitative predictions of TGD. That the time scales assignable to the CD of elementary particles correspond to biologically important time scales [8] forces to take this hypothesis very seriously.

The interpretation for T could be as a cosmic time quantized in powers of two. Even more general quantization is proposed to take place. The relative position of the second tip with respect to the first defines a point of the proper time constant hyperboloid of the future light cone. The hypothesis is that one must replace this hyperboloid with a lattice like structure. This implies very powerful cosmological predictions finding experimental support from the quantization of redshifts for instance [32]. For quite recent further empirical support see [100].

One should not take this argument without a grain of salt. Can one really realize zero energy ontology in this framework? The geometric picture is that translations correspond to translations of CDs. Translations should be done independently for the upper and lower tip of CD if one wants to speak about zero energy states but this is not possible if the proper time distance is quantized. If the relative M^4_+ coordinate is discrete, this pessimistic conclusion is strengthened further.

The manner to get rid of problem is to assume that translations are represented by quantum operators acting on states at the light-like boundaries. This is just what standard quantum theory assumes. An alternative- purely geometric- way out of difficulty is the Kac-Moody symmetry associated with light-like 3-surfaces meaning that local M^4 translations depending on the point of partonic 2-surface are gauge symmetries. For a given translation leading out of CD this gauge symmetry allows to make a compensating transformation which allows to satisfy the constraint.

This picture is roughly the recent view about WCW. What deserves to be emphasized is that a very concrete connection with basic structures of quantum field theory emerges already at the level of basic objects of the theory and GCI implies a strong form of holography and almost stringy picture.

5.3.2 Some Why's

In the following I try to summarize the basic motivations behind quantum TGD in form of various Why's.

1. Why WCW?

   Einstein's program has been extremely successful at the level of classical physics. Fusion of general relativity and quantum theory has however failed. The generalization of Einstein's geometrization program of physics from classical physics to quantum physics gives excellent hopes about the success in this project. Infinite-dimensional geometries are highly unique and this gives hopes about fixing the physics completetely from the uniqueness of the infinite-dimensional Kähler geometric existence.

2. Why spinor structure in WCW?

   Gamma matrices defining the Clifford algebra [54] of WCW are expressible in terms of fermionic oscillator operators. This is obviously something new as compared to the view about gamma matrices as bosonic objects. There is however no deep reason denying this kind of identification. As a consequence, a geometrization of fermionic oscillator operator algebra and fermionic statistics follows as also geometrization of super-conformal symmetries [64, 57] since gamma matrices define super-generators of the algebra of WCW isometries extended to a super-algebra.

3. Why Kähler geometry?

   Geometrization of the bosonic oscillator operators in terms of WCW vector fields and fermionic oscillator operators in terms of gamma matrices spanning Clifford algebra. Gamma matrices span hyper-finite factor of type II_1 and the extremely beautiful properties of these von Neuman algebras [71] (one of the three von Neuman algebras that von Neumann suggests as possible mathematical frameworks behind quantum theory) lead to a direct connection with the basic structures of modern physics (quantum groups, non-commutative geometries,.. [75]).

   A further reason why is the finiteness of the theory.

   (a) In standard QFTs there are two kinds of divergences. Action is a local functional of fields in 4-D sense and one performs path integral over **all** 4-surfaces to construct S-matrix. Mathematically path integration is a poorly defined procedure and one obtains diverging Gaussian determinants and divergences due to the local interaction vertices. Regularization provides the manner to get rid of the infinities but makes the theory very ugly.

(b) Kähler function defining the Kähler geometry is a expected to be non-local functional of the partonic 2-surface (Kähler action for a preferred extremal having as its ends the positive and negative energy 3-surfaces). Path integral is replaced with a functional integral which is mathematically well-defined procedure and one perfoms functional integral only over the partonic 2-surfaces rather than all 4-surfaces (holography). The exponent of Kähler function defines a unique vacuum functional. The local divergences of local quantum field theories of local quantum field theories since there are no local interaction vertices. Also the divergences associated with the Gaussian determinant and metric determinant cancel since these two determinants cancel each other in the integration over WCW. As a matter fact, symmetric space property suggest a much more elegant manner to perform the functional integral by reducing it to harmonic analysis in infinite-dimensional symmetric space [11].

(c) One can imagine also the possibility of divergences in fermionic degrees of freedom but it has turned out that the generalized Feynman diagrams in ZEO are manifestly finite. Even more: it is quite possible that only finite number of these diagrams give non-vanishing contributions to the scattering amplitude. This is essentially due to the new view about virtual particles, which are identified as bound states of on mass shell states assigned with the throats of wormhole contacts so that the integration over loop momenta of virtual particles is extremely restricted [11].

4. Why infinite-dimensional symmetries?

WCW must be a union of symmetric spaces in order that the Riemann connection exists (this generalizes the finding of Freed for loop groups [72]). Since the points of symmetric spaces are metrically equivalent, the geometrization becomes tractable although the dimension is infinite. A union of symmetric spaces is required because 3-surfaces with a size of galaxy and electron cannot be metrically equivalent. Zero modes distinguish these surfaces and can be regarded as purely classical degrees of freedom whereas the degrees of freedom contributing to the WCW line element are quantum fluctuating degrees of freedom.

One immediate implication of the symmetric space property is constant curvature space property meaning that the Ricci tensor proportional to metric tensor. Infinite-dimensionality means that Ricci scalar either vanishes or is infinite. This implies vanishing of Ricci tensor and vacuum Einstein equations for WCW.

5. Why $M^4 \times CP_2$?

This choice provides an explanation for standard model quantum numbers. The conjecture is that infinite-D geometry of 3-surfaces exists only for this choice. As noticed, the dimension of space-time surfaces and M^4 fixed by the requirement of generalized conformal invariance [63] making possible to achieve symmetric space property. If $M^4 \times CP_2$ is so special, there must be a good reason for this. Number theoretical vision [37] ,[47] indeed leads to the identification of this reason. One can assign the hierarchy of dimensions associated with partonic 2-surfaces, space-time surfaces and imbedding space to classical number fields and can assign to imbedding space what might be called hyper-octonionic structure. "Hyper" comes from the fact that the tangent space of H corresponds to the subspaces of complexified octonions with octonionic imaginary units multiplied by a commuting imaginary unit. The space-time reions would be either hyper-quaternionic or co-hyper-quaternionic so that associativity/co-associativity would become the basic dynamical principle at the level of space-time dynamics. Whether this dynamical principle is equivalent with the preferred extremal property of Kähler action remains an open conjecture.

6. Why zero energy ontology and why causal diamonds?

The consistency between Poincare invariance and GRT requires ZEO. In positive energy ontology only one of the infinite number of classical solutions is realized and partially fixed by the values of conserved quantum numbers so that the theory becomes obsolote. Even in quantum theory conservation laws mean that only those solutions of field equations with the quantum numbers of the initial state of the Universe are interesting and one faces the problem of understanding what the the initial state of the universe was. In ZEO these problems disappear. Everything is creatable from vacuum: if the physical state is mathematically realizable it is in principle reachable by a sequence of quantum jumps. There are no physically non-reachable entities in the theory. Zero energy ontology leads also to a fusion of thermodynamics with quantum theory. Zero energy states ae defined as entangled states of positive and negative energy states and entanglement coefficients define what I call M-matrix identified as "complex square root" of density matrix expressible as a product of diagonal real and positive density matrix and unitary S-matrix [6].

There are several good reasons why for causal diamonds. ZEO requires CDs, the generalized form of GCI and strong form of holography (light-like and space-like 3-surfaces are physically equivalent representations) require CDs, and also the view about light-like 3-surfaces as generalized Feynman diagrams requires CDs. Also the classical non-determinism of Kähler action can be understood using the hierarchy CDs and the addition of CDs inside CDs to obtain a fractal hierarchy of them provides an elegant manner to undersand radiative corrections and coupling constant evolution in TGD framework.

A strong physical argument in favor of CDs is the finding that the quantized proper time distance between the tips of CD fixed to be an octave of a fundamental time scale defined by CP_2 happens to define fundamental biological time scale for electron, u quark and d quark [8] : there would be a deep connection between elementary particle physics and living matter leading to testable predictions.

5.4 Modified Dirac action

The construction of the spinor structure for the world of classical worlds (WCW) leads to the vision that second quantized modified Dirac equation codes for the entire quantum TGD. Among other things this would mean that Dirac determinant would define the vacuum functional of the theory having interpretation as the exponent of Kähler function of WCW and Kähler function would reduce to Kähler action for a preferred extremal of Kähler action. In the following the recent view about the modified Dirac action are explained in more detail.

5.4.1 Identification of the modified Dirac action

The modified Dirac action action involves several terms. The first one is 4-dimensional assignable to Kähler action. Second term is instanton term reducible to an expression restricted to wormhole throats or any light-like 3-surfaces parallel to them in the slicing of space-time surface by light-like 3-surfaces. The third term is assignable to Chern-Simons term and has interpretation as a measurement interaction term linear in Cartan algebra of the isometry group of the imbedding space in order to obtain stringy propagators and also to realize coupling between the quantum numbers associated with super-conformal representations and space-time geometry required by quantum classical correspondence.

This means that 3-D light-like wormhole throats carry induced spinor field which can be regarded as independent degrees of freedom having the spinor fields at partonic 2-surfaces as sources and acting as 3-D sources for the 4-D induced spinor field. The most general measurement interaction would involve the corresponding coupling also for Kähler action but is not physically motivated. There are good arguments in favor of Chern-Simons Dirac action and corresponding measurement interaction.

1. A correlation between 4-D geometry of space-time sheet and quantum numbers is achieved by the identification of exponent of Kähler function as Dirac determinant making possible the entanglement of classical degrees of freedom in the interior of space-time sheet with quantum numbers.

2. Cartan algebra plays a key role not only at quantum level but also at the level of space-time geometry since quantum critical conserved currents vanish for Cartan algebra of isometries and the measurement interaction terms giving rise to conserved currents are possible only for Cartan algebras. Furthermore, modified Dirac equation makes sense only for eigen states of Cartan algebra generators. The hierarchy of Planck constants realized in terms of the book like structure of the generalized imbedding space assigns to each CD (causal diamond) preferred Cartan algebra: in case of Poincare algebra there are two of them corresponding to linear and cylindrical M^4 coordinates.

3. Quantum holography and dimensional reduction hierarchy in which partonic 2-surface defined fermionic sources for 3-D fermionic fields at light-like 3-surfaces Y_l^3 in turn defining fermionic sources for 4-D spinors find an elegant realization. Effective 2-dimensionality is achieved if the replacement of light-like wormhole throat X_l^3 with light-like 3-surface Y_l^3 "parallel" with it in the definition of Dirac determinant corresponds to the $U(1)$ gauge transformation $K \rightarrow K + f + \overline{f}$ for Kähler function of WCW so that WCW Kähler metric is not affected. Here f is holomorphic function of WCW ("world of classical worlds") complex coordinates and arbitrary function of zero mode coordinates.

4. An elegant description of the interaction between super-conformal representations realized at partonic 2-surfaces and dynamics of space-time surfaces is achieved since the values of Cartan charges are feeded to the 3-D Dirac equation which also receives mass term at the same time. Almost topological QFT at wormhole throats results at the limit when four-momenta vanish: this is in accordance with the original vision about TGD as almost topological QFT.

5. A detailed view about the physical role of quantum criticality results. Quantum criticality fixes the values of Kähler coupling strength as the analog of critical temperature. Quantum criticality implies that second variation of Kähler action vanishes for critical deformations and the existence of conserved current except in the case of Cartan algebra of isometries. Quantum criticality allows to fix the values of couplings appearing in the measurement interaction by using the condition $K \rightarrow K + f + \overline{f}$. p-Adic coupling constant evolution can be understood also and corresponds to scale hierarchy for the sizes of causal diamonds (CDs).

6. The inclusion of imaginary instanton term to the definition of the modified gamma matrices is not consistent with the conjugation of the induced spinor fields. Measurement interaction can be however assigned to both Kähler action and its instanton term. CP breaking, irreversibility and the space-time description of dissipation are closely related and the CP and T oddness of the instanton part of the measurement interaction term could provide first level description for dissipative effects. It must be however emphasized that the mere addition of instanton term to Kähler function could be enough.

7. A radically new view about matter antimatter asymmetry based on zero energy ontology emerges and one could understand the experimental absence of antimatter as being due to the fact antimatter corresponds to negative energy states. The identification of bosons as wormhole contacts is the only possible option in this framework.

8. Almost stringy propagators and a consistency with the identification of wormhole throats as lines of generalized Feynman diagrams is achieved. The notion of bosonic emergence leads to a long sought general master formula for the M-matrix elements. The counterpart for fermionic loop defining bosonic inverse propagator at QFT limit is wormhole contact with fermion and cutoffs in mass squared and hyperbolic angle for loop momenta of fermion and antifermion in the rest system of emitting boson have precise geometric counterpart.

5.4.2 Hyper-quaternionicity and quantum criticality

The conjecture that quantum critical space-time surfaces are hyper-quaternionic in the sense that the modified gamma matrices span a quaternionic subspace of complexified octonions at each point of the space-time surface is consistent with what is known about preferred extremals. The condition that both the modified gamma matrices and spinors are quaternionic at each point of the space-time surface leads to a precise ansatz for the general solution of the modified Dirac equation making sense also in the real context. The octonionic version of the modified Dirac equation is very simple since $SO(7,1)$ as vielbein group is replaced with G_2 acting as automorphisms of octonions so that only the neutral Abelian part of the classical electro-weak gauge fields survives the map.

Octonionic gamma matrices provide also a non-associative representation for the 8-D version of Pauli sigma matrices and encourage the identification of 8-D twistors as pairs of octonionic spinors conjectured to be highly relevant also for quantum TGD. Quaternionicity condition implies that octo-twistors reduce to something closely related to ordinary twistors.

5.4.3 The exponent of Kähler function as Dirac determinant for the modified Dirac action

Although quantum criticality in principle predicts the possible values of Kähler coupling strength, one might hope that there exists even more fundamental approach involving no coupling constants and predicting even quantum criticality and realizing quantum gravitational holography.

1. The Dirac determinant defined by the product of Dirac determinants associated with the light-like partonic 3-surfaces X_l^3 associated with a given space-time sheet X^4 is the simplest candidate for vacuum functional identifiable as the exponent of the Kähler function. Individual Dirac determinant is defined as the product of eigenvalues of the dimensionally reduced modified Dirac operator $D_{K,3}$ and there are good arguments suggesting that the number of eigenvalues is finite. p-Adicization requires that the eigenvalues belong to a given algebraic extension of rationals. This restriction would imply a hierarchy of physics corresponding to different extensions and could automatically imply the finiteness and algebraic number property of the Dirac determinants if only finite number of eigenvalues would contribute. The regularization would be performed by physics itself if this were the case.

2. It remains to be proven that the product of eigenvalues gives rise to the exponent of Kähler action for the preferred extremal of Kähler action. At this moment the only justification for the conjecture is that this the only thing that one can imagine.

3. A long-standing conjecture has been that the zeros of Riemann Zeta are somehow relevant for quantum TGD. Rieman zeta is however naturally replaced Dirac zeta defined by the eigenvalues of $D_{K,3}$ and closely related to Riemann Zeta since the spectrum consists essentially for the cyclotron energy spectra for localized solutions region of non-vanishing induced Kähler magnetic field and hence is in good approximation integer valued up to some cutoff integer. In zero energy ontology the Dirac zeta function associated with these eigenvalues defines "square root" of thermodynamics assuming that the energy levels of the system in question are expressible as logarithms of the eigenvalues of the modified Dirac operator defining kind of fundamental constants. Critical points correspond to approximate zeros of Dirac zeta and if Kähler function vanishes at criticality as it indeed should, the thermal energies at critical points are in first order approximation proportional to zeros themselves so that a connection between quantum criticality and approximate zeros of Dirac zeta emerges.

4. The discretization induced by the number theoretic braids reduces the world of classical worlds to effectively finite-dimensional space and configuration space Clifford algebra reduces to a finite-dimensional algebra. The interpretation is in terms of finite measurement resolution represented in terms of Jones inclusion $\mathcal{M} \subset \mathcal{N}$ of HFFs with $\mathcal{M}$ taking the role of complex numbers. The finite-D quantum Clifford algebra spanned by fermionic oscillator operators is identified as a representation for the coset space $\mathcal{N}/\mathcal{M}$ describing physical states modulo measurement resolution. In the sectors of generalized imbedding space corresponding to non-standard values of Planck constant quantum version of Clifford algebra is in question.

5.5 Three Dirac operators and their interpretation

The physical interpretation of Kähler Dirac equation is not at all straightforward. The following arguments inspired by effective 2-dimensionality suggest that the modified gamma matrices and corresponding effective metric could allow dual gravitational description of the physics associated with wormhole throats. This applies in particular to condensed matter physics.

5.5.1 Three Dirac equations

To begin with, Dirac equation appears in three forms in TGD.

1. The Dirac equation in world of classical worlds codes for the super Virasoro conditions for the super Kac-Moody and similar representations formed by the states of wormhole contacts forming the

counterpart of string like objects (throats correspond to the ends of the string. This Dirac generalizes the Dirac of 8-D imbedding space by bringing in vibrational degrees of freedom. This Dirac equation should gives as its solutions zero energy states and corresponding M-matrices generalizing S-matrix and their collection defining the unitary U-matrix whose natural application appears in consciousness theory as a coder of what Penrose calls U-process.

2. There is generalized eigenvalue equation for Chern-Simons Dirac operator at light-like wormhole throats. The generalized eigenvalue is $p^k\gamma_k$. The interpretation of pseudo-momentum p^k has been a problem but twistor Grassmannian approach suggests strongly that it can be interpreted as the counterpart of equally mysterious region momentum appearing in momentum twistor Grassmannian approach to $\mathcal{N}=4$ SYM. The pseudo-/region momentum p is quantized (this does not spoil the basics of Grasssmannian residues integral approach) and $1/p^k\gamma_k$ defines propagator in lines of generalized Feynman diagrams. The Yangian symmetry discovered generalizes in a very straightforward manner and leads alsoto the realization that TGD could allow also a twistorial formulation in terms of product $CP_3\times CP_3$ of two twistor spaces [42]. General arguments lead to a proposal for explicit form for the solutions of field equation represented identified as holomorphic 6-surfaces in this space subject to additional partial different equations for homogenenous functions of projective twistor coordinates suggesting strongly the quantal interpretation as analogs of partial waves. Therefore quantum-classical correspondence would be realize in beatiful manner.

3. There is Kähler Dirac equation in the interior of space-time. In this equation the gamma matrices are replaced with modified gamma matrices defined by the contractions of canonical momentum currents $T_k^\alpha=\partial L/\partial_\alpha h^k$ with imbedding space gamma matrices Γ_k. This replacement is required by internal consistency and by super-conformal symmetries.

Could Kähler Dirac equation provide a first principle justification for the light-hearted use of effective mass and the analog of Dirac equation in condensed manner physics? This would conform with the holographic philosophy. Partonic 2-surfaces with tangent space data and their light-like orbits would give hologram like representation of physics and the interior of space-time the 4-D representation of physics. Holography would have in the recent situation interpretation also as quantum classical correspondence between representations of physics in terms of quantized spinor fields at the light-like 3-surfaces on one hand and in terms of classical fields on the other hand.

The resulting dispersion relation for the square of the Kähler-Dirac operator assuming that induced like metric, Kähler field, etc. are very slowly varying contains quadratic and linear terms in momentum components plus a term corresponding to magnetic moment coupling. In general massive dispersion relation is obtained as is also clear from the fact that Kähler Dirac gamma matrices are combinations of M^4 and CP_2 gammas so that modified Dirac mixes different M^4 chiralities (basic signal for massivation). If one takes into account the dependence of the induced geometric quantities on space-time point dispersion relations become non-local.

5.5.2 Does energy metric provide the gravitational dual for condensed matter systems?

The modified gamma matrices define an effective metric via their anticommutators which are quadratic in components of energy momentum tensor (canonical momentum densities). This effective metric vanishes for vacuum extremals. Note that the use of modified gamma matrices guarantees among other things internal consistency and super-conformal symmetries of the theory. The physical interpretation has remained obscure hitherto although corresponding effective metric for Chern-Simons Dirac action has now a clear physical interpretation.

If the above argument is on the right track, this effective metric should have applications in condensed matter theory. In fact, energy metric has a natural interpretation in terms of effective light velocities which depend on direction of propagation. One can diagonalize the energy metric $g_e^{\alpha\beta}$ (contravariant form results from the anticommutators) and one can denote its eigenvalues by (v_0,v_i) in the case that the signature of the effective metric is $(1,-1,-1,-1)$. The 3-vector v_i/v_0has interpretation as components of effective light velocity in various directions as becomes clear by thinking the d'Alembert equation for the energy metric. This velocity field could be interpreted as that of hydrodynamic flow. The study of the extremals of Kähler action shows that if this flow is actually Beltrami flow so that the flow parameter associated with the flow lines extends to global coordinate, Kähler action reduces to a 3-D Chern-Simons action and one obtains effective topological QFT. The conserved fermion current $\overline{\Psi}\Gamma^{\alpha}_{e}\Psi$ has interpretation as incompressible hydrodynamical flow.

This would give also a nice analogy with AdS/CFT correspondence allowing to describe various kinds of physical systems in terms of higher-dimensional gravitation and black holes are introduced quite routinely to describe condensed matter systems. In TGD framework one would have an analogous situation but with 10-D space-time replaced with the interior of 4-D space-time and the boundary of AdS representing Minkowski space with the light-like 3-surfaces carrying matter. The effective gravitation would correspond to the "energy metric". One can associate with it curvature tensor, Ricci tensor and Einstein tensor using standard formulas and identify effective energy momentum tensor associated as Einstein tensor with effective Newton's constant appearing as constant of proportionality. Note however that the besides ordinary metric and "energy" metric one would have also the induced classical gauge fields having purely geometric interpretation and action would be Kähler action. This 4-D holography would provide a precise, dramatically simpler, and also a very concrete dual description. This cannot be said about model of graphene based on the introduction of 10-dimensional black holes, branes, and strings chosen in more or less ad hoc manner.

This raises questions. Does this give a general dual gravitational description of dissipative effects in terms of the "energy" metric and induced gauge fields? Does one obtain the counterparts of black holes? Do the general theorems of general relativity about the irreversible evolution leading to black holes generalize to describe analogous fate of condensed matter systems caused by dissipation? Can one describe non-equilibrium thermodynamics and self-organization in this manner?

One might argue that the incompressible Beltrami flow defined by the dynamics of the preferred extremals is dissipationless and viscosity must therefore vanish locally. The failure of complete non-determinism of Kähler action however means generation of entropy since the knowledge about the state decreases gradually. This in turn should have a phenomenological local description in terms of viscosity which characterizes the transfer of energy to shorter scales and eventually to radiation. The deeper description should be non-local and basically topological and might lead to quantization rules. For instance, one can imagine the quantization of the ratio η/s of the viscosity to entropy density as multiples of a basic unit defined by its lower bound (note that this would be analogous to Quantum Hall effect). For the first M-theory inspired derivation of the lower bound of η/s [99]. The lower bound for η/s is satisfied in good approximation by what should have been QCD plasma but found to be something different (RHIC and the first evidence for new physics from LHC [18]).

An encouraring sign comes from the observation that for so called massless extremals representing classically arbitrarily shaped pulses of radiation propagating without dissipation and dispersion along single direction the canonical momentum currents are light-like. The effective contravariant metric vanishes identically so that fermions cannot propate in the interior of massless extremals! This is of course the case also for vacuum extremals. Massless extremals are purely bosonic and represent bosonic radiation. Many-sheeted space-time decomposes into matter containing regions and radiation containing regions. Note that when wormhole contact (particle) is glued to a massless extremal, it is deformed so that CP_2 projection becomes 4-D guaranteing that the weak form of electric magnetic duality can be satisfied. Therefore massless extremals can be seen as asymptotic regions. Perhaps one could say that dissipation corresponds to a decoherence process creating space-time sheets consisting of matter and radiation. Those containing matter might be even seen as analogs blackholes as far as energy metric is considered.

5.5.3 Preferred extremals as perfect fluids

Almost perfect fluids seems to be abundant in Nature. For instance, QCD plasma was originally thought to behave like gas and therefore have a rather high viscosity to entropy density ratio $x = \eta/s$. Already RHIC found that it however behaves like almost perfect fluid with x near to the minimum predicted by AdS/CFT. The findings from LHC gave additional conform the discovery [94]. Also Fermi gas is predicted on basis of experimental observations to have at low temperatures a low viscosity roughly 5-6 times the minimal value [97]. In the following the argument that the preferred extremals of Kähler action are perfect fluids apart from the symmetry breaking to space-time sheets is developed. The argument requires some basic formulas summarized first.

The detailed definition of the viscous part of the stress energy tensor linear in velocity (oddness in velocity relates directly to second law) can be found in [96].

1. The symmetric part of the gradient of velocity gives the viscous part of the stress-energy tensor as a tensor linear in velocity. Velocity gardient decomposes to a term traceless tensor term and a term reducing to scalar.

$$\partial_i v_j + \partial_j v_i = \frac{2}{3}\partial_k v^k g_{ij} + (\partial_i v_j + \partial_j v_i - \frac{2}{3}\partial_k v^k g_{ij}) \ . \tag{5.1}$$

The viscous contribution to stress tensor is given in terms of this decomposition as

$$\sigma_{visc;ij} = \zeta\partial_k v^k g_{ij} + \eta(\partial_i v_j + \partial_j v_i - \frac{2}{3}\partial_k v^k g_{ij}) \ . \tag{5.2}$$

From $dF^i = T^{ij}S_j$ it is clear that bulk viscosity ζ gives to energy momentum tensor a pressure like contribution having interpretation in terms of friction opposing. Shear viscosity η corresponds to the traceless part of the velocity gradient often called just viscosity. This contribution to the stress tensor is non-diagonal and corresponds to momentum transfer in directions not parallel to momentum and makes the flow rotational. This termm is essential for the thermal conduction and thermal conductivity vanishes for ideal fluids.

2. The 3-D total stress tensor can be written as

$$\sigma_{ij} = \rho v_i v_j - p g_{ij} + \sigma_{visc;ij} \ . \tag{5.3}$$

The generalization to a 4-D relativistic situation is simple. One just adds terms corresponding to energy density and energy flow to obtain

$$T^{\alpha\beta} = (\rho - p)u^\alpha u^\beta + p g^{\alpha\beta} - \sigma^{\alpha\beta}_{visc} \ . \tag{5.4}$$

Here u^α denotes the local four-velocity satisfying $u^\alpha u_\alpha = 1$. The sign factors relate to the concentions in the definition of Minkowski metric ($(1,-1,-1,-1)$).

3. If the flow is such that the flow parameters associated with the flow lines integrate to a global flow parameter one can identify new time coordinate t as this flow parametger. This means a transition to a coordinate system in which fluid is at rest everywhere (comoving coordinates in cosmology) so that energy momentum tensor reduces to a diagonal term plus viscous term.

$$T^{\alpha\beta} = (\rho - p)g^{tt}\delta^\alpha_t\delta^\beta_t + p g^{\alpha\beta} - \sigma^{\alpha\beta}_{visc} \ . \tag{5.5}$$

In this case the vanishing of the viscous term means that one has perfect fluid in strong sense.

The existence of a global flow parameter means that one has

$$v_i = \Psi\partial_i\Phi \ . \tag{5.6}$$

Ψ and Φ depend on space-time point. The proportionality to a gradient of scalar Φ implies that Φ can be taken as a global time coordinate. If this condition is not satisfied, the perfect fluid property makes sense only locally.

AdS/CFT correspondence allows to deduce a lower limit for the coefficient of shear viscosity as

$$x = \frac{\eta}{s} \geq \frac{\hbar}{4\pi} . \tag{5.7}$$

This formula holds true in units in which one has $k_B = 1$ so that temperature has unit of energy.

What makes this interesting from TGD view is that in TGD framework perfect fluid property in approriately generalized sense indeed characterizes locally the preferred extremals of Kähler action defining space-time surface.

1. Kähler action is Maxwell action with U(1) gauge field replaced with the projection of CP_2 Kähler form so that the four CP_2 coordinates become the dynamical variables at QFT limit. This means enormous reduction in the number of degrees of freedom as compared to the ordinary unifications. The field equations for Kähler action define the dynamics of space-time surfaces and this dynamics reduces to conservation laws for the currents assignable to isometries. This means that the system has a hydrodynamic interpretation. This is a considerable difference to ordinary Maxwell equations. Notice however that the "topological" half of Maxwell's equations (Faraday's induction law and the statement that no non-topological magnetic are possible) is satisfied.

2. Even more, the resulting hydrodynamical system allows an interpretation in terms of a perfect fluid. The general ansatz for the preferred extremals of field equations assumes that various conserved currents are proportional to a vector field characterized by so called Beltrami property. The coefficient of proportionality depends on space-time point and the conserved current in question. Beltrami fields by definition is a vector field such that the time parameters assignable to its flow lines integrate to single global coordinate. This is highly non-trivial and one of the implications is almost topological QFT property due to the fact that Kähler action reduces to a boundary term assignable to wormhole throats which are light-like 3-surfaces at the boundaries of regions of space-time with Euclidian and Minkowskian signatures. The Euclidian regions (or wormhole throats, depends on one's tastes) define what I identify as generalized Feynman diagrams.

   Beltrami property means that if the time coordinate for a space-time sheet is chosen to be this global flow parameter, all conserved currents have only time component. In TGD framework energy momentum tensor is replaced with a collection of conserved currents assignable to various isometries and the analog of energy momentum tensor complex constructed in this manner has no counterparts of non-diagonal components. Hence the preferred extremals allow an interpretation in terms of perfect fluid without any viscosity.

This argument justifies the expectation that TGD Universe is characterized by the presence of low-viscosity fluids. Real fluids of course have a non-vanishing albeit small value of x. What causes the failure of the exact perfect fluid property?

1. Many-sheetedness of the space-time is the underlying reason. Space-time surface decomposes into finite-sized space-time sheets containing topologically condensed smaller space-time sheets containing.... Only within given sheet perfect fluid property holds true and fails at wormhole contacts and because the sheet has a finite size. As a consequence, the global flow parameter exists only in given length and time scale. At imbedding space level and in zero energy ontology the phrasing of the same would be in terms of hierarchy of causal diamonds (CDs).

2. The so called eddy viscosity is caused by eddies (vortices) of the flow. The space-time sheets glued to a larger one are indeed analogous to eddies so that the reduction of viscosity to eddy viscosity could make sense quite generally. Also the phase slippage phenomenon of super-conductivity meaning that the total phase increment of the super-conducting order parameter is reduced by a multiple of 2π in phase slippage so that the average velocity proportional to the increment of the phase along the channel divided by the length of the channel is reduced by a quantized amount.

   The standard arrangement for measuring viscosity involves a lipid layer flowing along plane. The velocity of flow with respect to the surface increases from $v = 0$ at the lower boundary to v_{upper}

at the upper boundary of the layer: this situation can be regarded as outcome of the dissipation process and prevails as long as energy is feeded into the system. The reduction of the velocity in direction orthogonal to the layer means that the flow becomes rotational during dissipation leading to this stationary situation.

This suggests that the elementary building block of dissipation process corresponds to a generation of vortex identifiable as cylindrical space-time sheets parallel to the plane of the flow and orthogonal to the velocity of flow and carrying quantized angular momentum. One expects that vortices have a spectrum labelled by quantum numbers like energy and angular momentum so that dissipation takes in discrete steps by the generation of vortices which transfer the energy and angular momentum to environment and in this manner generate the velocity gradient.

3. The quantization of the parameter x is suggestive in this framework. If entropy density and viscosity are both proportional to the density n of the eddies, the value of x would equal to the ratio of the quanta of entropy and kinematic viscosity η/n for single eddy if all eddies are identical. The quantum would be $\hbar/4\pi$ in the units used and the suggestive interpretation is in terms of the quantization of angular momentum. One of course expects a spectrum of eddies so that this simple prediction should hold true only at temperatures for which the excitation energies of vortices are above the thermal energy. The increase of the temperature would suggest that gradually more and more vortices come into play and that the ratio increases in a stepwise manner bringing in mind quantum Hall effect. In TGD Universe the value of $\hbar$ can be large in some situations so that the quantal character of dissipation could become visible even macroscopically. Whether this a situation with large $\hbar$ is encountered even in the case of QCD plasma is an interesting question.

The following poor man's argument tries to make the idea about quantization a little bit more concrete.

1. The vortices transfer momentum parallel to the plane from the flow. Therefore they must have momentum parallel to the flow given by the total cm momentum of the vortex. Before continuing some notations are needed. Let the densities of vortices and absorbed vortices be n and n_{abs} respectively. Denote by $v_{\|}$ *resp.* $v_{\perp}$ the components of cm momenta parallel to the main flow *resp.* perpendicular to the plane boundary plane. Let m be the mass of the vortex. Denote by S are parallel to the boundary plane.

2. The flow of momentum component parallel to the main flow due to the absorbed at S is

$$n_{abs} m v_{\|} v_{\perp} S \ .$$

This momentum flow must be equal to the viscous force

$$F_{visc} = \eta \frac{v_{\|}}{d} \times S \ .$$

From this one obtains

$$\eta = n_{abs} m v_{\perp} d \ .$$

If the entropy density is due to the vortices, it equals apart from possible numerical factors to

$$s = n$$

so that one has

$$\frac{\eta}{s} = m v_{\perp} d \ .$$

This quantity should have lower bound $x = \hbar/4\pi$ and perhaps even quantized in multiples of x, Angular momentum quantization suggests strongly itself as origin of the quantization.

3. Local momentum conservation requires that the comoving vortices are created in pairs with opposite momenta and thus propagating with opposite velocities $v_\perp$. Only one half of vortices is absorbed so that one has $n_{abs} = n/2$. Vortex has quantized angular momentum associated with its internal rotation. Angular momentum is generated to the flow since the vortices flowing downwards are absorbed at the boundary surface.

   Suppose that the distance of their center of mass lines parallel to plane is $D = \epsilon d$, ϵ a numerical constant not too far from unity. The vortices of the pair moving in opposite direction have same angular momentum $mv\ \ D/2$ relative to their center of mass line between them. Angular momentum conservation requires that the sum these relative angular momenta cancels the sum of the angular momenta associated with the vortices themselves. Quantization for the total angular momentum for the pair of vortices gives

$$\frac{\eta}{s} = \frac{n\hbar}{\epsilon}$$

   Quantization condition would give

$$\epsilon = 4\pi \ .$$

   One should understand why $D = 4\pi d$ - four times the circumference for the largest circle contained by the boundary layer- should define the minimal distance between the vortices of the pair. This distance is larger than the distance d for maximally sized vortices of radius $d/2$ just touching. This distance obviously increases as the thickness of the boundary layer increasess suggesting that also the radius of the vortices scales like d.

4. One cannot of course take this detailed model too literally. What is however remarkable that quantization of angular momentum and dissipation mechanism based on vortices identified as space-time sheets indeed could explain why the lower bound for the ratio η/s is so small.

5.5.4 Is the effective metric one- or two-dimensional?

The following argument suggests that the effective metric defined by the anti-commutators of the modified gamma matrices is effectively one- or two-dimensional. Effective one-dimensionality would conform with the observation that the solutions of the modified Dirac equations can be localized to one-dimensional world lines in accordance with the vision that finite measurement resolution implies discretization reducing partonic many-particle states to quantum superpositions of braids. This localization to 1-D curves occurs always at the 3-D orbits of the partonic 2-surfaces.

The argument is based on the following assumptions.

1. The modified gamma matrices for Kähler action are contractions of the canonical momentum densities T^α_k with the gamma matrices of H.

2. The strongest assumption is that the isometry currents

$$J^{A\alpha} = T^\alpha_k j^{Ak}$$

   for the preferred extremals of Kähler action are of form

$$J^{A\alpha} = \Psi^A (\nabla\Phi)^\alpha \qquad (5.8)$$

   with a common function Φ guaranteeing that the flow lines of the currents integrate to coordinate lines of single global coordinate variables (Beltrami property). Index raising is carried out by using the ordinary induced metric.

3. A weaker assumption is that one has two functions Φ_1 and Φ_2 assignable to the isometry currents of M^4 and CP_2 respectively.:

$$\begin{aligned} J_1^{A\alpha} &= \Psi_1^A(\nabla\Phi_1)^\alpha \ , \\ J_2^{A\alpha} &= \Psi_2^A(\nabla\Phi_2)^\alpha \ . \end{aligned} \quad (5.9)$$

The two functions Φ_1 and Φ_2 could define dual light-like curves spanning string world sheet. In this case one would have effective 2-dimensionality and decomposition to string world sheets [15]. Isometry invariance does not allow more that two independent scalar functions Φ_i.

Consider now the argument.

1. One can multiply both sides of this equation with j^{Ak} and sum over the index A labeling isometry currents for translations of M^4 and $SU(3)$ currents for CP_2. The tensor quantity $\sum_A j^{Ak}j^{Al}$ is invariant under isometries and must therefore satisfy

$$\sum_A \eta_{AB} j^{Ak} j^{Al} = h^{kl} \ , \quad (5.10)$$

where η_{AB} denotes the flat tangent space metric of H. In M^4 degrees of freedom this statement becomes obvious by using linear Minkowski coordinates. In the case of CP_2 one can first consider the simpler case $S^2 = CP_1 = SU(2)/U(1)$. The coset space property implies in standard complex coordinate transforming linearly under $U(1)$ that only the the isometry currents belonging to the complement of $U(1)$ in the sum contribute at the origin and the identity holds true at the origin and by the symmetric space property everywhere. Identity can be verified also directly in standard spherical coordinates. The argument generalizes to the case of $CP_2 = SU(3)/U(2)$ in an obvious manner.

2. In the most general case one obtains

$$\begin{aligned} T_1^{\alpha k} &= \sum_A \Psi_1^A j^{Ak} \times (\nabla\Phi_1)^\alpha \equiv f_1^k(\nabla\Phi_1)^\alpha \ , \\ T_2^{\alpha k} &= \sum_A \Psi_1^A j^{Ak} \times (\nabla\Phi_2)^\alpha \equiv f_2^k(\nabla\Phi_2)^\alpha \ . \end{aligned} \quad (5.11)$$

3. The effective metric given by the anti-commutator of the modified gamma matrices is in turn is given by

$$G^{\alpha\beta} = m_{kl} f_1^k f_1^l (\nabla\Phi_1)^\alpha (\nabla\Phi_1)^\beta + s_{kl} f_2^k f_2^l (\nabla\Phi_2)^\alpha (\nabla\Phi_2)^\beta \ . \quad (5.12)$$

The covariant form of the effective metric is effectively 1-dimensional for $\Phi_1 = \Phi_2$ in the sense that the only non-vanishing component of the covariant metric $G_{\alpha\beta}$ is diagonal component along the coordinate line defined by $\Phi \equiv \Phi_1 = \Phi_2$. Also the contravariant metric is effectively 1-dimensional since the index raising does not affect the rank of the tensor but depends on the other space-time coordinates. This would correspond to an effective reduction to a dynamics of point-like particles for given selection of braid points. For $\Phi_1 \neq \Phi_2$ the metric is effectively 2-dimensional and would correspond to stringy dynamics.

6 The role of twistors in quantum TGD

6.1 Could the Grassmannian program be realized in TGD framework?

In the following the TGD based modification of the approach based on zero energy ontology is discussed in some detail. It is found that pseudo-momenta are very much analogous to region momenta and the approach leading to discretization of pseudo-mass squared for virtual particles - and even the discretization of pseudo-momenta - is consistent with the Grassmannian approach in the simple case considered and allow to get rid of IR divergences. Also the possibility that the number of generalized Feynman diagrams contributing to a given scattering amplitude is finite so that the recursion formula for the scattering amplitudes would involve only a finite number of steps (maximum number of loops) is considered. One especially promising feature of the residue integral approach with discretized pseudo-momenta is that it makes sense also in the p-adic context in the simple special case discussed since residue integral reduces to momentum integral (summation) and lower-dimensional residue integral.

6.1.1 What Yangian symmetry could mean in TGD framework?

The loss of the Yangian symmetry in the integrations over the region momenta x^a ($p^a = x^{a+1} - x^a$) assigned to virtual momenta seems to be responsible for many ugly features. It is basically the source of IR divergences regulated by "moving out on the Coulomb branch theory" so that IR singularities remain the problem of the theory. This raises the question whether the loss of Yangian symmetry is the signature for the failure of QFT approach and whether the restriction of loop momentum integrations to avoid both kind of divergences might be a royal road beyond QFT. In TGD framework zero energy ontology indeed leads to to a concrete proposal based on the vision that virtual particles are something genuinely real.

The detailed picture is of course far from clear but to get an idea about what is involved one can look what kind of assumptions are needed if one wants to realize the dream that only a finite number of generalized Feynman diagrams contribute to a scattering amplitude which is Yangian invariant allowing a description using a generalization of the Grassmannian integrals.

1. Assume the bosonic emergence and its super-symmetric generalization holds true. This means that incoming and outgoing states are bound states of massless fermions assignable to wormhole throats but the fermions can opposite directions of three-momenta making them massive. Incoming and outgoing particles would consist of fermions associated with wormhole throats and would be characterized by a pair of twistors in the general situation and in general massive. This allows also string like mass squared spectrum for bound states having fermion and antifermion at the ends of the string as well as more general n-particle bound states. Hence one can speak also about the emergence of string like objects. For virtual particles the fermions would be massive and have discrete mass spectrum. Also super partners containing several collinear fermions and antifermions at a given throat are possible. Collinearity is required by the generalization of SUSY. The construction of these states bring strongly in mind the merge procedure involving the replacement $Z^{n+1} \rightarrow Z^n$.

2. The basic question is how the momentum twistor diagrams and the ordinary Feynman diagrams behind them are related to the generalized Feynman diagrams.

   (a) It is good to start from a common problem. In momentum twistor approach the relationship of region momenta to physical momenta remains somewhat mysterious. In TGD framework in turn the relationship of pseudo-momenta identified as generalized eigenvalues of the Chern-Simons Dirac operator at the lines of Feynman diagram (light-like wormhole throats) to the physical momenta has remained unclear. The identification of the pseudo-momentum as the TGD counterpart of the region momentum x looks therefore like a natural first guess.

   (b) The identification $x_{a+1} - x_a = p_a$ with p_a representing light-like physical four-momentum generalizes in obvious manner. Also the identification of the light-like momentum of the external parton as pseudo-momentum looks natural. What is important is that this does not require the identification of the pseudo-momenta propagating along internal lines of generalized Feynman diagram as actual physical momenta since pseudo-momentum just like x is fixed only apart from an overall shift. The identification allows the physical four-momenta associated with the wormhole throats to be always on mass shell and massless: if the sign of the physical energy can be also negative space-like momentum exchanges become possible.

(c) The pseudo-momenta and light-like physical massless momenta at the lines of generalized Feynman diagrams on one hand, and region momenta and the light-like momenta associated with the collinear singularities on the other hand would be in very similar mutual relationship. Partonic 2-surfaces can carry large number of collinear light-like fermions and bosons since super-symmetry is extended. Generalized Feynman diagrams would be analogous to momentum twistor diagrams if this picture is correct and one could hope that the recursion relations of the momentum twistor approach generalize.

3. The discrete mass spectrum for pseudo-momentum would in the momentum twistor approach mean the restriction of x to discrete mass shells, and the obvious reason for worry is that this might spoil the Grassmannian approach relying heavily on residue integrals and making sense also p-adically. It seems however that there is no need to worry. In [84] the $M_{6,4,l=0}(1234AB)$ the integration over twistor variables z_A and z_B using "entangled" integration contour leads to 1-loop MHV amplitude $N^pMHV, p=1$. The parametrization of the integration contour is $z_A=(\lambda_A, x\lambda_A)$, $z_B=(\lambda_B, x\lambda_B)$, where x is the M^4 coordinate representing the loop momentum. This boils down to an integral over $CP_1\times CP_1\times M^4$ [84]. The integrals over spheres CP_1s are contour integrals so that only an ordinary integral over M^4 remains. The reduction to this kind of sums occurs completely generally thanks to the recursion formula.

4. The obvious implication of the restriction of the pseudo-momenta x on massive mass shells is the absence of IR divergences and one might hope that under suitable assumptions one achieves Yangian invariance. The first question is of course whether the required restriction of x to mass shells in z_A and z_B or possibly even algebraic discretization of momenta is consistent with the Yangian invariance. This seems to be the case: the integration contour reduces to entangled integration contour in $CP_1\times CP_1$ not affected by the discretization and the resulting loop integral differs from the standard one by the discretization of masses and possibly also momenta with massless states excluded. Whether Yangian invariance poses also conditions on mass and momentum spectrum is an interesting question.

5. One can consider also the possibility that the incoming and outgoing particles - in general massive and to be distinguished from massless fermions appearing as their building blocks- have actually small masses presumably related to the IR cutoff defined by the size scale of the largest causal diamond involved. p-Adic thermodynamics could be responsible for this mass. Also the binding of the wormhole throats can give rise to a small contribution to vacuum conformal weight possibly responsible for gauge boson masses. This would imply that a given n-particle state can decay to N-particle states for which N is below some limit. The fermions inside loops would be also massive. This allows to circumvent the IR singularities due to integration over the phase space of the final states (say in Coulomb scattering).

6. The representation of the off mass shell particles as pairs of wormhole throats with non-parallel four-momenta (in the simplest case only the three-momenta need be in opposite directions) makes sense and that the particles in question are on mass shell with mass squared being proportional to inverse of a prime number as the number theoretic vision applied to the modified Dirac equation suggests. On mass shell property poses extremely powerful constraints on loops and when the number of the incoming momenta in the loop increases, the number of constraints becomes larger than the number of components of loop momentum for the generic values of the external momenta. Therefore there are excellent hopes of getting rid of UV divergences.

   A stronger assumption encouraged by the classical space-time picture about virtual particles is that the 3-momenta associated with throats of the same wormhole contact are always in same or opposite directions. Even this allows to have virtual momentum spectrum and non-trivial mass spectrum for them assuming that the three momenta are opposite.

7. The best that one can hope is that only a finite number of generalized Feynman diagrams contributes to a given reaction. This would guarantee that amplitudes belong to a finite-dimensional algebraic

extension of rational functions with rational coefficients since finite sums do not lead out from a finite algebraic extension of rationals. The first problem are self energy corrections. The assumption tht the mass non-renormalization theorems of SUSYs generalize to TGD framework would guarantee that the loops contributing to fermionic propagators (and their super-counterparts) do not affect them. Also the iteration of more complex amplitudes as analogs of ladder diagrams representing sequences of reactions $M \rightarrow M_1 \rightarrow M_2 \cdots . \rightarrow N$ such that at each M_n in the sequence can appear as on mass shell state could give a non-vanishing contribution to the scattering amplitude and would mean infinite number of Feynman diagrams unless these amplitudes vanish. If N appears as a virtual state the fermions must be however massive on mass shell fermions by the assumption about on-mass shell states and one can indeed imagine a situation in which the decay $M \rightarrow N$ is possible when N consists of states made of massless fermions is possible but not when the fermions have non-vanishing masses. This situation seems to be consistent with unitarity. The implication would be that the recursion formula for the all loop amplitudes for a given reaction would give vanishing result for some critical value of loops.

Already these assumptions give good hopes about a generalization of the momentum Grassmann approach to TGD framework. Twistors are doubled as are also the Grassmann variables and there are wave functions correlating the momenta of the the fermions associated with the opposite wormhole throats of the virtual particles as well as incoming gauge bosons which have suffered massivation. Also wave functions correlating the massless momenta at the ends of string like objects and more general many parton states are involved but do not affect the basic twistor formalism. The basic question is whether the hypothesis of unbroken Yangian symmetry could in fact imply something resembling this picture. The possibility to discretize integration contours without losing the representation as residue integral quite generally is basic prerequisite for this and should be shown to be true.

6.1.2 How to achieve Yangian invariance without trivial scattering amplitudes?

In $\mathcal{N} = 4$ SYM the Yangian invariance implies that the MHV amplitudes are constant as demonstrated in [84]. This would mean that the loop contributions to the scattering amplitudes are trivial. Therefore the breaking of the dual super-conformal invariance by IR singularities of the integrand is absolutely essential for the non-triviality of the theory. Could the situation be different in TGD framework? Could it be possible to have non-trivial scattering amplitudes which are Yangian invariants. Maybe! The following heuristic argument is formulated in the language of super-twistors.

1. The dual conformal super generators of the super-Lie algebra $U(2,2)$ acting as super vector fields reducing effectively to the general form $J = \eta_a^K \partial/\partial Z_a^J$ and the condition that they annihilate scattering amplitudes implies that they are constant as functions of twistor variables. When particles are replaced with pairs of wormhole throats the super generators are replaced by sums $J_1 + J_2$ of these generators for the two wormhole throats and it might be possible to achieve the condition

$$(J_1 + J_2)M = 0 \quad (6.1)$$

with a non-trivial dependence on the momenta if the super-components of the twistors associated with the wormhole throats are in a linear relationship. This should be the case for bound states.

2. This kind of condition indeed exists. The condition that the sum of the super-momenta expressed in terms of super-spinors λ reduces to the sum of real momenta alone is not usually posed but in the recent case it makes sense as an additional condition to the super-components of the the spinors λ associated with the bound state. This quadratic condition is exactly of the same general form as the one following from the requirement that the sum of all external momenta vanishes for scattering amplitude and reads as

$$X = \lambda_1\eta_1 + \lambda_2\eta_2 = 0 \ . \quad (6.2)$$

The action of the generators $\eta_1\partial_{\lambda_1} + \eta_2\partial_{\lambda_2}$ forming basic building blocks of the super generators on $p_1+p_2 = \lambda_1\tilde{\lambda}_1+\lambda_2\tilde{\lambda}_2$ appearing as argument in the scattering amplitude in the case of bound states gives just the quantity X, which vanishes so that one has super-symmetry. The generalization of this condition to n-parton bound state is obvious.

3. The argument does not apply to free fermions which have not suffered topological condensation and are therefore represented by CP_2 type vacuum extremal with single wormhole throat. If one accepts the weak form of electric-magnetic duality, one can circumvent this difficulty. The free fermions carry Kähler magnetic charge whereas physical fermions are accompanied by a bosonic wormhole throat carrying opposite Kähler magnetic charge and opposite electroweak isospin so that a ground state of string like object with size of order electroweak length scale is in question. In the case of quarks the Kähler magnetic charges need not be opposite since color confinement could involve Kähler magnetic confinement: electro-weak confinement holds however true also now. The above argument generalizes as such to the pairs formed by wormhole throats at the ends of string like object. One can of course imagine also more complex hybrids of these basic options but the general idea remains the same.

Note that the argument involves in an essential manner non-locality , which is indeed the defining property of the Yangian algebra and also the fact that physical particles are bound states. The massivation of the physical particles brings in the IR cutoff.

6.1.3 Number theoretical constraints on the pseudo-momenta

One can consider also further assumptions motivated by the recent view about the generalized eigenvalues of Chern-Simons Dirac operator having interpretation as pseudo-momentum. The details of this view need not of course be final.

1. Assume that the pseudo-momentum assigned to fermion lines by the modified Dirac equation [11] is the counterpart of region momentum as already explained and therefore does not directly correspond to the actual light-like four-momentum associated with partonic line of the generalized Feynman diagram. This assumption conforms with the assumption that incoming particles are built out of massless partonic fermions. It also implies that the propagators are massless propagators as required by twistorialization and Yangian generalization of super-conformal invariance.

2. Since (pseudo)-mass squared is number theoretically quantized as the length of a hyper-complex prime in preferred plane M^2 of pseudo-momentum space fermionic propagators are massless propagators with pseudo-masses restricted on discrete mass shells. Lorentz invariance suggests that M^2 cannot be common to all particles but corresponds to preferred reference frame for the virtual particle having interpretation as plane spanned by the quantization axes of energy and spin.

3. Hyper-complex primeness means also the quantization of pseudo-momentum components so that one has hyper-complex primes of form $\pm((p+1)/2, \pm(p-1)/1)$ corresponding to pseudo-mass squared $M^2 = p$ and hypercomplex primes $\pm(p, 0)$ with pseudo-mass squared $M2 = p^2$. Space-like fermionic momenta are not needed since for opposite signs of energy wormhole throats can have space-like net momenta. If space-like pseudo-momenta are allowed/needed for some reason, they could correspond to space-like hyper-complex primes $\pm((p-1)/2, \pm(p+1)/1)$ and $\pm(0,p)$ so that one would obtain also discretization of space-like mass shells also. The number theoretical mass squared is proportional to p, whereas p-adic mass squared is proportional to $1/p$. For p-adic mass calculations canonical identification $\sum x_n p^n$ maps p-adic mass squared to its real counterpart. The simplest mapping consistent with this would be $(p_0,p_1) \rightarrow (p_0,p_1)/p$. This could be assumed from the beginning in real context and would mean that the mass squared scale is proportional to $1/p$.

4. Lorentz invariance requires that the preferred coordinate system in which this holds must be analogous to the rest system of the virtual fermion and thus depends on the virtual particle. In accordance with the general vision discussed in [11] Lorentz invariance could correspond to a discrete algebraic subgroup of Lorentz group spanned by transformation matrices expressible in terms of roots of

unity. This would give a discrete version of mass shell and the preferred coordinate system would have a precise meaning also in the real context. Unless one allows algebraic extension of p-adic numbers p-adic mass shell reduces to the set of above number-theoretic momenta. For algebraic extensions of p-adic numbers the same algebraic mass shell is obtained as in real correspondence and is essential for the number theoretic universality. The interpretation for the algebraic discretization would be in terms of a finite measurement resolution. In real context this would mean discretization inducing a decomposition of the mass shell to cells. In the p-adic context each discrete point would be replaced with a p-adic continuum. As far as loop integrals are considered, this vision means that they make sense in both real and p-adic context and reduce to summations in p-adic context. This picture is discussed in detail in [11].

5. Concerning p-adicization the beautiful aspect of residue integral is that it makes sense also in p-adic context provided one can circumvent the problems related to the identification of p-adic counterpart of π requiring infinite-dimensional transcendental extension coming in powers of π. Together with the discretization of both real and virtual four-momenta this would allow to define also p-adic variants of the scattering amplitudes.

6.1.4 Could recursion formula allow interpretation in terms of zero energy ontology?

The identification of pseudo-momentum as a counterpart of region momentum suggests that generalized Feynman diagrams could be seen as a generalization of momentum twistor diagrams. Of course, the generalization from $\mathcal{N} = 4$ SYM to TGD is an enormous step in complexity and one must take all proposals in the following with a big grain of salt. For instance, the replacement of point-like particles with wormhole throats and the decomposition of gauge bosons to pairs of wormhole throats means that naive generalizations are dangerous.

With this in firmly in mind one can ask whether the recursion formula could allow interpretation in terms of zero energy states assigned to causal diamonds (CDs) containing CDs containing $\cdots$. In this framework loops could be assigned with sub-CDs.

The interpretation of the leading order singularities forming the basic building blocks of the twistor approach in zero ontology is the basic source of questions. Before posing these questions recall the basic proposal that partonic fermions are massless but opposite signs of energy are posssible for the opposite throats of wormhole contacts. Partons would be on mass shell but besides physical states identified as bound states formed from partons also more general intermediate states would be possible but restricted by momentum conservation and mass shell conditions for partons at vertices. Consider now the questions.

1. Suppose that the massivation of virtual fermions and their super partners allows only ladder diagrams in which the intermediate states contain on mass shell massless states. Should one allow this kind of ladder diagrams? Can one identify them in terms of leading order singularities? Could one construct the generalized Feynman diagrams from Yangian invariant tree diagrams associated with the hierarchy of sub-CDs and using BCFW bridges and entangled pairs of massless states having interpretation as box diagrams with on mass shell momenta at microscopic level? Could it make sense to say that scattering amplitudes are represented by tree diagrams inside CDs in various scales and that the fermionic momenta associated with throats and emerging from sub-CDs are always massless?

2. Could BCFW bridge generalizes as such and could the interpretation of BCFW bridge be in terms of a scattering in which the four on mass shell massless partonic states (partonic throats have arbitrary fermion number) are exchanged between four sub-CDs. This admittedly looks somewhat artificial.

3. Could the addition of 2-particle zero energy state responsible for addition of loop in the recursion relations and having interpretation in terms of the cutting of line carrying loop momentum correspond to an addition of sub-CD such that the 2-particle zero energy state has its positive and negative energy part on its past and future boundaries? Could this mean that one cuts a propagator line by adding CD and leaves only the portion of the line within CD. Could the reverse operation mean to the addition of zero energy "thermally entangled" states in shorter time and length scales

and assignable as a zero energy state to a sub-CD. Could one interpret the Cutkosky rule for propagator line in terms of this cutting or its reversal. Why only pairs would be needed in the recursion formula? Why not more general states? Does the recursion formula imply that they are included? Does this relate to the fact that these zero energy states have interpretation as single particle states in the positive energy ontology and that the basic building block of Feynman diagrams is single particle state? Could one regard the unitarity as an identity which states that the discontinuity of T-matrix characterizing zero energy state over cut is expressible in terms of $TT^\dagger$ and T matrix is the relevant quantity?

Maybe it is again dangerous to try to draw too detailed correspondences: after all, point like particles are replaced by partonic two-surfaces in TGD framework.

4. If I have understood correctly the genuine l-loop term results from $l-1$-loop term by the addition of the zero energy pair and integration over GL(2) as a representative of loop integral reducing $n+2$ to n and calculating the added loop at the same time [84]. The integrations over the two momentum twistor variables associated with a line in twistor space defining off mass shell four-momentum and integration over the lines represent the integration over loop momentum. The reduction to $GL(2)$ integration should result from the delta functions relating the additional momenta to $GL(2)$ variables (note that GL(2) performs linear transformations in the space spanned by the twistors Z_A and Z_B and means integral over the positions of Z_A an Z_B). The resulting object is formally Yangian invariant but IR divergences along some contours of integration breaks Yangian symmetry.

   The question is what happens in TGD framework. The previous arguments suggests that the reduction of the the loop momentum integral to integrals over discrete mass shells and possibly to a sum over their discrete subsets does not spoil the reduction to contour integrals for loop integrals in the example considered in [84]. Furthermore, the replacement of mass continuum with a discrete set of mass shells should eliminate IR divergences and might allow to preserve Yangian symmetry. One can however wonder whether the loop corrections with on mass shell massless fermions are needed. If so, one would have at most finite number of loop diagrams with on mass shell fermionic momenta and one of the TGD inspired dreams already forgotten would be realized.

6.1.5 What about unitarity?

The approach of Arkani-Hamed and collaborators means that loop integral over four-momenta are replaced with residue integrals around a small sphere $p^2 = \epsilon$. This is very much reminiscent of my own proposal for a few years ago based on the idea that the condition of twistorialization forces to accept only massless virtual states [41, 26]. I of course soon gave up this proposal as too childish.

This idea seems to however make a comeback in a modified form. At this time one would have only massive and quantized pseudo-momenta located at discrete mass shells. Can this picture be consistent with unitarity?

Before trying to answer this question one must make clear what one could assume in TGD framework.

1. Physical particles are in the general case massive and consist of collinear fermions at wormhole throats. External partons at wormhole throats must be massless to allow twistorial interpretation. Therefore massive states emerge. This applies also to stringy states.

2. The simplest assumption generalizing the childish idea is that on mass shell massless states for partons appear as both virtual particles and external particles. Space-like virtual momentum exchanges are possible if the virtual particles can consist of pairs of positive and negative energy fermions at opposite wormhole throats. Hence also partons at internal lines should be massless and this raises the question about the identification of propagators.

3. Generalized eigenvalue equation for Chern-Simons Dirac operator implies that virtual elementary fermions have massive and quantized pseudo-momenta whereas external elementary fermions are massless. The massive pseudo-momentum assigned with the Dirac propagator of a parton line cannot be identified with the massless real momentum assigned with the fermionic propagator line. The region momenta introduced in Grassmannian approach are something analogous.

As already explained, this brings in mind is the identification of this pseudo momentum as the counterpart of the region momentum of momentum twistor diagrams so that the external massless fermionic momenta would be differences of the pseudo-momenta. Indeed, since region momenta are determined apart from a common shift, they need not correspond to real momenta. Same applies to pseudo-momenta and one could assume that both internal and external fermion lines carry light-like pseudo-momenta and that external pseudo-momenta are equal to real momenta.

4. This picture has natural correspondence with twistor diagrams. For instance, the region momentum appearing in BCFW bridge defining effective propagator is in general massive although the underlying Feynman diagram would contain online massless momenta. In TGD framework massless lines of Feynman graphs associated with singularities would correspond to real momenta of massless fermions at wormhole throats. Also other canonical operations for Yangian invariants involve light-like momenta at the level of Feynman diagrams and would in TGD framework have a natural identification in terms of partonic momenta. Hence partonic picture would provide a microscopic description for the lines of twistor diagrams.

Let us assume being virtual particle means only that the discretized pseudo-momentum is on shell but massive whereas all real momenta of partons are light-like, and that negative partonic energies are possible. Can one formulate Cutkosky rules for unitarity in this framework? What could the unitarity condition

$$iDisc(T - T^\dagger) = -TT^\dagger$$

mean now? In particular, are the cuts associated with mass shells of physical particles or with mass shells of pseudo-momenta? Could these two assignments be equivalent?

1. The restriction of the partons to be massless but having both signs of energy means that the spectrum of intermediate states contains more states than the external states identified as bound states of partons with the same sign of energy. Therefore the summation over intermediate states does not reduce to a mere summation over physical states but involves a summation over states formed from massless partons with both signs of energy so that also space-like momentum exchanges become possible.

2. The understanding of the unitarity conditions in terms of Cutkosky rules would require that the cuts of the loop integrands correspond to mass shells for the virtual states which are also physical states. Therefore real momenta have a definite sign and should be massless. Besides this bound state conditions guaranteeing that the mass spectrum for physical states is discrete must be assumed. With these assumptions the unitary cuts would not be assigned with the partonic light-cones but with the mass shells associated of physical particles.

3. There is however a problem. The pseudo-momenta of partons associated with the external partons are assumed to be light-like and equal to the physical momenta.

   (a) If this holds true also for the intermediate physical states appearing in the unitarity conditions, the pseudo-momenta at the cuts are light-like and cuts must be assigned with pseudo-momentum light-cones. This could bring in IR singularities and spoil Yangian symmetry. The formation of bound states could eliminate them and the size scale of the largest CD involved would bring in a natural IR cutoff as the mass scale of the lightest particle. This assumption would however force to give up the assumption that only massive pseudo-momenta appear at the lines of the generalized Feynman diagrams.

   (b) On the other hand, if pseudo-momenta are not regarded as a property of physical state and are thus allowed to be massive for the real intermediate states in Cutkosky rules, the cuts at parton level correspond to on mass shell hyperboloids and IR divergences are absent.

6.2 Could TGD alllow formulation in terms of twistors

There are many questions to be asked. There would be in-numerable questions upwelling from my very incomplete understanding of the technical issues. In the following I restrict only to the questions which relate to the relationship of TGD approach to Witten's twistor string approach [91] and M-theory like frameworks. The arguments lead to an explicit proposal how the preferred extremals of Kähler action could correspond to holomorphic 4-surfaces in $CP_3 \times CP_3$. The basic motivation for this proposal comes from the observation that Kähler action is Maxwell action for the induced Kähler form and metric. Hence Penrose's original twistorial representation for the solutions of linear Maxwell's equations could have a generalization to TGD framework.

6.2.1 $M^4 \times CP_2$ from twistor approach

The first question which comes to mind relates to the origin of the Grassmannians. Do they have some deeper interpretation in TGD context. In twistor string theory Grassmannians relate to the moduli spaces of holomorphic surfaces defined by string world sheets in twistor space. Could partonic 2-surfaces have analogous interpretation and could one assign Grassmannians to their moduli spaces? If so, one could have rather direct connection with topological QFT defining twistor strings [91] and the almost topological QFT defining TGD. There are some hints to this direction which could be of course seen as figments of a too wild imagination.

1. The geometry of CD brings strongly in mind Penrose diagram for the conformally compactified Minkowski space [61], which indeed becomes CD when its points are replaced with spheres. This would suggest the information theoretic idea about interaction between observer and externals as a map in which M^4 is mapped to its conformal compactification represented by CD. Compactification means that the light-like points at the light-like boundaries of CD are identified and the physical counterpart for this in TGD framework is conformal invariance along light-rays along the boundaries of CD. The world of conscious observer for which CD is identified as a geometric correlate would be conformally compactified M^4 (plus CP_2 or course).

2. Since the points of the conformally compactified M^4 correspond to twistor pairs [88], which are unique only apart from opposite complex scalings, it would be natural to assign twistor space to CD and represent its points as pairs of twistors. This suggest an interpretation for the basic formulas of Grassmannian approach involving integration over twistors. The incoming and outgoing massless particles could be assigned at point-like limit light-like points at the lower and upper boundaries of CD and the lifting of the points of the light-cone boundary at partonic surfaces would give rise to the description in terms of ordinary twistors. The assumption that massless collinear fermions at partonic 2-surfaces are the basic building blocks of physical particles at partonic 2-surfaces defined as many particles states involving several partonic 2-surfaces would lead naturally to momentum twistor description in which massless momenta and described by twistors and virtual momenta in terms of twistor pairs. It is important to notice that in TGD framework string like objects would emerge from these massless fermions.

3. Partonic 2-surfaces are located at the upper and lower light-like boundaries of the causal diamond (CD) and carry energies of opposite sign in zero energy ontology. Quite generally, one can assign to the point of the conformally compactified Minkowski space a twistor pair using the standard description. The pair of twistors is determined apart from $Gl(2)$ rotation. At the light-cone boundary M^4 points are are light-like so that the two spinors of the two twistors differ from each other only by a complex scaling and single twistor is enough to characterize the space-time point this degenerate situation. The components of the twistor are related by the well known twistor equation $\mu^{a'} = -ix^{aa'}\lambda_a$. One can therefore lift each point of the partonic 2-surface to single twistor determined apart from opposite complex scalings of μ and λ so that the lift of the point would be 2-sphere. In the general case one must lift the point of CD to a twistor pair. The degeneracy of the points is given by $Gl(2)$ and each point corresponds to a 2-sphere in projective twistor space.

4. The new observation is that one can understand also CP_2 factor in twistor framework. The basic observation about which I learned in [88] (giving also a nice description of basics of twistor geometry)

is that a pair (X,Y) of twistors defines a point of CD on one hand and complex 2-planes of the dual twistor space -which is nothing but CP_2- by the equations

$$X_\alpha W^\alpha = 0 \ , \quad Y_\alpha W^\alpha = 0 \ .$$

The intersection of these planes is the complex line $CP_1 = S^2$. The action of $G(2)$ on the twistor pair affects the pair of surfaces CP_2 determined by these equations since it transforms the equations to their linear combination but not the the point of conformal CD resulting as projection of the sphere. Therefore twistor pair defines both a point of M^4 and assigns with it pair of CP_2:s represented as holomorphic surfaces of the projective dual twistor space. Hence the union over twistor pairs defines $M^4 \times CP_2$ via this assignment if it is possible to choose "the other" CP_2 in a unique manner for all points of M^4. The situation is similar to the assignment of a twistor to a point in the Grassmannian diagrams forming closed polygons with light-like edges. In this case one assigns to the the "region momenta" associated with the edge the twistor at the either end of the edge. One possible interpretation is that the two CP_2:s correspond to the opposite ends of the CD. My humble hunch is that this observation might be something very deep.

Recall that the assignment of CP_2 to M^4 point works also in another direction. $M^8 - H$ duality associates with so called hyper-quaternionic 4-surface of M^8 allowing preferred hyper-complex plane at each point 4-surfaces of $M^4 \times CP_2$. The basic observation behind this duality is that the hyper-quaternionic planes (copies of M^4) with preferred choices of hyper-complex plane M^2 are parameterized by points of CP_2. One can therefore assign to a point of CP_2 a copy of M^4. Maybe these both assignments indeed belong to the core of quantum TGD. There is also an interesting analogy with Uncertainty Principle: complete localization in M^4 implies maximal uncertainty of the point in CP_2 and vice versa.

6.2.2 Does twistor string theory generalize to TGD?

With this background the key speculative questions seem to be the following ones.

1. Could one relate twistor string theory to TGD framework? Partonic 2-surfaces at the boundaries of CD are lifted to 4-D sphere bundles in twistor space. Could they serve as a 4-D counterpart for Witten's holomorphic twistor strings assigned to point like particles? Could these surfaces be actually lifts of the holomorphic curves of twistor space replaced with the product $CP_3 \times CP_2$ to 4-D sphere bundles? If I have understood correctly, the Grassmannians $G(n,k)$ can be assigned to the moduli spaces of these holomorphic curves characterized by the degree of the polynomial expressible in terms of genus, number of negative helicity gluons, and the number of loops for twistor diagram.

   Could one interpret $G(n,k)$ as a moduli space for the δCD projections of n partonic 2-surfaces to which k negative helicity gluons and $n-k$ positive helicity gluons are assigned (or something more complex when one considers more general particle states)? Could quantum numbers be mapped to integer valued algebraic invariants? IF so, there would be a correlation between the geometry of the partonic 2-surface and quantum numbers in accordance with quantum classical correspondence.

2. Could one understand light-like orbits of partonic 2-surfaces and space-time surfaces in terms of twistors? To each point of the 2-surface one can assign a 2-sphere in twistor space CP_3 and CP_2 in its dual. These CP_2s can be identified. One should be able to assign to each sphere S^2 at least one point of corresponding CP_2s associated with its points in the dual twistor space and identified as single CP_2 union of CP_2:s in the dual twistor space a point of CP_2 or even several of them. One should be also able to continue this correspondence so that it applies to the light-like orbit of the partonic 2-surface and to the space-time surface defining a preferred extremal of Kähler action. For space-time sheets representable as graph of a map $M^4 \to CP_2$ locally one should select from a CP_2 assigned with a particular point of the space-time sheet a unique point of corresponding CP_2 in a manner consistent with field equations. For surfaces with lower dimensional M^4 projection one must assign a continuum of points of CP_2 to a given point of M^4. What kind equations-could allow to realize this assignment? Holomorphy is strongly favored also by the number theoretic considerations since in this case one has hopes of performing integrals using residue calculus.

(a) Could two holomorphic equations in $CP_3 \times CP_2$ defining 6-D surfaces as sphere bundles over $M^4 \times CP_2$ characterize the preferred extremals of Kähler action? Could partonic 2-surfaces be obtained by posing an additional holomorphic equation reducing twistors to null twistors and thus projecting to the boundaries of CD? A philosophical justification for this conjecture comes from effective 2-dimensionality stating that partonic 2-surfaces plus their 4-D tangent space data code for physics. That the dynamics would reduce to holomorphy would be an extremely beautiful result. Of course this is only an additional item in the list of general conjectures about the classical dynamics for the preferred extremals of Kähler action.

(b) One could also work in $CP_3 \times CP_3$. The first CP_3 would represent twistors endowed with a metric conformally equivalent to that of $M^{2,4}$ and having the covering of $SU(2,2)$ of $SO(2,4)$ as isometries. The second CP_3 defining its dual would have a metric consistent with the Calabi-Yau structure (having holonomy group $SU(3)$). Also the induced metric for canonically imbedded CP_2s should be the standard metric of CP_2 having $SU(3)$ as its isometries. In this situation the linear equations assigning to M^4 points twistor pairs and $CP_2 \subset CP_3$ as a complex plane would hold always true. Besides this two holomorphic equations coding for the dynamics would be needed.

(c) The issues related to the induced metric are important. The conformal equivalence class of M^4 metric emerges from the 5-D light-cone of $M^{2,4}$ under projective identification. The choice of a proper projective gauge would select M^4 metric locally. Twistors inherit the conformal metric with signature $(2,4)$ form the metric of 4+4 component spinors with metric having $(4,4)$ signature. One should be able to assign a conformal equivalence class of Minkowski metric with the orbits of pairs of twistors modulo $GL(2)$. The metric of conformally compactified M^4 would be obtained from this metric by dropping from the line element the contribution to the S^2 fiber associated with M^4 point.

(d) Witten related [91] the degree d of the algebraic curve describing twistor string, its genus g, the number k of negative helicity gluons, and the number l of loops by the following formula

$$d = k - 1 + l \ , \ g \leq l \ . \tag{6.3}$$

One should generalize the definition of the genus so that it applies to 6-D surfaces. For projective complex varieties of complex dimension n this definition indeed makes sense. Algebraic genus [52] is expressible in terms of the dimensions of the spaces of closed holomorphic forms known as Hodge numbers $h^{p,q}$ as

$$g = \sum (-1)^{n-k} h^{k,0} \ . \tag{6.4}$$

The first guess is that the formula of Witten generalizes by replacing genus with its algebraic counterpart . This requires that the allowed holomorphic surfaces are projective curves of twistori space, that is described in terms of homogenous polynomials of the 4+4 projective coordinates of $CP_3 \times CP_3$.

6.2.3 What is the relationship of TGD to M-theory and F-theory?

There are also questions relating to the possible relationship to M-theory and F-theory.

1. Calabi-Yau-manifolds [53, 68] are central for the compactification in super string theory and emerge from the condition that the super-symmetry breaks down to $\mathcal{N} = 1$ SUSY. The dual twistor space CP_3 with Euclidian signature of metric is a Calabi-Yau manifold [91]. Could one have in some sense two Calabi-Yaus! Twistorial CP_3 can be interpreted as a four-fold covering and conformal compactification of $M^{2,4}$. I do not know whether Calabi-Yau property has a generalization to the situation when Euclidian metric is replaced with a conformal equivalence class of flat metrics with Minkowskian signature and thus having a vanishing Ricci tensor. As far as differential forms (no

dependence on metric) are considered there should be no problems. Whether the replacement of the maximal holonomy group $SU(3)$ with its non-compact version $SU(1,2)$ makes sense is not clear to me.

2. The lift of the CD to projective twistor space would replace $CD \times CP_2$ with 10-dimensional space which inspires the familiar questions about connection between TGD and M-theory. If Calabi-Yau with a Minkowskian signature of metric makes sense then the Calabi-Yau of the standard M-theory would be replaced with its Minkowskian counterpart! Could it really be that M-theory like theory based on $CP_3 \times CP_2$ reduces to TGD in $CD \times CP_2$ if an additional symmetry mapping 2-spheres of CP_3 to points of CD is assumed? Could the formulation based on 12-D $CP_3 \times CP_3$ correspond to F-theory which also has two time-like dimensions. Of course, the additional conditions defined by the maps to M^4 and CP_2 would remove the second time-like dimension which is very difficult to justify on purely physical grounds.

3. One can actually challenge the assumption that the first CP_3 should have a conformal metric with signature $(2,4)$. Metric appears nowhere in the definition holomorphic functions and once the projections to M^4 and CP_2 are known, the metric of the space-time surface is obtained from the metric of $M^4 \times CP_2$. The previous argument for the necessity of the presence of the information about metric in the second order differential equation however suggests that the metric is needed.

4. The beginner might ask whether the 6-D 2-sphere bundles representing space-time sheets could have interpretation as Calabi-Yau manifolds. In fact, the Calabi-Yau manifolds defined as complete intersections in $CP_3 \times CP_3$ discovered by Tian and Yau are defined by three polynomials [68]. Two of them have degree 3 and depend on the coordinates of single CP_3 only whereas the third is bilinear in the coordinates of the CP_3:s. Obviously the number of these manifolds is quite too small (taking into account scaling the space defined by the coefficients is 6-dimensional). All these manifolds are deformation equivalent. These manifolds have Euler characteristic $\chi = \pm 18$ and a non-trivial fundamental group. By dividing this manifold by Z_3 one obtains $\chi = \pm 6$, which guarantees that the number of fermion generations is three in heterotic string theory. This manifold was the first one proposed to give rise to three generations and $\mathcal{N} = 1$ SUSY.

6.2.4 What could the field equations be in twistorial formulation?

The fascinating question is whether one can identify the equations determining the 3-D complex surfaces of $CP_3 \times CP_3$ in turn determining the space-time surfaces.

The first thing is to clarify in detail how space-time $M^4 \times CP_2$ results from $CP_3 \times CP_3$. Each point $CP_3 \times CP_3$ define a line in third CP_3 having interpretation as a point of conformally compactified M^4 obtained by sphere bundle projection. Each point of either CP_3 in turn defines CP_2 in in fourth CP_3 as a 2-plane. Therefore one has $(CP_3 \times CP_3) \times (CP_3 \times CP_3)$ but one can reduce the consideration to $CP_3 \times CP_3$ fixing $M^4 \times CP_2$. In the generic situation 6-D surface in 12-D $CP_3 \times CP_3$ defines 4-D surface in the dual $CP_3 \times CP_3$ and its sphere bundle projection defines a 4-D surface in $M^4 \times CP_2$.

1. The vanishing of three holomorphic functions f^i would characterize 3-D holomorphic surfaces of 6-D $CP_3 \times CP_3$. These are determined by three real functions of three real arguments just like a holomorphic function of single variable is dictated by its values on a one-dimensional curve of complex plane. This conforms with the idea that initial data are given at 3-D surface. Note that either the first or second CP_3 can determine the CP_2 image of the holomorphic 3-surface unless one assumes that the holomorphic functions are symmetric under the exchange of the coordinates of the two CP_3s. If symmetry is not assumed one has some kind of duality.

2. Effective 2-dimensionality means that 2-D partonic surfaces plus 4-D tangent space data are enough. This suggests that the 2 holomorphic functions determining the dynamics satisfy some second order differential equation with respect to their three complex arguments: the value of the function and its derivative would correspond to the initial values of the imbedding space coordinates and their normal derivatives at partonic 2-surface. Since the effective 2-dimensionality brings in dependence on the induced metric of the space-time surface, this equation should contain information about the induced metric.

3. The no-where vanishing holomorphic 3-form Ω, which can be regarded as a "complex square root" of volume form characterizes 6-D Calabi-Yau manifold [53, 68] , indeed contains this information albeit in a rather implicit manner but in spirit with TGD as almost topological QFT philosophy. Both CP_3:s are characterized by this kind of 3-form if Calabi-Yau with $(2,4)$ signature makes sense.

4. The simplest second order- and one might hope holomorphic- differential equation that one can imagine with these ingredients is of the form

$$\Omega_1^{i_1 j_1 k_1}\Omega_2^{i_2 j_2 k_2}\partial_{i_1 i_2}f^1\partial_{j_1 j_2}f^2\partial_{k_1 k_2}f^3 = 0 \ , \ \partial_{ij} \equiv \partial_i\partial_j \ . \tag{6.5}$$

   Since Ω_i is by its antisymmetry equal to $\Omega_i^{123}\epsilon^{ijk}$, one can divide Ω^{123}:s away from the equation so that one indeed obtains holomorphic solutions. Note also that one can replace ordinary derivatives in the equation with covariant derivatives without any effect so that the equations are general coordinate invariant.

   One can consider more complex equations obtained by taking instead of (f^1, f^2, f^3) arbitrary combinations (f^i, f^j, f^k) which results uniquely if one assumes anti-symmetrization in the labels $(1,2,3)$. In the sequel only this equation is considered.

5. The metric disappears completely from the equations and skeptic could argue that this is inconsistent with the fact that it appears in the equations defining the weak form of electric-magnetic duality as a Lagrange multiplier term in Chern-Simons action. Optimist would respond that the representation of the 6-surfaces as intersections of three hyper-surfaces is different from the representation as imbedding maps $X^4 \rightarrow H$ used in the usual formulation so that the argument does not bite, and continue by saying that the metric emerges in any case when one endows space-time with the induced metric given by projection to M^4.

6. These equations allow infinite families of obvious solutions. For instance, when some f^i depends on the coordinates of either CP_3 only, the equations are identically satisfied. As a special case one obtains solutions for which $f^1 = Z \cdot W$ and $(f^2, f^3) = (f^2(Z), f^3(W))$ This family contains also the Calabi-Yau manifold found by Yau and Tian, whose factor space was proposed as the first candidate for a compactification consistent with three fermion families.

7. One might hope that an infinite non-obvious solution family could be obtained from the ansatz expressible as products of exponential functions of Z and W. Exponentials are not consistent with the assumption that the functions f_i are homogenous polynomials of finite degree in projective coordinates so that the following argument is only for the purpose for learning something about the basic character of the equations.

$$\begin{aligned} & f^1 = E_{a_1,a_2,a_3}(Z)E_{\hat{a}_1,\hat{a}_2,\hat{a}_3}(W) \ , \qquad f^2 = E_{b_1,b_2,b_3}(Z)E_{\hat{b}_1,\hat{b}_2,\hat{b}_3}(W) \ , \\ & f^3 = E_{c_1,c_2,c_3}(Z)E_{\hat{c}_1,\hat{c}_2,\hat{c}_3}(W) \ , \\ & E_{a,b,c}(Z) = exp(az_1)exp(bz_2)exp(cz_3) \ . \end{aligned} \tag{6.6}$$

   The parameters a, b, c, and $\hat{a}, \hat{b}, \hat{c}$ can be arbitrary real numbers in real context. By the basic properties of exponential functions the field equations are algebraic. The conditions reduce to the vanishing of the products of determinants $det(a, b, c)$ and $det(\hat{a}, \hat{b}, \hat{c})$ so that the vanishing of either determinant is enough. Therefore the dependence can be arbitrary either in Z coordinates or in W coordinates. Linear superposition holds for the modes for which determinant vanishes which means that the vectors (a, b, c) or $(\hat{a}, \hat{b}, \hat{c})$ are in the same plane.

   Unfortunately, the vanishing conditions reduce to the conditions $f^i(W) = 0$ for case a) and to $f^i(Z) = 0$ for case b) so that the conditions are equivalent with those obtained by putting the "wave vector" to zero and the solutions reduce to obvious ones. The lesson is that the equations do not commute with the multiplication of the functions f^i with nowhere vanishing functions of W and Z. The equation selects a particular representation of the surfaces and one might argue that this should not be the case unless the hyper-surfaces defined by f^i contain some physically relevant information. One could consider the possibility that the vanishing conditions are replaced with conditions $f^i = c_i$ with $f^i(0) = 0$ in which case the information would be coded by a family of space-time surfaces obtained by varying c_i.

One might criticize the above equations since they are formulated directly in the product $CP_3 \times CP_3$ of projective twistor by choosing a specific projective gauge by putting$z^4 = 1, w^4 = 1$. The manifestly projectively invariant formulation for the equations is in full twistor space so that 12-D space would be replaced with 16-D space. In this case one would have 4-D complex permutation symbol giving for these spaces Calabi-Yau structure with flat metric. The product of functions $f = z^4 = constant$ and $g = w^4 = constant$ would define the fourth function $f_4 = fg$ fixing the projective gauge

$$\epsilon^{i_1 j_1 k_1 l_1} \epsilon^{i_2 j_2 k_2 l_2} \partial_{i_1 i_2} f^1 \partial_{j_1 j_2} f^2 \partial_{k_1 k_2} f^3 \partial_{l_1 l_2} f^4 = 0 \ , \ \partial_{ij} \equiv \partial_i \partial_j \ . \quad (6.7)$$

The functions f^i are homogenous polynomials of their twistor arguments to guarantee projective invariance. These equations are projectively invariant and reduce to the above form which means also loss of homogenous polynomial property. The undesirable feature is the loss of manifest projective invariance by the fixing of the projective gauge.

A more attractive ansatz is based on the idea that one must have one equation for each f^i to minimize the non-determinism of the equations obvious from the fact that there is single equation in 3-D lattice for three dynamical variables. The quartets (f^1, f^2, f^3, f^i), $i = 1, 2, 3$ would define a possible minimally non-linear generalization of the equation

$$\epsilon^{i_1 j_1 k_1 l_1} \epsilon^{i_2 j_2 k_2 l_2} \partial_{i_1 i_2} f^1 \partial_{j_1 j_2} f^2 \partial_{k_1 k_2} f^m \partial_{l_1 l_2} f^4 = 0 \ , \ \partial_{ij} \equiv \partial_i \partial_j \ , \ m = 1, 2, 3 \ . \quad (6.8)$$

Note that the functions are homogenous polynomials of their arguments and analogous to spherical harmonics suggesting that they can allow a nice interpretation in terms of quantum classical correspondence.

The minimal non-linearity of the equations also conforms with the non-linearity of the field equations associated with Kähler action. Note that also in this case one can solve the equations by diagonalizing the dynamical coefficient matrix associated with the quadratic term and by identifying the eigen-vectors of zero eigen values. One could consider also more complicated strongly non-linear ansätze such as (f^i, f^i, f^i, f^i), $i = 1, 2, 3$, but these do not seem plausible.

1. The explicit form of the equations using Taylor series expansion for multi-linear case

In this section the equations associated with (f_1, f_2, f_3) ansatz are discussed in order to obtain a perspective about the general structure of the equations by using simpler (multilinearity) albeit probably non-realistic case as starting point. This experience can be applied directly to the (f^1, f^2, f^3, f^i) ansatz, which is quadratic in f^i.

The explicit form of the equations is obtained as infinite number of conditions relating the coefficients of the Taylor series of f^1 and f^2. The treatment of the two variants for the equations is essentially identical and in the following only the manifestly projectively invariant form will be considered.

1. One can express the Taylor series as

$$\begin{aligned}
f^1(Z, W) &= \sum_{m,n} C_{m,n} M_m(Z) M_n(W) \ , \\
f^2(Z, W) &= \sum_{m,n} D_{m,n} M_m(Z) M_n(W) \ , \\
f^3(Z, W) &= \sum_{m,n} E_{m,n} M_m(Z) M_n(W) \ , \\
M_{m \equiv (m_1, m_2, m_3)}(Z) &= z_1^{m_1} z_2^{m_2} z_3^{m_3} \ .
\end{aligned} \quad (6.9)$$

2. The application of derivatives to the functions reduces to a simple algebraic operation

$$\partial_{ij}(M_m(Z)M_n(W)) = m_i n_j M_{m_1-e_i}(Z)M_{n-e_j}(W) . \quad (6.10)$$

Here e_i denotes i:th unit vector.

3. Using the product rule $M_m M_n = M_{m+n}$ one obtains

$$\begin{aligned} & \partial_{ij}(M_m(Z)M_n(W))\partial_{rs}(M_k(Z)M_l(W)) \\ = & m_i n_j k_r l_s \times M_{m-e_i}(Z)M_{n-e_j}(W) \times M_{k-e_r}(Z)M_{k-e_s}(W) \\ = & m_i n_j k_r l_s \times M_{m+k-e_i-e_r}(Z) \times M_{n+l-e_j-e_l}(W) . \end{aligned} \quad (6.11)$$

4. The equations reduce to the trilinear form

$$\begin{aligned} & \sum_{m,n,k,l,r,s} C_{m,n}D_{k,l}E_{r,s}(m,k,r)(n,l,s)M_{m+k+r-E}(Z)M_{n+l+s-E}(W) = 0 , \\ E = & e_1 + e_2 + e_3 , \quad (a,b,c) = \epsilon^{ijk}a_i b_j c_c . \end{aligned} \quad (6.12)$$

Here (a,b,c) denotes the determinant defined by the three index vectors involved. By introducing the summation indices

$$(M = m + k + r - E, k, r) , \quad (N = n + l + s - E, l, s)$$

one obtains an infinite number of conditions, one for each pair (M,N). The condition for a given pair (M,N) reads as

$$\sum_{k,l,r,s} C_{M-k-r+E,N-l-s+E}D_{k,l}E_{r,s} \times (M-k-r+E,k,r)(N-l-s+E,l,s) = 0 . \quad (6.13)$$

These equations can be regarded as linear equations by regarding any matrix selected from $\{C,D,E\}$ as a vector of linear space. The existence solutions requires that the determinant associated with the tensor product of other two matrices vanishes. This matrix is dynamical. Same applies to the tensor product of any of the matrices.

5. Hyper-determinant [80] is the generalization of the notion of determinant whose vanishing tells that multilinear equations have solutions. Now the vanishing of the hyper-determinant defined for the tensor product of the three-fold tensor power of the vector space defined by the coefficients of the Taylor expansion should provide the appropriate manner to characterize the conditions for the existence of the solutions. As already seen, solutions indeed exist so that the hyper-determinant must vanish. The elements of the hyper matrix are now products of determinants for the exponents of the monomials involved. The non-locality of the Kähler function as a functional of the partonic surface leads to the argument that the field equations of TGD for vanishing n:th variations of Kähler action are multilinear and that a vanishing of a generalized hyper-determinant characterizes this [11].

6. Since the differential operators are homogenous polynomials of partial derivatives, the total degrees of $M_m(Z)$ and $M_m(W)$ defined as a sum $D = \sum m_i$ is reduced by one unit by the action of both operators ∂_{ij}. For given value of M and N only the products

$$M_m(Z)M_n(W)M_k(Z)M_r(W)M_s(Z)M_l(W)$$

for which the vector valued degrees $D_1 = m + k + r$ and $D_2 = n + l + s$ have the same value are coupled. Since the degree is reduced by the operators appearing in the equation, polynomial solutions for which f^i contain monomials labelled by vectors m_i, n_i, r_i for which the components vary in a finite range $(0, n_{max})$ look like a natural solution ansatz. All the degrees $D_i \leq D_{i,max}$ appear in the solution ansatz so that quite a large number of conditions is obtained.

What is nice is that the equation can be interpreted as a difference equation in 3-D lattice with "time direction" defined by the direction of the diagonal.

1. The counterparts of time=constant slices are the planes $n_1 + n_2 + n_3 = n$ defining outer surfaces of simplices having E as a normal vector. The difference equation does not seem to say nothing about the behavior in the transversal directions. M and N vary in the simplex planes satisfying $\sum M_i = T_1$, $\sum N_i = T_2$. It seems natural to choose $T_1 = T_2 = T$ so that Z and W dynamics corresponds to the same "time". The number of points in the $T = constant$ simplex plane increases with T which is analogous to cosmic expansion.

2. The "time evolution" with respect to T can be solved iteratively by increasing the value of $\sum M_i = N_i = T$ by one unit at each step. Suppose that the values of coefficients are known and satisfy the conditions for (m, k, r) and (n, l, s) up to the maximum value T for the sum of the components of each of these six vectors. The region of known coefficients -"past"- obviously corresponds to the interior of the simplex bounded by the plane $\sum M_i = \sum N_i = T$ having E as a normal. Let (m_{min}, n_{min}), (k_{min}, l_{min}) and (r_{min}, s_{min}) correspond to the smallest values of 3-indices for which the coefficients are non-vanishing- this could be called the moment of "Big Bang". The simplest but not necessary assumption is that these indices correspond zero vectors $(0, 0, 0)$ analogous to the tip of light-cone.

3. For given values of M and N corresponding to same value of "cosmic time" T one can separate from the formula the terms which correspond to the un-known coefficients as the sum $C_{M+E,N+E}D_{0,0}E_{0,0}+ D_{M+E,N+E}D_{0,0}C_{0,0}+E_{M+E,N+E}C_{0,0}D_{0,0}$. The remaining terms are by assumption already known. One can fix the normalization by choosing $C_{0,0} = D_{0,0} = E_{0,0} = 1$. With these assumptions the equation reduces at each point of the outer boundary of the simplex to the form

$$C_{M+E,N+E} + D_{M+E,N+E} + E_{M+E,N+E} = X$$

where X is something already known and contain only data about points in the plane $m+k+r = M$ and $n+r+s = N$. Note that these planes have one "time like direction" unlike the simplex plane so that one could speak about a discrete analog of string world sheet in 3+3+3-D lattice space defined by a 2-plane with one time-like direction.

4. For each point of the simplex plane one has equation of the above form. The equation is non-deterministic since only constrain only the sum $C_{M+E,N+E} + D_{M+E,N+E} + E_{M+E,N+E}$ at each point of the simplex plane to a plane in the complex 3-D space defined by them. Hence the number of solutions is very large. The condition that the solutions reduce to polynomials poses conditions on the coefficients since the quantities X associated with the plane $T = T_{max}$ must vanish for each point of the simplex plane in this case. In fact, projective invariance means that the functions involved are homogenous functions in projective coordinates and thus polynomials and therefore reduce to polynomials of finite degree in 3-D treatment. This obviously gives additional condition to the equations.

2. The minimally non-linear option

The simple equation just discussed should be taken with a caution since the non-determinism seems to be too large if one takes seriously the analogy with classical dynamics. By the vacuum degeneracy also the time evolution associated with Kähler action breaks determinism in the standard sense of the word. The non-determinism is however not so strong and removed completely in local sense for non-vacuum extremals. One could also try to see the non-determinism as the analog for non-deterministic time evolution by quantum jumps.

One can however consider the already mentioned possibility of increasing the number of equations so that one would have three equations corresponding to the three unknown functions f^i so that the determinism associated with each step would be reduced. The equations in question would be of the same general form but with (f^1, f^2, f^3) replaced with some some other combination.

1. In the genuinely projective situation where one can consider the (f^1, f^2, f^3, f^i), $i = 1, 2, 3$ as a unique generalization of the equation. This would make the equations quadratic in f_i and reduce the non-determinism at given step of the time evolution. The new element is that now only monomials $M_m(z)$ associated with the f^i with same degree of homegenity defined by $d = \sum m_i$ are consistent with projective invariance. Therefore the solutions are characterized by six integers $(d_{i,1}, d_{i,2})$ having interpretation as analogs of conformal weights since they correspond to eigenvalues of scaling operators. That homogenous polynomials are in question gives hopes that a generalization of Witten's approach might make sense. The indices m vary at the outer surfaces of the six 3-simplices defined by $(d_{i,1}, d_{i,2})$ and looking like tedrahedrons in 3-D space. The functions f^i are highly analogous to the homogenous functions appearing in group representations and quantum classical correspondence could be realized through the representation of the space-time surfaces in this manner.

2. The 3-determinants (a, b, c) appearing in the equations would be replaced by 4-determinants and the equations would have the same general form. One has

$$\begin{aligned} &\sum_{k,l,r,s,t,u} C_{M-k-r-t+E,N-l-s-u+E} D_{k,l} E_{r,s} C_{t,u} \times \\ &\times (M-k-r-t+E, k, r, t)(N-l-s-u+E, l, s, u) = 0 \ , \\ &E = e_1 + e_2 + e_3 + e_4 \ , \quad (a, b, c, d) = \epsilon^{ijkl} a_i b_j c_k d_l \ . \end{aligned} \tag{6.14}$$

   and its variants in which D and E appear quadratically. The values of M and N are restricted to the tedrahedrons $\sum M_i = \sum d_{k,1} + d_{1,i}$ and $\sum N_i = \sum d_{i,2} + d_{i,2}$, $i = 1, 2, 3$. Therefore the dynamics in the index space is 3-dimensional. Since the index space is in a well-defined sense dual to CP_3 as is also the CP_3 in which the solutions are represented as counterparts of 3-surfaces, one could say that the 3-dimensionality of the dynamics corresponds to the dynamics of Chern-Simons action at space-like at the ends of CD and at light-like 3-surfaces.

3. The view based on 4-D time evolution is not useful since the solutions are restricted to time=constant plane in 4-D sense. The elimination of one of the projective coordinates would lead however to the analog of the above describe time evolution. In four-D context a more appropriate form of the equations is

$$\sum_{m,n,k,l,r,s} C_{m,n} D_{k,l} E_{r,s} C_{t,u} (m, k, r, t)(n, l, s, u) M_{m+k+r-E}(Z) M_{n+l+s-E}(W) = 0 \tag{6.15}$$

   with similar equations for f^2 and f^3. If one assumes that the CP_2 image of the holomorphic 3-surface is unique (it can correspond to either CP_3) the homogenous polynomials f^i must be symmetric under the exchange of Z and W so that the matrices C, D, and E are symmetric. This is equivalent to a replacement of the product of determinants with a sum of 16 products of determinants obtained by permuting the indices of each index pair (m, n), (k, l),(r, s) and (t, u).

4. The number N_{cond} of conditions is given by the product $N_{cond} = N(d_M)N(d_N)$ of numbers of points in the two tedrahedrons defined by the total conformal weights

$$\sum M_r = d_M = \sum_k d_{k,1} + d_{i,1} \text{ and } \sum N_r = d_N = \sum_k d_{k,2} + d_{i,2} \ , \ i = 1,2,3.$$

The number N_{coeff} of coefficients is

$$N_{coeff} = \sum_k n(d_{k,1}) + \sum_k n(d_{k,2}) \ ,$$

where $n(d_{k,i})$ is the number points associated with the tedrahedron with conformal weight $d_{k,i}$. Since one has $n(d) \propto d^3$, N_{cond} scales as

$$N_{cond} \propto d_M^3 d_N^3 = (\sum_k d_{k,1} + d_{1,i})^3 \times (\sum_k d_{k,2} + d_{i,2})^3$$

whereas the number N_{coeff} of coefficients scales as

$$N_{coeff} \propto \sum_k (d_{k,1}^3 + d_{k,2}^3) \ .$$

N_{cond} is clearly much larger than N_{coeff} so the solutions are analogous to partial waves and that the reduction of the rank for the matrices involved is an essential aspect of being a solution. The reduction of the rank for the coefficient matrices should reduce the effective number of coefficients so that solutions can be found. An interesting question is whether the coefficients are rationals with a suitable normalization allowed by independent conformal scalings. An analogy for the dynamics is quantum entanglement for 3+3 systems respecting the conservation of conformal weights and quantum classical correspondence taken to extreme suggests something like this.

5. One can interpret these equations as linear equations for the coefficients of the either linear term or as quadratic equations for the non-linear term. Also in the case of quadratic term one can apply general linear methods to identify the vanishing eigen values of the matrix of the quadratic form involved and to find the zero modes as solutions. The rank of the dynamically determined multiplier matrix must be non-maximal for the solutions to exist. One can imagine that the rank changes at critical surfaces in the space of Taylor coefficients meaning a multi-furcation in the space determined by the coefficients of the polynomials. Also the degree of the polynomial can change at the critical point.

   Solutions for which either determinant vanishes for all terms present in the solution exist. This is is achieved if either the index vectors (m,l,r,t) or (n,l,s,u) in their respective parallel 3-planes are also in a 3-plane going through the origin. These solutions might seen as the analogs of vacuum extremals of Chern-Simons action for which the CP_2 projection is at most 2-D Lagrangian manifold.

   Quantum classical correspondence requires that the space-time surface carries also information about the momenta of partons. This information is quasi-continuous. Also information about zero modes should have representation in terms of the coefficients of the polynomials. Is this really possible if only products of polynomials of fixed conformal weights with strong restrictions on coefficients can be used? The counterpart for the vacuum degeneracy of Kähler action might resolve the problem. The analog for the construction of space-time surfaces as deformations of vacuum extremals would be starting from a trivial solution and adding to the building blocks of f^i some terms of same degree for which the wave vectors are not in the intersection of a 3-plane and simplex planes. The still existing "vacuum part" of the solution could carry the needed information.

6. One can take "obvious solutions" characterized by different common 3-planes for the "wave vectors" characterizing the 8 monomials $M_a(Z)$ and $M_b(W)$, $a \in \{m,k,r,t\}$ and $b \in \{n,l,s,u\}$. The coefficient matrices C, D, E, F are completely free. For the sum of these solutions the equations contain

interaction terms for which at least two "wave vectors" belong to different 3-planes so that the corresponding 4-determinants are non-vanishing. The coefficients are not anymore free. Could the "obvious solutions" have interpretation in terms of different space-time sheets interacting via wormhole contacts? Or can one equate "obvious" with "vacuum" so that interaction between different vacuum space-time sheets via wormhole contact with 3-D CP_2 projection would deform vacuum extremals to non-vacuum extremals? Quantum classical correspondence inspires the question whether the products for functions f_i associated with an obvious solution associated with a particular plane correspond to a tensor products for quantum states associated with a particular partonic 2-surface or space-time sheet.

7. Effective 2-dimensionality realized in terms of the extremals of Chern-Simons actions with Lagrange multiplier term coming from the weak form of electric magnetic duality should also have a concrete counterpart if one takes the analogy with the extremals of Kähler action seriously. The equations can be transformed to 3-D ones by the elimination of the fourth coordinate but the interpretation in terms of discrete time evolution seems to be impossible since all points are coupled. The total conformal weights of the monomials vary in the range $[0, d_{1,i}]$ and $[0, d_{2,i}]$ so that the non-vanishing coefficients are in the interior of 3-simplex. The information about the fourth coordinate is preserved being visible via the four-determinants.

8. It should be possible to relate the hierarchy with respect to conformal weights would to the geometrization of loop integrals if a generalization of twistor strings is in question. One could hope that there exists a hierarchy of solutions with levels characterized by the rank of the matrices appearing in the linear representation. There is a temptation to associate this hierarchy with the hierarchy of deformations of vacuum extremals of Kähler action forming also a hierarchy. If this is the case the obvious solutions would correspond to vacuum exremals. At each step when the rank of the matrices involved decreases the solution becomes nearer to vacuum extremal and there should exist vanishing second variation of Kähler action. This structural similarity gives hopes that the proposed ansatz might work. Also the fact that a generalization of the Penrose's twistorial description for the solutions of Maxwell's equations to the situation when Maxwell field is induced from the Kähler form of CP_2 raises hopes. One must however remember that the consistency with other proposed solution ansätze and with what is believed to be known about the preferred extremals is an enormously powerful constraint and a mathematical miracle would be required.

7 Finiteness of generalized Feynman diagrams zero energy ontology

By effective 2-dimensionality partonic 2-surfaces plus the 4-D tangent space data at them code for quantum physics. The light-like orbits of partonic 2-surfaces in turn have interpretation as analogs of Feynman diagrams which correspond to 3-surfaces defining the regions at which the signature of the induced metric changes and 4-metric becomes degenerate. One could also identify the space-like regions of the space-time surfaces (deformed CP_2 type vacuum extremals, in particular wormhole throats) as the counterparts of generalized Feynman diagrams. The regions with Minkowskian signature of the induced metric would in turn correspond to the many-sheeted version of external space-time in which the particles move. A very concrete connection between particle and space-time geometry and topology is clearly in question.

Zero energy ontology has already led to the idea of interpreting the virtual particles as pairs of positive and negative energy wormhole throats. Hitherto I have taken it as granted that ordinary Feynman diagrammatics generalizes more or less as such. It is however far from clear what really happens in the verties of the generalized Feynmann diagrams. The safest approach relies on the requirement that unitarity realized in terms of Cutkosky rules in ordinary Feynman diagrammatics allows a generalization. This requires loop diagrams. In particular, photon-photon scattering can take place only via a fermionic square loop so that it seems that loops must be present at least in the topological sense.

One must be however ready for the possibility that something unexpectedly simple might emerge. For instance, the vision about algebraic physics allows naturally only finite sums for diagrams and does not favor infinite perturbative expansions. Hence the true believer on algebraic physics might dream about finite number of diagrams for a given reaction type. For simplicity generalized Feynman diagrams without the complications brought by the magnetic confinement since by the previous arguments the generalization need not bring in anything essentially new.

The basic idea of duality in early hadronic models was that the lines of the dual diagram representing particles are only re-arranged in the vertices. This however does not allow to get rid of off mass shell momenta. Zero energy ontology encourages to consider a stronger form of this principle in the sense that the virtual momenta of particles could correspond to pairs of on mass shell momenta of particles. If also interacting fermions are pairs of positive and negative energy throats in the interaction region the idea about reducing the construction of Feynman diagrams to some kind of lego rules might work.

7.1 Virtual particles as pairs of on mass shell particles in ZEO

The first thing is to try to define more precisely what generalized Feynman diagrams are. The direct generalization of Feynman diagrams implies that both wormhole throats and wormhole contacts join at vertices.

1. A simple intuitive picture about what happens is provided by diagrams obtained by replacing the points of Feynman diagrams (wormhole contacts) with short lines and imagining that the throats correspond to the ends of the line. At vertices where the lines meet the incoming on mass shell quantum numbers would sum up to zero. This approach leads to a straightforward generalization of Feynman diagrams with virtual particles replaced with pairs of on mass shell throat states of type $++$, $--$, and $+-$. Incoming lines correspond to $++$ type lines and outgoing ones to $--$ type lines. The first two line pairs allow only time like net momenta whereas $+-$ line pairs allow also space-like virtual momenta. The sign assigned to a given throat is dictated by the the sign of the on mass shell momentum on the line. The condition that Cutkosky rules generalize as such requires $++$ and $--$ type virtual lines since the cut of the diagram in Cutkosky rules corresponds to on mass shell outgoing or incoming states and must therefore correspond to $++$ or $--$ type lines.

2. The basic difference as compared to the ordinary Feynman diagrammatics is that loop integrals are integrals over mass shell momenta and that all throats carry on mass shell momenta. In each vertex of the loop mass incoming on mass shell momenta must sum up to on mass shell momentum. These constraints improve the behavior of loop integrals dramatically and give excellent hopes about finiteness. It does not however seem that only a finite number of diagrams contribute to the scattering amplitude besides tree diagrams. The point is that if a the reactions $N_1 \rightarrow N_2$ and $N_2 \rightarrow N_3$,, where N_i denote particle numbers, are possible in a common kinematical region for N_2-particle states then also the diagrams $N_1 \rightarrow N_2 \rightarrow N_2 \rightarrow N_3$ are possible. The virtual states N_2 include all all states in the intersection of kinematically allow regions for $N_1 \rightarrow N_2$ and $N_2 \rightarrow N_3$. Hence the dream about finite number possible diagrams is not fulfilled if one allows massless particles. If all particles are massive then the particle number N_2 for given N_1 is limited from above and the dream is realized.

3. For instance, loops are not possible in the massless case or are highly singular (bringing in mind twistor diagrams) since the conservation laws at vertices imply that the momenta are parallel. In the massive case and allowing mass spectrum the situation is not so simple. As a first example one can consider a loop with three vertices and thus three internal lines. Three on mass shell conditions are present so that the four-momentum can vary in 1-D subspace only. For a loop involving four vertices there are four internal lines and four mass shell conditions so that loop integrals would reduce to discrete sums. Loops involving more than four vertices are expected to be impossible.

4. The proposed replacement of the elementary fermions with bound states of elementary fermions and monopoles $X_\pm$ brings in the analog of stringy diagrammatics. The 2-particle wave functions in the momentum degrees of freedom of fermiona and $X_\pm$ migh allow more flexibility and allow more loops. Note however that there are excellent hopes about the finiteness of the theory also in this case.

7.2 Loop integrals are manifestly finite

One can make also more detailed observations about loops.

1. The simplest situation is obtained if only 3-vertices are allowed. In this case conservation of momentum however allows only collinear momenta although the signs of energy need not be the same. Particle creation and annihilation is possible and momentum exchange is possible but is always light-like in the massless case. The scattering matrices of supersymmetric YM theories would suggest something less trivial and this raises the question whether something is missing. Magnetic monopoles are an essential element of also these theories as also massivation and symmetry breaking and this encourages to think that the formation of massive states as fermion $X_{\pm}$ pairs is needed. Of course, in TGD framework one has also high mass excitations of the massless states making the scattering matrix non-trivial.

2. In YM theories on mass shell lines would be singular. In TGD framework this is not the case since the propagator is defined as the inverse of the 3-D dimensional reduction of the modified Dirac operator D containing also coupling to four-momentum (this is required by quantum classical correspondence and guarantees stringy propagators),

$$\begin{aligned} D &= i\hat{\Gamma}^{\alpha}p_{\alpha} + \hat{\Gamma}^{\alpha}D_{\alpha} \ , \\ p_{\alpha} &= p_k\partial_{\alpha}h^k \ . \end{aligned} \tag{7.1}$$

   The propagator does not diverge for on mass shell massless momenta and the propagator lines are well-defined. This is of course of essential importance also in general case. Only for the incoming lines one can consider the possibility that 3-D Dirac operator annihilates the induced spinor fields. All lines correspond to generalized eigenstates of the propagator in the sense that one has $D_3\Psi = \lambda\gamma\Psi$, where γ is modified gamma matrix in the direction of the stringy coordinate emanating from light-like surface and D_3 is the 3-dimensional dimensional reduction of the 4-D modified Dirac operator. The eigenvalue λ is analogous to energy. Note that the eigenvalue spectrum depends on 4-momentum as a parameter.

3. Massless incoming momenta can decay to massless momenta with both signs of energy. The integration measure $d^2k/2E$ reduces to dx/x where $x \geq 0$ is the scaling factor of massless momentum. Only light-like momentum exchanges are however possible and scattering matrix is essentially trivial. The loop integrals are finite apart from the possible delicacies related to poles since the loop integrands for given massless wormhole contact are proportional to dx/x^3 for large values of x.

4. Irrrespective of whether the particles are massless or not, the divergences are obtained only if one allows too high vertices as self energy loops for which the number of momentum degrees of freedom is $3N-4$ for N-vertex. The construction of SUSY limit of TGD in [12] led to the conclusion that the parallelly propagating N fermions for given wormhole throat correspond to a product of N fermion propagators with same four-momentum so that for fermions and ordinary bosons one has the standard behavior but for $N>2$ non-standard so that these excitations are not seen as ordinary particles. Higher vertices are finite only if the total number N_F of fermions propagating in the loop satisfies $N_F > 3N-4$. For instance, a 4-vertex from which $N=2$ states emanate is finite.

7.3 Taking into account magnetic confinement

What has been said above is not quite enough. As shown in the accompanying article and in [11] the weak form of electric-magnetic duality [81] leads to the picture about elementary particles as pairs of magnetic monopoles inspiring the notions of weak confinement based on magnetic monopole force. Also color confinement would have magnetic counterpart. This means that elementary particles would behave like string like objects in weak boson length scale. Therefore one must also consider the stringy case with wormhole throats replaced with fermion-$X_{\pm}$ pairs ($X_{\pm}$ is electromagnetically neutral and $\pm$ refers to the sign of the weak isospin opposite to that of fermion) and their super partners.

1. The simplest assumption in the stringy case is that fermion-$X_\pm$ pairs behave as coherent objects, that is scatter elastically. In more general case only their higher excitations identifiable in terms of stringy degrees of freedom would be created in vertices. The massivation of these states makes possible non-collinear vertices. An open question is how the massivation fermion-$X_\pm$ pairs relates to the existing TGD based description of massivation in terms of Higgs mechanism and modified Dirac operator.

2. Mass renormalization could come from self energy loops with negative energy lines as also vertex normalization. By very general arguments supersymmetry implies the cancellation of the self energy loops but would allow non-trivial vertex renormalization [12].

3. If only 3-vertices are allowed, the loops containing only positive energy lines are possible if on mass shell fermion-$X_\pm$ pair (or its superpartner) can decay to a pair of positive energy pair particles of same kind. Whether this is possible depends on the masses involved. For ordinary particles these decays are not kinematically possible below intermediate boson mass scale (the decays $F_1 \rightarrow F_2 + \gamma$ are forbidden kinematically or by the absence of flavor changing neutral currents whereas intermediate gauge bosons can decay to on mass shell fermion-antifermion pair).

4. The introduction of IR cutoff for 3-momentum in the rest system associated with the largest CD (causal diamond) looks natural as scale parameter of coupling constant evolution and p-adic length scale hypothesis favors the inverse of the size scale of CD coming in powers of two. This parameter would define the momentum resolution as a discrete parameter of the p-adic coupling constant evolution. This scale does not have any counterpart in standard physics. For electron, d quark, and u quark the proper time distance between the tips of CD corresponds to frequency of 10 Hz, 1280 Hz, and 160 Hz: all these frequencies define fundamental bio-rhythms [8].

These considerations have left completely untouched one important aspect of generalized Feynman diagrams: the necessity to perform a functional integral over the deformations of the partonic 2-surfaces at the ends of the lines- that is integration over WCW. Number theoretical universality requires that WCW and these integrals make sense also p-adically and in the following these aspects of generalized Feynman diagrams are discussed.

References

Books related to TGD

[1] M. Pitkänen. Basic Properties of CP_2 and Elementary Facts about p-Adic Numbers. In *Towards M-matrix*. Onlinebook. `http://tgd.wippiespace.com/public_html/pdfpool/append.pdf`, 2006.

[2] M. Pitkänen. About the New Physics Behind Qualia. In *Quantum Hardware of Living Matter*. Onlinebook. `http://tgd.wippiespace.com/public_html/bioware/bioware.html#newphys`, 2006.

[3] M. Pitkänen. Basic Extremals of Kähler Action. In *Physics in Many-Sheeted Space-Time*. Online-book. `http://tgd.wippiespace.com/public_html/tgdclass/tgdclass.html#class`, 2006.

[4] M. Pitkänen. Configuration Space Spinor Structure. In *Quantum Physics as Infinite-Dimensional Geometry*. Onlinebook. `http://tgd.wippiespace.com/public_html/tgdgeom/tgdgeom.html#cspin`, 2006.

[5] M. Pitkänen. Construction of Configuration Space Kähler Geometry from Symmetry Principles. In *Quantum Physics as Infinite-Dimensional Geometry*. Onlinebook. `http://tgd.wippiespace.com/public_html/tgdgeom/tgdgeom.html#compl1`, 2006.

[6] M. Pitkänen. Construction of Quantum Theory: M-matrix. In *Towards M-Matrix*. Onlinebook. `http://tgd.wippiespace.com/public_html/tgdquant/tgdquant.html#towards`, 2006.

[7] M. Pitkänen. Construction of Quantum Theory: Symmetries. In *Towards M-Matrix*. Onlinebook. http://tgd.wippiespace.com/public_html/tgdquant/tgdquant.html#quthe, 2006.

[8] M. Pitkänen. Dark Matter Hierarchy and Hierarchy of EEGs. In *TGD and EEG*. Onlinebook. http://tgd.wippiespace.com/public_html/tgdeeg/tgdeeg.html#eegdark, 2006.

[9] M. Pitkänen. DNA as Topological Quantum Computer. In *Genes and Memes*. Onlinebook. http://tgd.wippiespace.com/public_html/genememe/genememe.html#dnatqc, 2006.

[10] M. Pitkänen. Does TGD Predict the Spectrum of Planck Constants? In *Towards M-Matrix*. Onlinebook. http://tgd.wippiespace.com/public_html/tgdquant/tgdquant.html#Planck, 2006.

[11] M. Pitkänen. Does the Modified Dirac Equation Define the Fundamental Action Principle? In *Quantum Physics as Infinite-Dimensional Geometry*. Onlinebook. http://tgd.wippiespace.com/public_html/tgdgeom/tgdgeom.html#Dirac, 2006.

[12] M. Pitkänen. Does the QFT Limit of TGD Have Space-Time Super-Symmetry? In *Towards M-Matrix*. Onlinebook. http://tgd.wippiespace.com/public_html/tgdquant/tgdquant.html#susy, 2006.

[13] M. Pitkänen. Evolution in Many-Sheeted Space-Time. In *Genes and Memes*. Onlinebook. http://tgd.wippiespace.com/public_html/genememe/genememe.html#prebio, 2006.

[14] M. Pitkänen. Identification of the Configuration Space Kähler Function. In *Quantum Physics as Infinite-Dimensional Geometry*. Onlinebook. http://tgd.wippiespace.com/public_html/tgdgeom/tgdgeom.html#kahler, 2006.

[15] M. Pitkänen. Knots and TGD. In *Quantum Physics as Infinite-Dimensional Geometry*. Onlinebook. http://tgd.wippiespace.com/public_html/tgdgeom/tgdgeom.html#knotstgd, 2006.

[16] M. Pitkänen. Massless states and particle massivation. In *p-Adic Length Scale Hypothesis and Dark Matter Hierarchy*. Onlinebook. http://tgd.wippiespace.com/public_html/paddark/paddark.html#mless, 2006.

[17] M. Pitkänen. Negentropy Maximization Principle. In *TGD Inspired Theory of Consciousness*. Onlinebook. http://tgd.wippiespace.com/public_html/tgdconsc/tgdconsc.html#nmpc, 2006.

[18] M. Pitkänen. New Particle Physics Predicted by TGD: Part I. In *p-Adic Length Scale Hypothesis and Dark Matter Hierarchy*. Onlinebook. http://tgd.wippiespace.com/public_html/paddark/paddark.html#mass4, 2006.

[19] M. Pitkänen. *p-Adic length Scale Hypothesis and Dark Matter Hierarchy*. Onlinebook. http://tgd.wippiespace.com/public_html/paddark/paddark.html, 2006.

[20] M. Pitkänen. p-Adic Numbers and Generalization of Number Concept. In *TGD as a Generalized Number Theory*. Onlinebook. http://tgd.wippiespace.com/public_html/tgdnumber/tgdnumber.html#padmat, 2006.

[21] M. Pitkänen. p-Adic Particle Massivation: Elementary Particle Masses. In *p-Adic Length Scale Hypothesis and Dark Matter Hierarchy*. Onlinebook. http://tgd.wippiespace.com/public_html/paddark/paddark.html#mass2, 2006.

[22] M. Pitkänen. p-Adic Physics as Physics of Cognition and Intention. In *TGD Inspired Theory of Consciousness*. Onlinebook. http://tgd.wippiespace.com/public_html/tgdconsc/tgdconsc.html#cognic, 2006.

[23] M. Pitkänen. *Physics in Many-Sheeted Space-Time*. Onlinebook. http://tgd.wippiespace.com/public_html/tgdclass/tgdclass.html, 2006.

[24] M. Pitkänen. Quantum Antenna Hypothesis. In *Quantum Hardware of Living Matter*. Onlinebook. http://tgd.wippiespace.com/public_html/bioware/bioware.html#tubuc, 2006.

[25] M. Pitkänen. Quantum Astrophysics. In *Physics in Many-Sheeted Space-Time*. Onlinebook. http://tgd.wippiespace.com/public_html/tgdclass/tgdclass.html#qastro, 2006.

[26] M. Pitkänen. Quantum Field Theory Limit of TGD from Bosonic Emergence. In *Towards M-Matrix*. Onlinebook. http://tgd.wippiespace.com/public_html/tgdquant/tgdquant.html#emergence, 2006.

[27] M. Pitkänen. Quantum Hall effect and Hierarchy of Planck Constants. In *Towards M-Matrix*. Onlinebook. http://tgd.wippiespace.com/public_html/tgdquant/tgdquant.html#anyontgd, 2006.

[28] M. Pitkänen. Quantum Model for Nerve Pulse. In *TGD and EEG*. Onlinebook. http://tgd.wippiespace.com/public_html//tgdeeg/tgdeeg/tgdeeg.html#pulse, 2006.

[29] M. Pitkänen. *Quantum Physics as Infinite-Dimensional Geometry*. Onlinebook.http://tgd.wippiespace.com/public_html/tgdgeom/tgdgeom.html, 2006.

[30] M. Pitkänen. *Quantum TGD*. Onlinebook. http://tgd.wippiespace.com/public_html/tgdquant/tgdquant.html, 2006.

[31] M. Pitkänen. TGD and Astrophysics. In *Physics in Many-Sheeted Space-Time*. Onlinebook. http://tgd.wippiespace.com/public_html/tgdclass/tgdclass.html#astro, 2006.

[32] M. Pitkänen. TGD and Cosmology. In *Physics in Many-Sheeted Space-Time*. Onlinebook. http://tgd.wippiespace.com/public_html/tgdclass/tgdclass.html#cosmo, 2006.

[33] M. Pitkänen. *TGD and Fringe Physics*. Onlinebook. http://tgd.wippiespace.com/public_html/freenergy/freenergy.html, 2006.

[34] M. Pitkänen. *TGD as a Generalized Number Theory*. Onlinebook. http://tgd.wippiespace.com/public_html/tgdnumber/tgdnumber.html, 2006.

[35] M. Pitkänen. TGD as a Generalized Number Theory: Infinite Primes. In *TGD as a Generalized Number Theory*. Onlinebook. http://tgd.wippiespace.com/public_html/tgdnumber/tgdnumber.html#visionc, 2006.

[36] M. Pitkänen. TGD as a Generalized Number Theory: p-Adicization Program. In *TGD as a Generalized Number Theory*. Onlinebook. http://tgd.wippiespace.com/public_html/tgdnumber/tgdnumber.html#visiona, 2006.

[37] M. Pitkänen. TGD as a Generalized Number Theory: Quaternions, Octonions, and their Hyper Counterparts. In *TGD as a Generalized Number Theory*. Onlinebook. http://tgd.wippiespace.com/public_html/tgdnumber/tgdnumber.html#visionb, 2006.

[38] M. Pitkänen. The Notion of Free Energy and Many-Sheeted Space-Time Concept. In *TGD and Fringe Physics*. Onlinebook. http://tgd.wippiespace.com/public_html/freenergy/freenergy.html#freenergy, 2006.

[39] M. Pitkänen. The Relationship Between TGD and GRT. In *Physics in Many-Sheeted Space-Time*. Onlinebook. http://tgd.wippiespace.com/public_html/tgdclass/tgdclass.html#tgdgrt, 2006.

[40] M. Pitkänen. *Topological Geometrodynamics: an Overview*. Onlinebook.http://tgd.wippiespace.com/public_html/tgdview/tgdview.html, 2006.

[41] M. Pitkänen. Twistors, N=4 Super-Conformal Symmetry, and Quantum TGD. In *Towards M-Matrix*. Onlinebook. http://tgd.wippiespace.com/public_html/tgdquant/tgdquant.html#twistor, 2006.

[42] M. Pitkänen. Yangian Symmetry, Twistors, and TGD. In *Towards M-Matrix*. Onlinebook. http://tgd.wippiespace.com/public_html/tgdquant/tgdquant.html#Yangian, 2006.

Articles about TGD

[43] M. Pitkänen. Basic Properties of CP_2 and Elementary Facts about p-Adic Numbers. http://tgd.wippiespace.com/public_html/pdfpool/append.pdf, 2006.

[44] M. Pitkänen. Physics as Generalized Number Theory II: Classical Number Fields. https://www.createspace.com/3569411, July 2010.

[45] M. Pitkänen. Physics as Infinite-dimensional Geometry I: Identification of the Configuration Space Kähler Function. https://www.createspace.com/3569411, July 2010.

[46] M. Pitkänen. Physics as Infinite-dimensional Geometry II: Configuration Space Kähler Geometry from Symmetry Principles. https://www.createspace.com/3569411, July 2010.

[47] M. Pitkänen. Physics as Generalized Number Theory I: p-Adic Physics and Number Theoretic Universality. https://www.createspace.com/3569411, July 2010.

[48] M. Pitkänen. Physics as Generalized Number Theory III: Infinite Primes. https://www.createspace.com/3569411, July 2010.

[49] M. Pitkänen. Physics as Infinite-dimensional Geometry III: Configuration Space Spinor Structure. https://www.createspace.com/3569411, July 2010.

[50] M. Pitkänen. Physics as Infinite-dimensional Geometry IV: Weak Form of Electric-Magnetic Duality and Its Implications. https://www.createspace.com/3569411, July 2010.

[51] M. Pitkänen. The Geometry of CP_2 and its Relationship to Standard Model. https://www.createspace.com/3569411, July 2010.

Mathematics

[52] Algebraic genus. http://en.wikipedia.org/wiki/Arithmetic_genus.

[53] Calabi-Yau manifold. http://en.wikipedia.org/wiki/CalabiYau_manifold.

[54] Clifford algebra. http://en.wikipedia.org/wiki/Clifford_algebra.

[55] Fields. http://en.wikipedia.org/wiki/Field_(mathematics).

[56] Harmonic analysis. http://en.wikipedia.org/wiki/Harmonic_analysis.

[57] Kac-Moody algebra. http://en.wikipedia.org/wiki/KacMoody_algebra.

[58] Kähler manifold. http://en.wikipedia.org/wiki/Khler_manifold.

[59] Number theory. http://en.wikipedia.org/wiki/Number_theory.

[60] Octonions. http://en.wikipedia.org/wiki/Octonions.

[61] Penrose diagram. http://en.wikipedia.org/wiki/Penrose_diagram.

[62] Quaternions. http://en.wikipedia.org/wiki/Quaternion.

[63] Scale invariance vs. conformal invariance. http://en.wikipedia.org/wiki/Conformal_field_theory#Scale_invariance_vs._conformal_invariance.

[64] Super Virasoro algebra. http://en.wikipedia.org/wiki/Super_Virasoro_algebra.

[65] Symmetric space. http://en.wikipedia.org/wiki/Symmetric_space.

[66] Symplectic geometry. http://en.wikipedia.org/wiki/Symplectic_geometry.

[67] Symplectic manifold. http://en.wikipedia.org/wiki/Symplectic_manifold.

[68] V. Bouchard. Lectures on complex geometry, Calabi-Yau manifolds and toric geometry. http://www.ulb.ac.be/sciences/ptm/pmif/Rencontres/ModaveI/CGL.ps, 2005.

[69] L. Brekke and P. G. O. Freund. p-Adic Numbers in Physics. *Phys. Rep.*, 233(1), 1993.

[70] A. C. da Silva. Symplectic geometry. http://www.math.princeton.edu/~acannas/symplectic.pdf, 2004.

[71] J. Dixmier. *Von Neumann Algebras.* North-Holland, Amsterdam, 1981.

[72] D. S. Freed. The Geometry of Loop Groups , 1985.

[73] C. N. Pope G. W. Gibbons. CP_2 as gravitational instanton. *Comm. Math. Phys.*, 55, 1977.

[74] W. Hawking, S. and N. Pope, C. Generalized Spin Structures in Quantum Gravity. *Phys. Lett.*, (1), 1978.

[75] C. Kassel. *Quantum Groups.* Springer Verlag, 1995.

[76] A. Kent P. Goddard and D. Olive. Unitary representations of the Virasoro and super-Virasoro algebras. *Comm. Math. Phys.*, 103(1), 1986.

[77] N. Pope, C. Eigenfunctions and $Spin^c$ Structures on CP_2, 1980.

[78] J. Hanson T. Eguchi, B. Gilkey. *Phys. Rep.*, 66:1980, 1980.

Theoretical Physics

[79] Chern-Simons theory. http://en.wikipedia.org/wiki/ChernSimons_theory.

[80] Hyperdeterminant. http://en.wikipedia.org/wiki/Hyperdeterminant.

[81] Montonen Olive Duality. http://en.wikipedia.org/wiki/Montonen-Olive_duality.

[82] Super symmetry. http://en.wikipedia.org/wiki/SUSY.

[83] R. Boels. On BCFW shifts of integrands and integrals. http://arxiv.org/abs/1008.3101, 2010.

[84] Nima Arkani-Hamed et al. The All-Loop Integrand For Scattering Amplitudes in Planar N=4 SYM. http://arxiv.org/find/hep-th/1/au:+Bourjaily_J/0/1/0/all/0/1, 2010.

[85] P. Svrcek F. Cachazo and E. Witten. MHV Vertices and Tree Amplitudes In Gauge Theory. http://arxiv.org/abs/hep-th/0403047, 2004.

[86] J. Plefka J. Drummond, J. Henn. Yangian symmetry of scattering amplitudes in $\mathcal{N} = 4$ super Yang-Mills theory. http://cdsweb.cern.ch/record/1162372/files/jhep052009046.pdf, 2009.

[87] E. Witten L. Dolan, C. R. Nappi. Yangian Symmetry in $D = 4$ superconformal Yang-Mills theory. http://arxiv.org/abs/hep-th/0401243, 2004.

[88] L. Mason and D. Skinner. Dual Superconformal Invariance, Momentum Twistors and Grassmannians. http://arxiv.org/pdf/0909.0250v2.

[89] S. Parke and T. Taylor. An Amplitude for N gluon Scattering. *Phys. Rev.*, 56, 1986.

[90] B. Feng R. Britto, F. Cachazo and E. Witten. Direct Proof of Tree-level Recursion Relation in Yang-Mills Theory. http://arxiv.org/PS_cache/hep-th/pdf/0501/0501052v2.pdf, 2005.

[91] E. Witten. Perturbative Gauge Theory As a String Theory In Twistor Space. http://arxiv.org/abs/hep-th/0312171, 2003.

Particle and Nuclear Physics

[92] Lamb shift. http://en.wikipedia.org/wiki/Lamb_shift.

[93] J. D. Bjorken. *Acta Phys. Polonica B*, 28:2773, 1997.

[94] Alice Collaboration. Charged-particle multiplicity density at mid-rapidity in central Pb-Pb collisions at $\sqrt{s_{NN}}$= 2.76 TeV. http://arxiv.org/abs/1011.3916, 2010.

Condensed Matter Physics

[95] Fractional quantum Hall Effect. http://en.wikipedia.org/wiki/Fractional_quantum_Hall_effect.

[96] Viscosity. http://en.wikipedia.org/wiki/Viscosity.

[97] C. Cao et al. Universal Quantum Viscosity in a Unitary Fermi Gas. http://www.sciencemag.org/content/early/2010/12/08/science.1195219, 2010.

[98] D. J. P. Morris et al. Dirac Strings and Magnetic Monopoles in Spin Ice Dy2Ti2O7. *Physics World*, 326(5951):411–414, 2009.

[99] P. K. Kotvun et al. Viscosity in Strongly Interacting Quantum Field Theories from Black Hole Physics. http://arxiv.org/abs/hep-th/0405231, 2010.

Cosmology and Astro-Physics

[100] M. R. S. Hawkins. On time dilation in quasar light curves. *Monthly Notices of the Royal Astronomical Society*, 2010.

[101] D. Da Roacha and L. Nottale. Gravitational Structure Formation in Scale Relativity. http://arxiv.org/abs/astro-ph/0310036, 2003.

Biology

[102] High energy phosphate. http://en.wikipedia.org/wiki/High-energy_phosphate.

[103] F. A. Popp et al. Emission of Visible and Ultraviolet Radiation by Active Biological Systems. *Collective Phenomena*, 3, 1981.

Neuroscience and Consciousness

[104] C. F. Blackman. *Effect of Electrical and Magnetic Fields on the Nervous System*, pages 331–355. Plenum, New York, 1994.

Article

New Particle Physics Predicted by TGD: Part I

Matti Pitkänen [1]

Abstract

TGD predicts a lot of new physics and it is quite possible that this new physics becomes visible at LHC. Although the calculational formalism is still lacking, p-adic length scale hypothesis allows to make precise quantitative predictions for particle masses by using simple scaling arguments.

The basic elements of quantum TGD responsible for new physics are following.

- The new view about particles relies on their identification as partonic 2-surfaces (plus 4-D tangent space data to be precise). This effective metric 2-dimensionality implies generalizaton of the notion of Feynman diagram and holography in strong sense. One implication is the notion of field identity or field body making sense also for elementary particles and the Lamb shift anomaly of muonic hydrogen could be explained in terms of field bodies of quarks.
- The topological explanation for family replication phenomenon implies genus generation correspondence and predicts in principle infinite number of fermion families. One can however develop a rather general argument based on the notion of conformal symmetry known as hyper-ellipticity stating that only the genera $g = 0, 1, 2$ are light. What "light" means is however an open question. If light means something below CP_2 mass there is no hope of observing new fermion families at LHC. If it means weak mass scale situation changes.

  For bosons the implications of family replication phenomenon can be understood from the fact that they can be regarded as pairs of fermion and antifermion assignable to the opposite wormhole throats of wormhole throat. This means that bosons formally belong to octet and singlet representations of dynamical SU(3) for which 3 fermion families define 3-D representation. Singlet would correspond to ordinary gauge bosons. Also interacting fermions suffer topological condensation and correspond to wormhole contact. One can either assume that the resulting wormhole throat has the topology of sphere or that the genus is same for both throats.
- The view about space-time supersymmetry differs from the standard view in many respects. First of all, the super symmetries are not associated with Majorana spinors. Super generators correspond to the fermionic oscillator operators assignable to leptonic and quark-like induced spinors and there is in principle infinite number of them so that formally one would have $\mathcal{N} = \infty$ SUSY. I have discussed the required modification of the formalism of SUSY theories and it turns out that effectively one obtains just $\mathcal{N} = 1$ SUSY required by experimental constraints. The reason is that the fermion states with higher fermion number define only short range interactions analogous to van der Waals forces. Right handed neutrino generates this super-symmetry broken by the mixing of the M^4 chiralities implied by the mixing of M^4 and CP_2 gamma matrices for induced gamma matrices. The simplest assumption is that particles and their superpartners obey the same mass formula but that the p-adic length scale can be different for them.
- One of the basic distinctions between TGD and standard model is the new view about color.
  - The first implication is separate conservation of quark and lepton quantum numbers implying the stability of proton against the decay via the channels predicted by GUTs. This does not mean that proton would be absolutely stable. p-Adic and dark length scale hierarchies indeed predict the existence of scale variants of quarks and leptons and proton could decay to hadons of some zoomed up copy of hadrons physics. These decays should be slow and presumably they would involve phase transition changing the value of Planck constant characterizing proton. It might be that the simultaneous increase of Planck constant for all quarks occurs with very low rate.
  - Also color excitations of leptons and quarks are in principle possible. Detailed calculations would be required to see whether their mass scale is given by CP_2 mass scale. The so called leptohadron physics proposed to explain certain anomalies associated with both electron, muon, and τ lepton could be understood in terms of color octet excitations of leptons.

[1] Correspondence: E-mail:matpitka@luukku.com

- The new view about particle massivation involves besides p-adic thermodynamics also Higgs but there is no need to assume that Higgs vacuum expectation plays any role. The most natural option favored by the assumption that elementary bosons are bound states of massless elementary fermions, by twistorial considerations, and by the fact that both gauge bosons and Higgs form SU(2) triplet and singlet, predicts that also photon and other massless gauge bosons develop small mass so that all Higgs particles and their colored variants would disappear from spectrum. Same could happen for Higgsinos.
- Fractal hierarchies of weak and hadronic physics labelled by p-adic primes and by the levels of dark matter hierarchy are highly suggestive. Ordinary hadron physics corresponds to $M_{107} = 2^{107} - 1$. One especially interesting candidate would be scaled up hadronic physics which would correspond to $M_{89} = 2^{89} - 1$ defining the p-adic prime of weak bosons. The corresponding string tension is about 512 GeV and it might be possible to see the first signatures of this physics at LHC. Nuclear string model in turn predicts that nuclei correspond to nuclear strings of nucleons connected by colored flux tubes having light quarks at their ends. The interpretation might be in terms of M_{127} hadron physics. In biologically most interesting length scale range 10 nm-2.5 μm there are four Gaussian Mersennes and the conjecture is that these and other Gaussian Mersennes are associated with zoomed up variants of hadron physics relevant for living matter. Cosmic rays might also reveal copies of hadron physics corresponding to M_{61} and M_{31}
- Weak form of electric magnetic duality implies that the fermions and antifermions associated with both leptons and bosons are Kähler magnetic monopoles accompanied by monopoles of opposite magnetic charge and with opposite weak isospin. For quarks Kähler magnetic charge need not cancel and cancellation might occur only in hadronic length scale. The magnetic flux tubes behave like string like objects and if the string tension is determined by weak length scale, these string aspects should become visible at LHC. If the string tension is 512 GeV the situation becomes less promising.

In this article and the following article the predicted new physics and possible indications for it are discussed.

Contents

1 Introduction

TGD predicts a lot of new physics and it is quite possible that this new physics becomes visible at LHC. Although calculational formalism is still lacking, p-adic length scale hypothesis allows to make precise quantitative predictions for particle masses by using simple scaling arguments. Actually there is already now evidence for effects providing further support for TGD based view about QCD and first rumors about super-symmetric particles have appeared.

Before detailed discussion it is good to summarize what elements of quantum TGD are responsible for new physics.

1. The new view about particles relies on their identification as partonic 2-surfaces (plus 4-D tangent space data to be precise). This effective metric 2-dimensionality implies generalizaton of the notion of Feynman diagram and holography in strong sense. One implication is the notion of field identity or field body making sense also for elementary particles and the Lamb shift anomaly of muonic hydrogen could be explained in terms of field bodies of quarks.

2. The topological explanation for family replication phenomenon implies genus generation correspondence and predicts in principle infinite number of fermion families. One can however develop a rather general argument based on the notion of conformal symmetry known as hyper-ellipticity stating that only the genera $g = 0, 1, 2$ are light [3]. What "light" means is however an open question. If light means something below CP_2 mass there is no hope of observing new fermion families at LHC. If it means weak mass scale situation changes.

   For bosons the implications of family replication phenomenon can be understood from the fact that they can be regarded as pairs of fermion and antifermion assignable to the opposite wormhole throats of wormhole throat. This means that bosons formally belong to octet and singlet representations of dynamical SU(3) for which 3 fermion families define 3-D representation. Singlet would correspond to ordinary gauge bosons. Also interacting fermions suffer topological condensation and correspond to wormhole contact. One can either assume that the resulting wormhole throat has the topology of sphere or that the genus is same for both throats.

3. The view about space-time supersymmetry differs from the standard view in many respects. First of all, the super symmetries are not associated with Majorana spinors. Super generators correspond to the fermionic oscillator operators assignable to leptonic and quark-like induced spinors and there is in principle infinite number of them so that formally one would have $\mathcal{N} = \infty$ SUSY. I have discussed the required modification of the formalism of SUSY theories in [11] and it turns out that effectively one obtains just $\mathcal{N} = 1$ SUSY required by experimental constraints. The reason is that the fermion states with higher fermion number define only short range interactions analogous to van der Waals forces. Right handed neutrino generates this super-symmetry broken by the mixing of the M^4 chiralities implied by the mixing of M^4 and CP_2 gamma matrices for induced gamma matrices. The simplest assumption is that particles and their superpartners obey the same mass formula but that the p-adic length scale can be different for them.

4. The new view about particle massivation involves besides p-adic thermodynamics also Higgs but there is no need to assume that Higgs vacuum expectation plays any role. The most natural option favored by the assumption that elementary bosons are bound states of massless elementary fermions, by twistorial considerations, and by the fact that both gauge bosons and Higgs form SU(2) triplet and singlet, predicts that also photon and other massless gauge bosons develop small mass so that all Higgs particles and their colored variants would disappear from spectrum. Also Higgsinos could be eaten by gauginos so that only massive gauginos would be seen at LHC.

5 One of the basic distinctions between TGD and standard model is the new view about color.

   (a) The first implication is separate conservation of quark and lepton quantum numbers implying the stability of proton against the decay via the channels predicted by GUTs. This does not mean that proton would be absolutely stable. p-Adic and dark length scale hierarchies indeed predict the existence of scale variants of quarks and leptons and proton could decay to hadons of some zoomed up copy of hadrons physics. These decays should be slow and presumably they would involve phase transition changing the value of Planck constant characterizing proton. It might be that the simultaneous increase of Planck constant for all quarks occurs with very low rate.

   (b) Also color excitations of leptons and quarks are in principle possible. Detailed calculations would be required to see whether their mass scale is given by CP_2 mass scale. The so called leptohadron physics proposed to explain certain anomalies associated with both electron, muon, and τ lepton could be understood in terms of color octet excitations of leptons [26].

5 Fractal hierarchies of weak and hadronic physics labelled by p-adic primes and by the levels of dark matter hierarchy are highly suggestive. Ordinary hadron physics corresponds to $M_{107} = 2^{107} - 1$. One especially interesting candidate would be scaled up hadronic physics which would correspond to $M_{89} = 2^{89} - 1$ defining the p-adic prime of weak bosons. The corresponding string tension is about 512 GeV and it might be possible to see the first signatures of this physics at LHC. Nuclear string model in turn predicts that nuclei correspond to nuclear strings of nucleons connected by colored flux tubes having light quarks at their ends. The interpretation might be in terms of M_{127} hadron physics. In biologically most interesting length scale range 10 nm-2.5 μm there are four Gaussian Mersennes and the conjecture is that these and other Gaussian Mersennes are associated with zoomed up variants of hadron physics relevant for living matter. Cosmic rays might also reveal copies of hadron physics corresponding to M_{61} and M_{31}

5 Weak form of electric magnetic duality implies that the fermions and antifermions associated with both leptons and bosons are Kähler magnetic monopoles accompanied by monopoles of opposite magnetic charge and with opposite weak isospin. For quarks Kähler magnetic charge need not cancel and cancellation might occur only in hadronic length scale. The magnetic flux tubes behave like string like objects and if the string tension is determined by weak length scale, these string aspects should become visible at LHC. If the string tension is 512 GeV the situation becomes less promising.

In this article the predicted new elementry particle physics and possible indications for it are discussed. Subsequent article is devoted to new hadron physics and scaled up variants of hardon physics in both quark and lepton sector.

2 Higgs or no Higgs?

The question whether TGD predicts Higgs or not has been one of the longstanding issues of TGD. The view about bosons as wormhole throats carrying fermionic quantum numbers are the opposite light-like wormhole throats of the contact makes it very difficult to assume that scalar particles would not exist. On the other hand, twistorial considerations force to assume that the wormhole throat as basic building bricks of particles are massless and that even virtual particles correspond to composites of on mass shell massless states with both signs of energy allowed. This forces the conclusion that spin 1 particles are necessary massive so that all Higgs like particles are eaten by gauge bosons. This makes sense since Higgs and gauge bosons belong to same representations of electroweak symmetries. Even the super-partners of Higgs bosons would experience the same fate. LHC will soon solve this issue.

In TGD framework p-adic thermodynamics gives the dominating contribution to fermion masses, which is something completely new. In the case of gauge bosons thermodynamic contribution is small since the inverse integer valued p-adic temperature is $T = 1/2$ for bosons or even lower: for fermions one has $T = 1$.

Whether Higgs can contribute to the masses is not completely clear. In TGD framework Mexican hat potential however looks like trick. One must however keep in mind that any other mechanism must explain the ratio of W and Z^0 masses and how these bosons receive their longitudinal polarizations. One must also consider seriously the possibility that all components for the TGD counterpart of Higgs boson are transformed to the longitudinal polarizations of the gauge bosons. Twistorial approach to TGD indeed strongly suggests that also the gauge bosons regarded usually as massless have a small mass guaranteing cancellation of IR singularities. As I started write to write this piece of text I believed that photon does not eat Higgs but had to challenge my beliefs. Maybe there is no Higgs to be found at LHC! Only pseudo-scalar partner of Higgs would remain to be discovered.

The weak form of electric magnetic duality implying that each wormhole throat carrying fermionic quantum numbers is accompanied by a second wormhole throat carrying opposite magnetic charge and neutrino pair screening weak isospin and making gauge bosons massive. Concerning the implications the following view looks the most plausible one at this moment.

1. Neutral Higgs-if not eaten by photon- could develop a coherent state meaning vacuum expectation value and this is naturally proportional to the inverse of the p-adic length scale as are boson masses. This contribution can be assigned to the magnetic flux tube mentioned above since it screens weak force - or equivalently - makes them massive. Higgs expectation would not cause boson massivation. Rather, massivation and Higgs vacuum expectation would be caused by the presence of the magnetic flux tubes. Standard model would suffer from a causal illusion. Even a worse illusion is possible if the photon eats the neutral Higgs.

2. The "stringy" magnetic flux tube connecting fermion wormhole throat and the wormhole throat containing neutrino pair would give to the vacuum conformal weight a small contribution and therefore to the mass squared of both fermions and gauge bosons (dominating one for the latter). This contribution would be small in the p-adic sense (proportional $1/p^2$ rather than $1/p$). I cannot calculate this "stringy" contribution but stringy formula in weak scale is very suggestive.

3. In the case of light fermions and massless gauge bosons the stringy contribution must vanish and therefore must correspond to $n = 0$ string excitation (string does not vibrate at all) : otherwise the mass of fermion would be of order weak boson mass. For weak bosons $n = 1$ would look like a natural identification but also $n = 0$ makes sense since $h \pm 1$ states corresponds opposite three-momenta for massless fermion and antifermion so that the state is massive. The mechanism bringing in the $h = 0$ helicity of gauge boson would be the TGD counterpart for the transformation of Higgs component to a longitudinal polarization. $n \geq 0$ excited states of fermions and $n \geq 1$ excitations of bosons having masses above weak boson masses are predicted and would mean new physics becoming possibly visible at LHC.

2.1 The identification of Higgs

Consider now the identification of Higgs in TGD framework.

1. In TGD framework Higgs particles do not correspond to complex $SU(2)$ doublets but to triplet and singlet having same quantum numbers as gauge bosons. Therefore the idea that photon eats neutral Higgs is suggestive. Also a pseudo-scalar variant of Higgs is predicted. Let us see how these states emerge from weak strings.

2. The two kinds of massive states corresponding to $n = 0$ and $n = 1$ give rise to massive spin 1 and spin 2 particles. First of all, the helicity doublet $(1, -1)$ is necessarily massive since the 3-momenta for massless fermion and anti-fermion are opposite. For $n = L = 0$ this gives two states but helicity zero component is lacking. For $n = L = 1$ one has tensor product of doublet $(1, -1)$ and angular momentum triplet formed by $L = 1$ rotational state of the weak string. This gives 2×3 states corresponding to $J = 0$ and $J = 2$ multiplets. Note however than in spin degrees of freedom the Higgs candidate is not a genuine elementary scalar particle.

3. Fermion and antifermion can have parallel three momenta summing up to a massless 4-momentum. Spin vanishes so that one has Higgs like particle also now. This particle is however pseudo-scalar being group theoretically analogous to meson formed as a pair of quark and antiquark. p-Adic thermodynamics gives a contribution to the mass squared. By taking a tensor product with rotational

states of strings one would obtain Regge trajectory containing pseudoscalar Higgs as the lowest state.

2.2 Do all gauge bosons possess small mass?

Consider now the problem how the gauge bosons can eat the Higgs boson to get their longitudinal component.

1. $(J = 0, n = 1)$ Higgs state can be combined with $n = 0$ $h = \pm 1$ doublet to give spin 1 massive triplet provided the masses of the two states are same. This will be discussed below.

2. Also gauge bosons usually regarded as massless can eat the scalar Higgs so that Higgs like particle could disappear completely. There would be no Higgs to be discovered at LHC! But is this a real prediction? Could it be that it is not possible to have exactly massless photons and gluons? The mixing of M^4 chiralities for Chern-Simons Dirac equation implies that also collinear massless fermion and antifermion can have helicity ± 1. The problem is that the mixing of the chiralities is a signature of massivation!

   Could it really be that even the gauge bosons regarded as massless have a small mass characterized by the length scale of the causal diamond defining the physical IR cutoff and that the remaining Higgs component would correspond to the longitudinal component of photon? This would mean the number of particles in the final states for a particle reaction with a fixed initial state is always bounded from above. This is important for the twistorial aesthetics of generalized Feynman diagrammatics implied by zero energy ontology. Also the vanishing of IR divergences is guaranteed by a small physical mass [28]. Maybe internal consistency allows only pseudo-scalar Higgs.

2.3 Weak Regge trajectories

The weak form of electric-magnetic duality suggests strongly the existence of weak Regge trajectories.

1. The most general mass squared formula with spin-orbit interaction term $M^2_{L-S} L \cdot S$ reads as

$$M^2 = nM_1^2 + M_0^2 + M^2_{L-S} L \cdot S \ , \ n = 0, 2, 4 \text{ or } n = 1, 3, 5, ..., \ . \tag{2.1}$$

   M_1^2 corresponds to string tension and M_0^2 corresponds to the thermodynamical mass squared and possible other contributions. For a given trajectory even (odd) values of n have same parity and can correspond to excitations of same ground state. From ancient books written about hadronic string model one vaguely recalls that one can have several trajectories (satellites) and if one has something called exchange degeneracy, the even and odd trajectories define single line in $M^2 - J$ plane. As already noticed TGD variant of Higgs mechanism combines together $n = 0$ states and $n = 1$ states to form massive gauge bosons so that the trajectories are not independent.

2. For fermions, possible Higgs, and pseudo-scalar Higgs and their super partners also p-adic thermodynamical contributions are present. M_0^2 must be non-vanishing also for gauge bosons and be equal to the mass squared for the $n = L = 1$ spin singlet. By applying the formula to $h = \pm 1$ states one obtains

$$M_0^2 = M^2(boson) \ . \tag{2.2}$$

   The mass squared for transversal polarizations with $(h, n, L) = (\pm 1, n = L = 0, S = 1)$ should be same as for the longitudinal polarization with $(h = 0, n = L = 1, S = 1, J = 0)$ state. This gives

$$M_1^2 + M_0^2 + M^2_{L-S} L \cdot S = M_0^2 \ . \tag{2.3}$$

From $L \cdot S = [J(J+1) - L(L+1) - S(S+1)]/2 = -2$ for $J = 0, L = S = 1$ one has

$$M_{L-S}^2 = -\frac{M_1^2}{2} . \quad (2.4)$$

Only the value of weak string tension M_1^2 remains open.

3. If one applies this formula to arbitrary $n = L$ one obtains total spins $J = L+1$ and $L-1$ from the tensor product. For $J = L-1$ one obtains

$$M^2 = (2n+1)M_1^2 + M_0^2 .$$

For $J = L+1$ only M_0^2 contribution remains so that one would have infinite degeneracy of the lightest states. Therefore stringy mass formula must contain a non-linear term making Regge trajectory curved. The simplest possible generalization which does not affect n=0 and n=1 states is of from

$$M^2 = n(n-1)M_2^2 + (n - \frac{L \cdot S}{2})M_1^2 + M_0^2 . \quad (2.5)$$

The challenge is to understand the ratio of W and Z^0 masses, which is purely group theoretic and provides a strong support for the massivation by Higgs mechanism.

1. The above formula and empirical facts require

$$\frac{M_0^2(W)}{M_0^2(Z)} = \frac{M^2(W)}{M^2(Z)} = cos^2(\theta_W) . \quad (2.6)$$

in excellent approximation. Since this parameter measures the interaction energy of the fermion and antifermion decomposing the gauge boson depending on the net quantum numbers of the pair, it would look very natural that one would have

$$M_0^2(W) = g_W^2 M_{SU(2)}^2 , \quad M_0^2(Z) = g_Z^2 M_{SU(2)}^2 . \quad (2.7)$$

Here $M_{SU(2)}^2$ would be the fundamental mass squared parameter for $SU(2)$ gauge bosons. p-Adic thermodynamics of course gives additional contribution which is vanishing or very small for gauge bosons.

2. The required mass ratio would result in an excellent approximation if one assumes that the mass scales associated with $SU(2)$ and $U(1)$ factors suffer a mixing completely analogous to the mixing of $U(1)$ gauge boson and neutral $SU(2)$ gauge boson W_3 leading to γ and Z^0. Also Higgs, which consists of $SU(2)$ triplet and singlet in TGD Universe, would very naturally suffer similar mixing. Hence $M_0(B)$ for gauge boson B would be analogous to the vacuum expectation of corresponding mixed Higgs component. More precisely, one would have

$$\begin{aligned} M_0(W) &= M_{SU(2)} , \\ M_0(Z) &= cos(\theta_W)M_{SU(2)} + sin(\theta_W)M_{U(1)} , \\ M_0(\gamma) &= -sin(\theta_W)M_{SU(2)} + cos(\theta_W)M_{U(1)} . \end{aligned} \quad (2.8)$$

The condition that photon mass is very small and corresponds to IR cutoff mass scale gives $M_0(\gamma) = \epsilon cos(\theta_W)M_{SU(2)}$, where ϵ is very small number, and implies

$$\frac{M_{U(1)}}{M(W)} = tan(\theta_W) + \epsilon \ ,$$
$$\frac{M(\gamma)}{M(W)} = \epsilon \times cos(\theta_W) \ ,$$
$$\frac{M(Z)}{M(W)} = \frac{1 + \epsilon \times sin(\theta_W)cos(\theta_W)}{cos(\theta_W)} \ . \tag{2.9}$$

There is a small deviation from the prediction of the standard model for W/Z mass ratio but by the smallness of photon mass the deviation is so small that there is no hope of measuring it. One can of course keep mind open for $\epsilon = 0$. The formulas allow also an interpretation in terms of Higgs vacuum expectations as it must. The vacuum expectation would most naturally correspond to interaction energy between the massless fermion and antifermion with opposite 3-momenta at the throats of the wormhole contact and the challenge is to show that the proposed formulas characterize this interaction energy. Since CP_2 geometry codes for standard model symmetries and their breaking, it woul not be surprising if this would happen. One cannot exclude the possibility that p-adic thermodynamics contributes to $M_0^2(boson)$. For instance, ϵ might characterize the p-adic thermal mass of photon.

If the mixing applies to the entire Regge trajectories, the above formulas would apply also to weak string tensions, and also photons would belong to Regge trajectories containing high spin excitations.

3. What one can one say about the value of the weak string tension M_1^2? The naive order of magnitude estimate is $M_1^2 \simeq m_W^2 \simeq 10^4$ GeV^2 is by a factor 1/25 smaller than the direct scaling up of the hadronic string tension about 1 GeV^2 scaled up by a factor 2^{18}. The above argument however allows also the identification as the scaled up variant of hadronic string tension in which case the higher states at weak Regge trajectories would not be easy to discover since the mass scale defined by string tension would be 512 GeV to be compared with the recent beam energy 7 TeV. Weak string tension need of course not be equal to the scaled up hadronic string tension. Weak string tension - unlike its hadronic counterpart- could also depend on the electromagnetic charge and other characteristics of the particle.

2.4 Is the earlier conjectured pseudoscalar Higgs there at all?

Spin 1 gauge bosons and Higgs differ only by different spin direction of fermions at opposite wormhole throats. For spin 1 gauge bosons the 3-momenta at two wormhole throats cannot be parallel if if one wants non-vanishing spin component in the direction of moment. 3-momenta are most naturally opposite for the massless states at throats. This forces massivation for all gauge bosons and even graviton and this in turn requires Higgs even in the case of gluons.

This inspires the question whether the parity properties of the couplings of gauge boson and corresponding Higgs transforming like 3+1 under SU(2) (this is due to the special character of imbedding space spinors) could be exactly the same? Higgs would couple like a mixture of scalar and pseudoscalar to fermions just as weak gauge bosons couple and the mixture would be just the same. If there are no axial variants of vector gauge bosons there should exist no pseudoscalar Higgs. The nonexistence of axial variants of vector gauge bosons is suggested by quantum classical correspondence: only gauge bosons having classical space-time correlates as induced gauge potentials should be allowed, nothing else. Note that color variant of Higgs would exist and would be eaten by gluons to get mass.

The similarity of the construction of gauge bosons and Higgsinos as pairs of wormhole throats containing fermion and antifermion encourages to think that Higgs mechanism is invariant under supersymmetries. If so, also Higgsinos would be eaten and one would have massive super-symmetric gauge theory with fermions with photon and other massless particle possessing a tiny mass. This looks very simple. The testable implication would be that only weak gauginos should contribute to muon g-2 anomaly.

2.5 Higgs issue after Europhysics 2011

The general feeling at the Eve of Europhysics 2011 conference was that this meeting might become one of the key events in the history of physics. This might turn out to be the case. CDF and D0 were the groups representing the data from p-pbar collisions at Tevatron whereas ATLAS and CMS represented the data about p-p collisions at LHC. The blog participation transformed the conference from a closed meeting of specialists to a media event inspiring intense blog discussions and viXra log became the most interesting discussion forum thanks to the excellent postings of Phil Gibbs giving focused summaries of various reports about SUSY and Higgs.

The hope was that two basic questions would receive a unique answer. Does Higgs exist and if so what is its mass? Is the standard view about SUSY correct: in other words do the super-partners exist with masses below TeV scale? It was clear that negative answer to even the Higgs issue would force a thorough reconsideration of the status of not only MSSM but also that of super string theory and M-theory because of the general role of Higgs mechanism in the description massivation and symmetry breaking for the QFT limits of these theories. The implications are far reaching also for the inflationary cosmology where scalar fields and Higgs mechanism are taken as granted. Actually the non-existence of Higgs forces to reconsider the entire quantum field theoretic description of particle massivation.

Already before the conference several anomalies had emerged and the question was whether LHC data gives a support for these anomalies.

- A 145 GeV bump with 4 sigma significance in the mass distribution of jet pairs jj in Wjj final states was reported by CDF [38] but not confirmed by D0 [72]. The interpretation as Higgs was excluded and some of the proposed identifications of 145 GeV bump was as decay products of leptophobic Z' boson or of technicolor pion. There were also indications for 300 GeV bump in the mass distribution of Wjj states themselves suggesting cascade like decay.

- Both CDF and D0 had reported two bumps at almost same mass about 325 GeV [49, 50] having no obvious interpretation in standard model framework. Technicolor approach and also TGD suggests an interpretation as pionlike state.

- CDF and D0 had also reported anomalous forward-backward asymmetry in top-pair production in p-pbar collisions suggesting the existence of new kind of flavor changing colored neutral currents [62, 88]. TGD based explanation of family replication phenomenon combined with bosonic emergence predicts that gauge bosons should appear as flavor singlets and octets. Octets would indeed induce flavor changing currents and asymmetry. Also many other indications for new physics such as anomalously large CP breaking in BBbar system had been reported and one should not forget long list of forgotten anomalies from previous years, say the two and half year old CDF anomaly which D0 failed to observe. Recall also that proton has shown no signs of decaying.

What did we learn during these days? Already before the conference it was clear that standard SUSY had transformed from the healer of the standard model to a patient. The parameter space for MSSM (minimal supersymmetric extension of standard model predicting two Higgs multiplets) had been narrowed down by strong lower limits on squark and sgluon masses to the extent that the original basic motivation for MSSM (stability of Higgs mass against radiative corrections) had been lost as well as the explanation for the anomaly of g-2 of muon. During the conference the bounds on SUSY parameters were tightened further and the rough conclusion is that squark and gluinos masses must be above 1 TeV. Even Lubos Motl was forced to conclude that the probability that LCH discovers standard SUSY is 50 per cent instead of 90 per cent or more of 2008 blog posting. In TGD framework simple p-adic scaling arguments lead to the proposal that the only sfermions with mass below 1 TeV are selectron and sneutrinos with selectron having mass equal to 262 GeV. Low sneutrino masses allow in principle to understand g-2 of muon. Selectron could decay to electron plus neutralino for which mass must be larger than 46 GeV neutralino would eventually decay to photon or virtual Z plus neutrino.

The Higgs issue became the central theme of the conference and the three days from Thursday to Sunday were loaded with excitement. After many twists, the final conclusion was that there is 2.5 sigma evidence from ATLAS for a state in the mass range 140-150 GeV, which might be Higgs or something else. The press release of Fermi lab at Friday announced that they have confined Higgs to the interval 120-137

GeV. After the announcement of ATLAS both D0 and CDF discovered suddenly evidence for Higgs in 140-150 GeV mass range. The evidence for this mass range emerged from the decays of a might-be Higgs to WW pairs decaying in turn to lepton pairs. The proponent of technicolor would of course see this as evidence for an off mass shell state of a neutral pion like state explaining also the jj bump in Wjj system and at 145 GeV mass and not allowing an interpretation as Higgs. In TGD framework the experience with earlier anomalies such as two year old CDF anomaly encouraged the interpretation in terms p-adic mass octaves of the pion of p-adically scaled up variant of hadron physics with mass scale 512 times higher than that of the ordinary hadron physics. Somewhat frustratingly, the final conclusion about the Higgs issue was promised to emerge only towards the end of the next year but it is clear that already now standard model might well be inconsistent with all data irrespective of the mass of Higgs. MSSM would allow additional flexibility but is also in difficulties.

The surprise of the first conference day was additional evidence for the bump at 327 GeV reported already earlier by CDF. This state is a complete mystery in standard model framework and therefore extremely interesting. The proponents of technicolor would probably suggest interpretation as exotic ρ or ω meson. in TGD framework both 145 GeV pion and 325 GeV ρ and ω appear as mesons of M_{89} hadron physics if one assumes that the u and d quarks of M_{89} physics have masses corresponding to the p-adic length scale $k = 93$ (mass is 102 GeV and should be visible as a preferred quark jet mass). I would not be surprised if technicolor models would experience a brief renessaince but fail experimentally since a lot of new states and elementary particles is implied by the extension of the color gauge group. The mere p-adic scaling does not imply anything like this.

Also super string inspired predictions of various exotics such as microscopic black holes, strong gravity, large extra dimensions, Randall-Sundrum gravitons, split supersymmetry, and whatever were tested. No evidence was found. Neither there was evidence for lepto-quarks, heavier partners of intermediate gauge bosons, and various other exotics.

To my view, the results of the conference force to re-consider the basic assumptions of the approach followed during last more than three decades. Is it possible be find a more realistic physical interpretation of the mathematically extremely attractive supersymmetry? Unitarity requires new physics in TeV scale: is this new physics technicolor or its TGD analog without gauge group extenssion or something else? To me however the mother of all questions concerns the microscopic description of massivation. The description in terms of Higgs is after all a phenomenological description borrowed from condensed matter physics. It does not work for extended objects like strings but require quantum field theory limit. p-Adic thermodynamics for conformal weight (to which mass squared is proportional to) should be an essential element of the microscopic approach too since it is a description working for the fundamental objects and in presence of super-conformal invariance.

What actually happens in the massivation: could it be that all components of Higgs, of its super partners, and of its higher spin generalizations are eaten in a process in which massless multiplets with various spins combine to form only massive multiplets? Here twistor approach might provide the guideline since its applicability requires that massive particles should allow an interpretation as bound states of massless ones. Perhaps the simple observation that spin one bound states of massless fermion and anti-fermion are automatically massive might help to get to the deeper waters.

What next? Standard model Higgs is more or less excluded and the same fate is very probably waiting the SUSY Higgs. I would not be surprised if technicolor models would experience a brief renaissance but fail experimentally since very many new hadronlike states and new elementary particles are implied by the extension of the color gauge group. Sooner or later the simple p-adic scaling of the ordinary hadron physics probably turns out to be the only realistic option. If technicolor becomes in fashion, the hadrons of M_{89} hadron physics will be however found as a side product of this search.

Eventually this requires giving up the Planck length scale reductionism as the basic philosophy and replacing it with p-adic fractality as the basic guiding principle tying together physics at very short and at very long length scales making possible the long sough for ultraviolet completion of known physics. This led to the landscape catastrophe in M-theory since very many physics in long length scales had the same UV completion. Some general principle fixing the long range physics is obviously missing. p-Adic smoothness for which infinite in real sense is infinitesimal selects the unique long length scale physics among infinitely many alternatives. The real problems are really much much deeper than finding proper parameters for SUSY and it would be a high time for theoreticians to finally realize this.

3 Family replication phenomenon

3.1 Family replication phenomenon for bosons

TGD predicts that also gauge bosons, with gravitons included, should be characterized by family replication phenomenon but not quite in the expected manner. The first expectation was that these gauge bosons would have at least 3 light generations just like quarks and leptons.

Only within last years it has become clear that there is a deep difference between fermions and gauge bosons. Elementary fermions and particles super-conformally related to elementary fermions correspond to single throat of a wormhole contact assignable to a topologically condensed CP_2 type vacuum extremal whereas gauge bosons would correspond to a wormhole throat pair assignable to wormhole contact connecting two space-time sheets. Wormhole throats correspond to light-like partonic 3-surfaces at which the signature of the induced metric changes.

In the case of 3 generations gauge bosons can be arranged to octet and singlet representations of a dynamical SU(3) and octet bosons for which wormhole throats have different genus could be massive and effectively absent from the spectrum.

Exotic gauge boson octet would induce particle reactions in which conserved handle number would be exchanged between incoming particles such that total handle number of boson would be difference of the handle numbers of positive and negative energy throat. These gauge bosons would induce flavor changing but genus conserving neutral current. There is no evidence for this kind of currents at low energies which suggests that octet mesons are heavy. Typical reaction would be $\mu + e \rightarrow e + \mu$ scattering by exchange of $\Delta g = 1$ exotic photon.

3.2 Higher gauge boson families

TGD predicts that also gauge bosons, with gravitons included, should be characterized by family replication phenomenon but not quite in the expected manner. The first expectation was that these gauge bosons would have at least 3 light generations just like quarks and leptons.

Only within last two years it has become clear that there is a deep difference between fermions and gauge bosons. Elementary fermions and particles super-conformally related to elementary fermions correspond to single throat of a wormhole contact assignable to a topologically condensed CP_2 type vacuum extremal whereas gauge bosons would correspond to a wormhole throat pair assignable to wormhole contact connecting two space-time sheets. Wormhole throats correspond to light-like partonic 3-surfaces at which the signature of the induced metric changes.

In the case of 3 generations gauge bosons can be arranged to octet and singlet representations of a dynamical SU(3) and octet bosons for which wormhole throats have different genus could be massive and effectively absent from the spectrum.

Exotic gauge boson octet would induce particle reactions in which conserved handle number would be exchanged between incoming particles such that total handle number of boson would be difference of the handle numbers of positive and negative energy throat. These gauge bosons would induce flavor changing but genus conserving neutral current. There is no evidence for this kind of currents at low energies which suggests that octet mesons are heavy. Typical reaction would be $\mu + e \rightarrow e + \mu$ scattering by exchange of $\Delta g = 1$ exotic photon.

3.2.1 New view about interaction vertices and bosons

There are two options for the identification of particle vertices as topological vertices.

1. *Option a)*

The original assumption was that one can assign also to bosons a partonic 2-surface X^2 with more or less well defined genus g. The hypothesis is consistent with the view that particle reactions are described by smooth 4-surfaces with vertices being singular 3-surfaces intermediate between two three-topologies. The basic objection against this option is that it can induce too high rates for flavor changing currents. In particular $g > 0$ gluons could induce these currents. Second counter argument is that stable $n > 4$-particle vertices are not possible.

2. Option b)

According to the new vision (option 2)), particle decays correspond to branchings of the partonic 2-surfaces in the same sense as the vertices of the ordinary Feynman diagrams do correspond to branchings of lines. The basic mathematical justification for this vision is the enormous simplification caused by the fact that vertices correspond to non-singular 2-manifolds. This option allows also $n > 3$-vertices as stable vertices.

A consistency with the experimental facts is achieved if the observed gauge bosons have each value of $g(X^2)$ with the same probability. Hence the general boson state would correspond to a phase $exp(in2\pi g/3)$, $n = 0, 1, 2$, in the discrete space of 3 lowest topologies $g = 0, 1, 2$. The observed bosons would correspond to $n = 0$ state and exotic higher states to $n = 1, 2$.

The nice feature of this option is that no flavor changing neutral electro-weak or color currents are predicted. This conforms with the fact that CKM mixing can be understood as electro-weak phenomenon described most naturally by causal determinants X_l^3 (appearing as lines of generalized Feynman diagram) connecting fermionic 2-surfaces of different genus.

Consider now objections against this scenario.

1. Since the modular contribution does not depend on the gradient of the elementary particle vacuum functional but only on its logarithm, all three boson states should have mass squared which is the average of the mass squared values $M^2(g)$ associated with three generations. The fact that modular contribution to the mass squared is due to the super-symplectic thermodynamics allows to circumvent this objection. If the super-symplectic p-adic temperature is small, say $T_p = 1/2$, then the modular contribution to the mass squared is completely negligible also for $g > 0$ and photon, graviton, and gluons could remain massless. The wiggling of the elementary particle vacuum functionals at the boundaries of the moduli spaces $\mathcal{M}_g$ corresponding to 2-surfaces intermediate between different 2-topologies (say pinched torus and self-touching sphere) caused by the change of overall phase might relate to the higher p-adic temperature T_p for exotic bosons.

2. If photon states had a 3-fold degeneracy, the energy density of black body radiation would be three times higher than it is. This problem is avoided if the the super-symplectic temperature for $n = 1, 2$ states is higher than for $n = 0$ states, and same as for fermions, say $T_p = 1$. In this case two mass degenerate bosons would be predicted with mass squared being the average over the three genera. In this kind of situation the factor 1/3 could make the real mass squared very large, or order CP_2 mass squared, unless the sum of the modular contributions to the mass squared values $M^2_{mod}(g) \propto n(g)$ is divisible by 3. This would make also photon, graviton, and gluons massive. Fortunately, $n(g)$ is divisible by 3 as is clear form $n(0) = 0$, $n(1) = 9$, $n(2) = 60$.

3.2.2 Masses of genus-octet bosons

For option 1) ordinary bosons are accompanied by $g > 0$ massive partners. For option 2) both ordinary gauge bosons and their exotic partners have suffered maximal topological mixing in the case that they are singlets with respect to the dynamical SU(3). There are good reasons to expect that Higgs mechanism for ordinary gauge bosons generalizes as such and that $1/T_p > 1$ means that the contribution of p-adic thermodynamics to the mass is negligible. The scale of Higgs boson expectation would be given by p-adic length scale and mass degeneracy of octet is expected. A good guess is obtained by scaling the masses of electro-weak bosons by the factor $2^{(k-89)/2}$. Also the masses of genus-octet of gluons and photon should be non-vanishing and induced by a vacuum expectation of Higgs particle which is electro-weak singlet but genus-octet.

3.2.3 Indications for genus-generation correspondence for gauge bosons

Tommaso Dorigo is a highly inspiring blogger since he writes from the point of view of experimental physicist without the burden of theoretical dogmas. I share with him also the symptons of splitting of personality to fluctuation-enthusiast and die-hard skeptic. This makes life interesting but not easy. This time Tommaso told about the evidence for new neutral gauge boson states in $p\overline{p}$ collisions. The title of the posting was "A New Z' Boson at 240 GeV? No, Wait, at 720!?" [79].

1. The findings

The title tells that the tentative interpretation of these states are as excited states of Z^0 boson and that the masses of the states are around 240 GeV and 720 GeV. The evidence for the new states comes from electron-positron pairs in relatively narrow energy interval produced by the decays of the might-be-there gauge boson. This kind of decay is an especially clean signature since strong interaction effects are not present and it appears at sharp energy.

240 GeV bump was reported by CDF last year [60] CDF last year in $p\overline{p}$ collisions at $\sqrt{s} = 1.96$ TeV. The probability that it is a fluctuation is .6 per cent. What is encouraging that also D0 found the same bump. If the particle in question is analogous to Z^0, it should decay also to muons. CDF checked this and found a negative result. This made Tommaso rather skeptic.

Also indications for 720 GeV resonance (720 GeV is just a nominal value, the mass could be somewhere between 700-800 GeV) was reported by D0 collaboration: the report is titled as "Search for high-mass narrow resonances in the di-electron channel at D0" [69]. There are just 2 events above 700 GeV but background is small: just three events above 600 GeV. It is easy to guess what skeptic would say.

Before continuing I want to make clear that I refuse to be blind believer or die-hard skeptic and that I am unable to say anything serious about the experimental side. I am just interested to see whether these events might be interpreted in TGD framework. TGD indeed predicts -or should I say strongly suggests- a lot of new physics above intermediate boson length scale.

2. Are exotic Z^0 bosons p-adically scaled up variants of ordinary Z^0 boson?

p-Adic length scale hypothesis allows the p-adic length scale characterized by prime $p \simeq 2^k$ vary since k can have several integer values. The TGD counterpart of Gell-Mann-Okubo mass formula involves varying value of k for quark masses. Several anomalies reported by Tommaso during years could be resolved if k can have several values. Last anomaly was the discovery that Ω_b baryon containing two strange quarks and bottom quark seems to appear with two masses differing by about 100 MeV. TGD explains the mass difference correctly by assuming that strange quark can have besides ordinary mass scale mass differing by factor of 2. The prediction is 105 MeV.

One can look whether p-adic length scale hypothesis could explains the masses of exotic Z^0 candidates as multiples of half octaves of Z^0 mass which is 91 GeV. k=3 would give 257 GeV, not too far from 240 GeV. k=6 would give 728 GeV consistent with the nominal value of the mass. Also other masses are predicted and this could serve as a test for the theory. This option does not however explain why muon pairs are not produced in the case of 240 GeV resonance.

3. Support for topological explanation of family replication phenomenon?

The improved explanation is based on TGD based view about family replication phenomenon [3].

1. In TGD the explanation of family replication is in terms of genus of 2-dimensional partonic surface representing fermion. Fermions correspond to SU(3) triplet of a dynamical symmetry assignable to the three lowest genera (sphere, torus, sphere with two handles). Bosons as wormhole contacts have two wormhole throats carrying fermion numbers and correspond to SU(3) singlet and octet. Sooner or later the members of the octet - presumably heavier than singlet- should be observed (maybe this has been done now;-)).

2. The exchange of these particles predicts also charged flavor changing currents respecting conservation of corresponding "isospin" and "hypercharge." For instance, lepton quark scattering $e + s \rightarrow \mu + d$ would be possible. The most dramatic signature of these states is production of muon-positron pairs (for instance) via decays.

3. Since the Z^0 or photon like boson in question has vanishing "isospin" and "hypercharge", it must be orthogonal to the ordinary Z^0 which couples identically to all families. There are two states of this kind and they correspond to superpositions of fermion pairs of different generations in TGD framework. The two bosons - very optimistically identified as 240 GeV and 720 GeV Z^0, must be orthogonal to the ordinary Z^0. This requires that the phase factors in superposition of pairs adjust themselves properly. Also mixing effects breaking color symmetry are possible and expected to occur since the SU(3) in question is not an exact symmetry. Hence the exotic Z^0 bosons *could*

couple preferentially to some fermion generation. This kind of mixing might be used to explain the absence of muon pair signal in the case of 240 GeV resonance.

4. The prediction for the masses is same as for the first option if the octet and singlet bosons have identical masses for same p-adic mass scale so that mass splitting between different representations would take place via the choice of the mass scale alone.

4. Could scaled up copy of hadron physics involved?

One can also ask whether these particles could come from the decays of hadrons of a scaled up copy of hadron physics strongly suggested by p-Adic length scale hypothesis.

1. Various hadron physics would correspond to Mersenne primes: standard hadron physics to M_{107} and new hadron physics to Mersenne prime $M_{89} = 2^{89} - 1$. The first guess for the mass scale of "light" M^{89} hadrons would be $2^{(107-89)/2} = 512$ times that for ordinary hadrons. The electron pairs might result in a decay of scaled up variant of pseudoscalar mesons π , η, or of η' or spin one ρ and ω mesons with nearly the same mass. Only scaled up ρ and ω mesons remains under consideration if one assumes spin 1.

2. The scaling of pion mass about 140 MeV gives 72 GeV. This is three times smaller than 240 GeV but this is extremely rough estimate. Actually it is the p-adic mass scale of quarks involved which matters rather than that of hadronic space-time sheet characterized by M_{89}. The naive scaling of the mass of η meson with mass 548 MeV would give about 281 GeV. η' would give 490 GeV. ρ meson with mass would give 396 GeV. The estimates are just order of magnitude estimates since the mass splitting between pseudoscalar and corresponding vector meson is sensitive to quark mass scale.

3. This option does not provide any explanation for the lack of muon pairs in decays of 240 GeV resonance.

To conclude, family replication phenomenon for gauge bosons is consistent with the claimed masses and also absence of muon pairs might be understood and it remains to be seen whether only statistical fluctuations are in question.

3.2.4 A slight indication for the exotic octet of gauge bosons from forward-backward asymmetry in top pair production

CDF has reported two anomalies related to the production of top quark pairs. The production rate for the pairs is too high and the forward backward asymmetry is also anomalously high. Both these anomalies could be understood as support for the octet of gauge bosons associated predicted by TGD based explanation of family replication phenomenon [3]. The exchange of both gauge bosons would induce both charged and neutral flavour changing electroweak and color currents.

1. Two high production rate for top quark pairs

Both Jester and Lubos tell about top quark related anomaly in proton-antiproton collisions at Tevatron reported by CDF collaboration. The anomaly has been actually reported already last summer but has gone un-noticed. For more detailed data see this [42].

What has been found is that the production rate for jet pairs with jet mass around 170 GeV, which happens to correspond to top quark mass, the production cross section is about 3 times higher higher than QCD simulations predict. 3.44 sigma deviation is in question meaning that its probability is same as for the normalized random variable x/σ to be larger than 3.44 for Gaussian distribution $exp(-(x/\sigma)^2/2)/(2\pi\sigma^2)^{1/2}$. Recall that 5 sigma is regarded as so unprobable fluctuation that one speaks about discovery. If top pairs are produced by some new particle, this deviation should be seen also when second top decays leptonically meaning a large missing energy of neutrino. There is however a slight deficit rather than excess of these events.

One can consider three interpretations.

1. The effect is a statistical fluke. But why just at the top quark mass?

2. The hadronic signal is real but there is a downwards fluctuation reducing the number of leptonic events slightly from the expected one. In the leptonic sector the measurement resolution is poorer so that this interpretation looks reasonable. In this case the decay of some exotic boson to top quark pair could explain the signal. Below this option will be considered in more detail in TGD framework and the nice thing is that it can be connected to the anomalously high forward backward asymmetry in top quark pair production reported by CDF for few weeks ago [62].

3. Both effects are real and the signal is due to R-parity violating 3-particle decays of gluinos with mass near to top quark mass. This is the explanation proposed in the paper of Perez and collabators.

2. Too high forward backward asymmetry in the production rate for top quark pairs

There is also a second anomaly involved with top pair production. Jester reports new data [88] about the strange top-pair forward-backward asymmetry in top pair production in p-pbar collisions already mentioned [62]. In Europhysics 2011 conference D0 collaboration reported the same result. CMS collaboration found however no evidence for the asymmetry in p-p collisions at LHC [66]. For top pairs with invariant mass above 450 GeV the asymmetry is claimed by CDF to be stunningly large 48+/-11 per cent. 3 times more often top quarks produced in qqbar annihilation prefer to move in the direction of quark. Note that this experiment would have reduced the situation from the level of ppbar collisions to the level of quark-antiquark collisions and the negative result suggests that valence quarks might play an essential role in the anomaly.

The TGD based explanation for the finding would relation on the flavor octet of gluons and the new view about Feynman diagrams.

1. The identification of family replication phenomenon in terms of genus of the wormhole throats (see this) predicts that family replication corresponds to a dynamical SU(3) symmetry (having nothing to do with color SU(3)or Gell-Mann's SU(3)) with gauge bosons belonging to the octet and singlet representations. Ordinary gauge bosons would correspond besides the familar singlet representation also to exotic octet representation for which the exchanges induce neutral flavor changing currents in the case of gluons and neutral weak bosons and charge changing ones in the case of charged gauge bosons. The exchanges of the octet representation for gluons could explain both the anomalously high production rate of top quark pairs and the anomalously large forward backward asymmetry! Also electroweak octet could of course contribute.

2. This argument requires a more detailed explanation for what happens in the exchange of gauge boson changing the genus. Particles correspond to wormhole contacts. For topologically condensed fermions the genus of the second throat is that of sphere created when the fermionic CP_2 vacuum extremal touches background space sheet. For bosons both wormhole throats are dynamical and the topologiies of both throats matter. The exchange diagram corresponds to a situation in which $g = g_i$ fermionic wormhole throat from past turns back in time spending some time as second throat of virtual boson wormhole contact and $g = g_f$ from future turns back in time and defines the second throat of virtual boson wormhole contact. The turning corresponds to gauge boson exchange represented by a wormhole contact with $g = g_i$ and $g = g_f$ wormhole throats. Ordinary gauge bosons are quantum superpositions of (g, g) pairs transforming as SU(3) singlets and SU(3) charged octet bosons are of pairs (g_1, g_2) with $g_1 \neq g_2$. In the absence of topological mixing of fermions inducing CKM mixing the exchange is possible only between fermions of same generation. The mixing is however large and changes the situation.

3. One could say that top quark from the geometric future transforms at exchange line to space-like t-quark (genus $g = 2$) and returns to future. The quark from the geometric past does the same and returns back to the past as antiquark of antiproton. In the exchange line this quark combines with t-quark to form a virtual color octet gluon.

This mechanism could also give additional contributions to the mechanism generating CP breaking since new box diagrams involving two exchanges of flavor octet bosons contribute to the mixings of quark pairs in mesons. The exchanges giving rise to an intermediate state of two top quarks are expected to give the largest contribution to the mixing of the quark pairs making up the meson. This might relate to the observed anomalous like sign muon asymmetry in BBbar decay [70] suggesting that the CP breaking in this system is roughly 50 times larger than predicted by CKM matrix. It might be that there the new diagrams only amplify the CP breaking associated wiht CKM matrix rather than bringing in any new source of CP breaking.

3.3 The physics of $M - \overline{M}$ systems forces the identification of vertices as branchings of partonic 2-surfaces

For option 2) gluons are superpositions of $g = 0, 1, 2$ states with identical probabilities and vertices correspond to branchings of partonic 2-surfaces. Exotic gluons do not induce mixing of quark families and genus changing transitions correspond to light like 3-surfaces connecting partonic 2-surfaces with different genera. CKM mixing is induced by this topological mixing. The basic testable predictions relate to the physics of $M\overline{M}$ systems and are due to the contribution of exotic gluons and large direct CP breaking effects in $K - \overline{K}$ favor this option.

For option 1) vertices correspond to fusions rather than branchings of the partonic 2-surfaces. The prediction that quarks can exchange handle number by exchanging $g > 0$ gluons (to be denoted by G_g in the sequel) could be in conflict with the experimental facts.

1. CP breaking in $K - \bar{K}$ as a basic test

CP breaking physics in kaon-antikaon and other neutral pseudoscalar meson systems is very sensitive to the new physics. What makes the situation especially interesting, is the recently reported high precision value for the parameter ϵ'/ϵ describing direct CP breaking in kaon-antikaon system [81]. The value is almost by an order of magnitude larger than the standard model expectation. $K - \bar{K}$ mass difference predicted by perturbative standard model is 30 per cent smaller than the the experimental value and one cannot exclude the possibility that new physics instead of/besides non-perturbative QCD might be involved.

In standard model the low energy effective action is determined by box and penguin diagrams. $\Delta S = 2$ piece of the effective weak Lagrangian, which describes processes like $s\bar{d} \rightarrow d\bar{s}$, determines the value of the $K - \bar{K}$ mass difference Δm_K and since this piece determines $K \rightarrow \bar{K}$ amplitude it also contributes to the parameter ϵ characterizing indirect CP breaking. $\Delta S = 2$ part of the weak effective action corresponds to box diagrams involving two W boson exchanges.

2. Δm_K kills option a

For option 1) box diagrams involving Z and $g > 0$ exchanges are allowed provided exchanges correspond to exchange of both Z and $g > 0$ gluon. The most obvious objection is that the exchanges of $g > 0$ gluons make strong $\Delta S > 0$ decays of mesons possible: $K_S \rightarrow \pi\pi$ is a good example of this kind of decay. The enhancement of the decay rate would be of order $(\alpha_s(g = 1)/\alpha_{em})^2(m_W/m_G(g = 1)^2 \sim 10^3$. Also other $\Delta S = 1$ decay rates would be enhanced by this factor. The real killer prediction is a gigantic value of Δm_K for kaon-antikaon system resulting from the possibility of $\bar{s}d \rightarrow \bar{d}s$ decay by single $g = 1$ gluon exchange. This prediction alone excludes option 1).

3. Option 2) could explain direct CP breaking

For option 2) box diagrams are not affected in the lowest order by exotic gluons. The standard model contributions to Δm_K and indirect CP breaking are correct for the observed value of the top quark mass which results if top corresponds to a secondary padic length scale $L(2, k)$ associated with $k = 47$ (Appendix). Higher order gluonic contribution could increase the value of Δm_K predicted to be about 30 per cent too small by the standard model.

In standard model penguin diagrams contribute to $\Delta S = 1$ piece of the weak Lagrangian, which determines the direct CP breaking characterized by the parameter ϵ'/ϵ. Penguin diagrams, which describe processes like $s\bar{d} \rightarrow d\bar{d}$, are characterized by effective vertices dsB, where B denotes photon, gluon or Z boson. dsB vertices give the dominant contribution to direct CP breaking in standard model. The new penguin diagrams are obtained from ordinary penguin diagrams by replacing ordinary gluons with exotic gluons.

For option 2) the contributions predicted by the standard model are multiplied by a factor 3 in the approximation that exotic gluon mass is negligible in the mass scale of intermediate gauge boson. These diagrams affect the value of the parameter ϵ'/ϵ characterizing direct CP breaking in $K-\bar{K}$ system found experimentally to be almost order of magnitude larger than standard model expectation [81].

4 Super-symmetry

TGD based vision about supersymmetry has developed rather slowly.

1. From the beginning it was clear that super-conformal symmetry is realized in TGD but differs in many respects from the more standard realizations such as $\mathcal{N}=\infty$ SUSY realized in MSSM [33] involving Majorana spinors in an essential manner. The covariantly constant right-handed neutrino generates the super-symmetry at the level of CP_2 geometry and the construction of super-partners would be more or less equivalent with the addition of right-handed covariantly constant right-handed neutrino and antineutrinos. It was however not clear whether space-time supersymmetry is realized at all since one could argue that that these states are just gauge degrees of freedom.

2. A more general general SUSY algebra is generated by the generalized eigen modes of the modified Dirac operator at partonic 2-surface. The value of $\mathcal{N}$ can be very large-even infinite- for this algebra and the SUSY is badly broken: this picture leads to a construction of QFT limit of TGD [11], which seems to be crucial also for the understanding of TGD itself. Right-handed neutrinos represent the sub-SUSY with minimal breaking induced by the mixing of right- and left handed neutrinos caused by the properties of the modified gamma matrices for which mixing between M^4 and CP_2 gamma matrices takes place induced breaking of M^4 chirality serving as a signature for massivation.

3. R-parity conservation leading to strong predictions in the case of MSSM is broken and since super-particles can decay to neutrino and particles the life-times of super-partners are finite and there is no lightest sparticle. One cannot of course take this view as the final one: only the experimental input allows to fill in the details correctly. It is however clear that TGD based SUSY differs dramatically from the SUSY as it is usually understood and that LHC could allow to decide which of these views is nearer to truth.

4.1 A concise view about SUSY phenomenology in TGD inspired Universe

The results from SUSY have been very inspiring and forced me to learn the basics of MSSM phenomenology and abstract what it shares with TGD view. As always this kind of process has bee very useful. In the following I summarize the big ideas which distinguish TGD SUSY from MSSM and try to formulate the resulting general phenomenological picture. Of course, one can argue that this kind of approach assuming that QFT calculations can be interpreted in TGD framework is adhoc since twistor approach applied in TGD means deviation from QFT at the fundamental level. I just assume that QFT approach is a good approximation to TGD and look for consequences. Also the experimental constraints on SUSY parameters deduced in the framework of MSSM model with R-symmetry are discussed in TGD framework and shown to be relaxed considerably so that it is possible to avoid problems plaguing MSSM approach: mention only little hierarchy problem and conficting demands from g-2 anomaly of muon and from squark mass limits.

4.1.1 Super-conformal invariance and generalized space-time supersymmetry

Super-conformal symmetry is behind the space-time supersymmetry in TGD framework. It took long time to get convinced that one obtains space-time supersymmetry in some sense.

1. The basic new idea is that the fermionic oscillator operators assignable to partonic 2-surfaces defined the SUSY algebra analagous to space-time SUSY algebra. Without length scale cutoff the number of oscillator operators is infinite and one would have $\mathcal{N}=\infty$ SUSY. Finite measurement resolution

realized automatically by the dynamics of the modified Dirac action however effectively replaces the partonic two-surface with a discrete set of points which is expected to be finite and one obtains reduction finite value of $\mathcal{N}$. Braids become basic objects in finite measurement resolution at light-like 3-D wormhole throats. String world sheets become the basic objects at the level of 4-D space-time surfaces and it is now possible to identify them uniquely and a connection with the theory of 1-knots, their cobordisms and 2-knots emerges.

The supersymmetry involves several algebras for which the fermionic oscillator operator algebra serves as a building block. Also the gamma matrices of the world of classical worlds (WCW) are expressible in terms of the fermionic oscillator operators so that fermionic anticommutations have purely geometric interpretation. The presence of SUSY in the sense of conservation of fermionic supra currents requires a consistency condition. For induced gamma matrices the surface must be minimal surfaces (extremal of volume action). For Kähler action and Chern-Simons action one must replace induced gammas by modified ones defined by the contractions of canonical momentum densities with the imbedding space gamma matrices.

The super-symmetry in TGD framework differs from that in standard approach in that Majorana spinors are not involved. One has 8-D imbedding space spinors with interpretation as quark and lepton spinors. This makes sense because color corresponds to color partial waves and quarks move in triality $t = \pm 1$ and leptons in $t = 0$ waves. Baryon and lepton number are conserved exactly.

2. A highly non-trivial aspect of TGD based SUSY is bosonic emergence meaning that bosons can be constructed from fermions. Zero energy ontology makes this construction extremely elegant since both massive states and virtual states are composites of massless states. The mass shell constraints on vertices reduce dramatically the momentum space for virtual momenta and resolves UV divergences. General arguments support the existence of pseudoscalar Higgs but it is not quite clear whether its existence is somehow forbidden by symmetries. Scalar and pseudoscalar Higgs transforming according to 3+1 decomposition under weak SU(2) replace two complex doublets of MSSM if this is not the case. This difference is essentially due to the fact that spinors are not M^4 spinors but $M^4 \times CP_2$ spinors. A proper notation would be B, h_B for the gauge bosons and corresponding Higgs particles and one expects that electroweak mixing characterized by Weinberg angle takes place for neutral Higgs particles and also their super-counterparts.

   The sfermions associated with left and right handed fermions should couple to fermions via $P_\pm = 1 \pm \gamma_5$ so that one can speak about left- and right-handed scalars. Maximal mixing between them leads to scalar and pseudoscalar. This observation raises a question about Higgs and its pseudoscalar variant. Could one also assume that the initial states are right and left handed Higgs and that maximal mixing leads to scalar and pseudoscalar with scalar eaten by gauge bosons?

3. Also spin one particles regarded usually as massless must have small mass and this means that Higgs scalar is completely eaten by gauge bosons. Also scalar gluons are predicted and would be eaten by gluons to develop a small mass. This resolves the IR difficulties of massless gauge theories and conjecture to make possible exact Yangian symmetry in the twistor approach to TGD. The disappearance of Higgs means that corresponding limits on the parameters of SUSY are lost. For instance the limits in $(tan(\beta), M_{SUSY})$ coming from Higgs mass do not hold anymore. The disappearence of Higgs means also the disappearance of the little hierarchy problem, which is one of the worst headaches of MSSM SUSY: no Higgs- no Higgs mass to be stabilized.

4.1.2 Induced spinor structure and purely geometric breaking of SUSY

Particle massivation and the breaking of SUSY and R-symmetry are the basic problems of both QFT and stringy approach to SUSY. Just by looking the arguments related to how MSSM could emerge from string models make clear how hopelessly ad hoc the constructions are. The notion of modified gamma matrix provides a purely geometric approach to these symmetric breakings involving no free parameters. To my view the failure to realize partially explains the recent situation in the forefront of theoretical physics and LHC findings are now making clear that something is badly wrong.

1. The space-time supersymmetry is broken. The reason is that the modified gamma matrices are superpositions of M^4 and CP_2 gamma matrices. This implies mixing of M^4 chiralities, which is a direct symptom of massivation and is responsible for Higgs like aspects of massivation: p-adic thermodynamics is second and completely new aspect. There is hierarchy of supersymmetries according to the strength of breaking. Right handed neutrino generates supersymmetry which is broken only by the mixing of right handed neutrino with left handed one and induced by the mixing of gamma matrices. This corresponds to the supersymmetry analogous to that of MSSM. The supersymmetries generated by other fermionic oscillator operators with electroweak quantum numbers break the super-symmetry in much more dramatic manner but the basic algebra remains and should allow an elegant formulation of TGD in terms of generalized super fields [11].

2. One important implication is R-parity breaking due to the transformation of right handed neutrino of superpartner to left-handed neutrino. If this takes place fast enough, the process $\tilde{P} \rightarrow P + \nu$ becomes possible. The universal decay signature would be lonely neutrino representing missing energy without accompanying charged lepton. This means that the experimental limits on sparticle masses deduced assuming R-symmetry do not hold anymore. For instance, the masses of charginos and neutralinos can be considerably lower than weak mass scale as suggested by strange 1995 event [96]. The recent very high lower bounds of squark masses putting them above 800 GeV assume also R-symmetry and therefore need not hold true if the decays $\tilde{h} \rightarrow q + \nu$ and $\tilde{g} \rightarrow g + \nu$ take place fast enough. A further implication is that the scale of SUSY can be weak mass scale- say the p-adic mass scale of 105 GeV corresponding to Mersenne prime M_{89}.

4.1.3 p-Adic length scale hypothesis and breaking of SUSY by a selection of p-adic length scale

p-Adic length scale symmetry leads to a completely new view about SUSY breaking which leads to extremely strong prediction. The basic conjecture is that if the p-adic length scales associated with particles and sparticle are same, their masses are identical. The basic aspect of SUSY breaking is therefore different p-adic mass scales of particle and sparticle. By p-adic length scale hypothesis the masses of particle and superpartner therefore differ by a power of $\sqrt{2}$. This is extremely powerful prediction and given only minimal kinematical constraints on the event suggesting super-symmetry allows to deduce the mass of super-partner. Some examples give an idea about what is involved.

1. The idea about M_{89} as the prime characterizing both electroweak scale and SUSY mass scale leads to the proposal that all superpartners correspond to the same p-adic mass scale characterized by $k = 89$. For instance, the masses of sfermions would be given in first order p-adic thermodynamics calculation by

$$\begin{aligned} \frac{m_{\tilde{L}}}{GeV} &= (262, 439, 945) \ , \\ \frac{m_{\tilde{\nu}}}{GeV} &= (235, 423, 938) \ , \\ \frac{m_{\tilde{U}}}{GeV} &= (262, 287, 893) \ , \\ \frac{m_{\tilde{D}}}{GeV} &= (235, 287, 900) \ . \end{aligned} \tag{4.1}$$

   The recent results from LHC force imply that in absence of R-parity breaking the masses of squarks must be in the region 800-1000 GeV, which forces to conclude that neutrino mixing must be fast enough to allow direct decays $\tilde{P} \rightarrow P + \nu$.

2. It should be noticed that the single strange event reported 1995 [96] discussed in [16] gives for the mass of selectron the estimate 131 GeV, which corresponds to M_{91} instead of M_{89} and is thus one half of the selectron mass for Mersenne option. This event allowed also to estimate the masses of Zino and corresponding Higgsino. The results are summarized by the following table:

$$m(\tilde{e}) = 131\ GeV\ ,\quad m(\tilde{Z}^0) = 91.2\ GeV\ ,\quad m(\tilde{h}) = 45.6\ GeV\ . \tag{4.2}$$

If one takes these results at face value one must conclude either that M_{89} hypothesis is too strong or M_{SUSY} corresponds to M_{91} or that M_{89} is correct identification but also sfermions can appear in several p-adic mass scales.

3. In case of the mixing of gaugino and corresponding Higgsino the hypothesis means that the mass matrix is of such form that its eigenvalues have same magnitude but opposite sign. For instance, for the mixing of wino and h_W one would have

$$M = \begin{pmatrix} M_2 & \sqrt{2}M_W cos(\beta) \\ \sqrt{2}M_W cos(\beta) & \mu \end{pmatrix}$$

so that the hypothesis implies

$$M_2 = -\mu\ .$$

The masses of the resulting two states would be same but they could correspond to different p-adic primes so that mass scales would differ by a power of $\sqrt{2}$. This formula applies also zino and $\tilde{h}_Z$ and photino and $\tilde{h}_\gamma$. One possibility is that heavier weak gaugino corresponds to intermediate gauge boson mass scale and light gaugino to 1/2 of this scale: lighter mass scales are forbidden by the decay widths of weak gauge bosons. The exotic event of 1995 suggest that heavier zino has very nearly the same mass as Z^0 and the lighter one mass which in good approximation equal to one half of Z^0 mass. This would mean $M_2 = -\mu << M_Z$.

4. If one accepts the SUSY formula

$$m_{\tilde{\nu}}^2 = m_{\tilde{L}}^2 + \frac{1}{2}M_{\tilde{Z}}^2 cos(2\beta)$$

relating neutrino and sneutrino masses, one can conclude that $cos(\beta) = 1/\sqrt{2}$ is the only possible option so that $tan(\beta) = 1$ is obtained. This value is excluded by R-parity conserving MSSM but could be consistent with the explanation of g-2 anomaly of muon in terms of loops involving weak gauginos and corresponding higgsinos.

5. One must distinguish between right- and left handed sfermions $\tilde{F}_R$ and $\tilde{F}_L$. These states couple via $1 \pm \gamma_5$ to fermions and are not therefore either pure scalars or pure pseudoscalars. One expects that a maximal mixing of left and right handed sfermions occurs and leads to scalar and pseudoscalar. Mass formula is naturally same for the states for the same value of p-adic prime and also same value of p-adic prime is suggestive. It might however happen that a p-adic mass scales are different for scalars and pseudoscalars. This would allow to have light and heavy variants of squarks and sleptons with scalars probably being the lighter ones.

To sum up, TGD leads to a very predictive model for SUSY and its breaking.

1. Since there is no Higgs, there are no bounds to parameters from Higgs mass and little hierarchy problem is avoided.

2. The basic element is R-parity breaking reflected in the possibility of decays of sparticles to particle and lonely neutrino not balanced by charge lepton. This together with the absence of Higgs would allow to circumvent various mass limits deduced from LHC and its predecessors and only the limits from the decay width of intermediate gauge bosons on charginos and neutralinos would remain.

3. The assumption that the masses of particles and sparticles are same for same p-adic length scale and that the choice of p-adic length scale breaks SUSY means that sparticle masses can be deduce from those of particles apart from a scaling by a power of $\sqrt{2}$. This is a powerful and directly testable prediction and predicts the 2×2 mixing matrices for charginos and neutralinos completely apart from the parameter $M = -\mu$. One must however take it very cautiously in TGD framework since in TGD framework higgsino might disappear from the spectrum just like Higgs and the situaton becomes degenerate. An attractive idea is that m_{SUSY} corresponds to p-adic mass scale associated with Mersenne prime M_{89} characterizing electro-weak length scale and that Mesrenne hypothesis poses conditions on sfermion masses so that all sparticles so that the only light particles are selectron and sneutrinos and other sfermions have masses about 13 TeV.

4. Muon g-2 anomaly could be understood from the predictions $tan(\beta) = 1$, $m_{SUSY} \simeq 105$ GeV if one assumes that sneutrinos are light. Scalar and pseudoscalar sfermions could in principle have different mass scales but there seems to be no recently accepted anomalies requiring this.

4.1.4 Do also Higgsinos disappear from the spectrum?

In the following still some comments about TGD based view about symmetry breaking, Higgs, electroweak symmetry breaking, and SUSY. There are several unclear issues at the level of detais. This is thanks to my unforgivable laziness in writing down the details. The results from LHC are however so fascinating that they force me to win my laziness. In the following I try to clarify my thoughs.

1. Is the earlier conjectured pseudoscalar Higgs there at all?

Spin 1 gauge bosons and Higgs differ only by different spin direction of fermions at opposite wormhole throats. For spin 1 gauge bosons the 3-momenta at two wormhole throats cannot be parallel if if one wants non-vanishing spin component in the direction of moment. 3-momenta are most naturally opposite for the massless states at throats. This forces massivation for all gauge bosons and even graviton and this in turn requires Higgs even in the case of gluons.

The question is whether the parity properties of the couplings of gauge boson and corresponding Higgs transforming like 3+1 under SU(2) (this is due to the special character of imbedding space spinors) be exactly the same? Higgs would couple like a mixture of scalar and pseudoscalar to fermions just as weak gauge bosons couple and the mixture would be just the same. If there are no axial variants of vector gauge bosons there should exist no pseudoscalar Higgs. The nonexistence of axial variants of vector gauge bosons is suggested by quantum classical correspondence: only gauge bosons having classical space-time correlates as induced gauge potentials should be allowed, nothing else. Note that color variant of Higgs would exist and would be eaten by gluons to get mass.

2. Could Higgs mechanism lead to the disappearance of also Higgsinos?

The similarity of the construction of gauge bosons and Higgsinos as pairs of wormhole throats containing fermion and antifermion encourages to think that Higgs mechanism is invariant under supersymmetries. If so, also Higgsinos would be eaten and one would have massive super-symmetric gauge theory with fermions with photon and other massless particle possessing a tiny mass. This looks very simple. The testable implication would be that only weak gauginos should contribute to muon g-2 anomaly.

3. Electroweak symmetry breaking

The recent view about electroweak symmetry breaking is less than year old. The basic realization was that wormhole throats carrying elementary particle quantum numbers possess Kähler magnetic charge (in homological sense- CP_2 has non-trivial second homology). This magnetic charge must be compensated and this is achieved if the particle wormhole throat is connected to a second wormhole throat by a magnetic flux tube. The second wormhole would carry a weak charge of neutrino pair compensating the weak isospin of the particle so that weak interactions would be screened above the weak length scale. For colored states the compensation could also occur in longer length scale and corresponds to color confinement.

This does not actually require the length scale of flux tubes associated with all elementary particles to be the weak length scale as I have thought. Rather, the flux tube length for a particle at rest could correspond to the Compton length of the particle. For instance, for electron the maximal flux tube length would be about 10^{-13} meters. For particles not at rest the length would get shorter by length contraction. For very light but massive particles such as photon and graviton the maximum length of flux tube would be very long. The interaction of very low energy photons and gravitons would be essentially classical and induced by the classical oscillations of induced gauge fields induced by a long flux tube connecting the interacting systems. For high energy quanta this interaction would be essentially quantal and realized as absorption of quanta with flux tube length -essentially wave length of quantum- much shorter than the distance between the interacting systems. Gravitational waves would interact essentially classically even when absorbed since absorption would mean that the flux tube would connects two parts of the measurement apparatus. For large hbar gravitons the length of flux tube could correspond to the distance between interaction systems.

A fascinating possibility is that electronic Cooper pairs of superconductors with large value of hbar, could correspond to long flux tubes with electron's quantum numbers at both ends. Maybe this takes place in high T_c super conductors.

4. Some details of the SUSY predictions

TGD SUSY differs from the standard SUSY in many respects.

1. All fermionic oscillator operators assignable to the wormhole throats generate supersymmetries. These oscillator operators differ from ordinary ones in that they do not have momentum label and momentum can be only assigned to the entire state. Therefore the interpretation of all states assignable to wormhole throats as large SUSY multiplet is possible. This SUSY is badly broken and there is hierarchy of breakings defined by the interactions inducing the breaking in turn define by the quantum numbers of SUSY generators. For quark generators the breaking is largest and the smallest breaking is associated with the oscillator operators assignable to right-handed neutrinos since they have only gravitational interactions.

2. The symmetry generators are not Majorana spinors and this does not lead to any difficulties as has been found. Only if one would try stringy quantization trying to define stringy diagrams in terms of stringy propagators defined by stringy form of super-conformal algebra, one would end up with difficulties. Majorana property is also excluded by the separate conservation of baryon and lepton number.

   For single wormhole throat one can see the situation in terms of N=2 SUSY with right handed neutrino and its antiparticle appearing as SUSY generators carrying conserved fermion number. One can classify the superpartners by their right-handed neutrino number which is +/-1. For instance, for single wormhole throat one obtains fermion and its partner containing ν_R pair, and fermion number 0 and fermion number 2 sfermions. In the case of gauge bosons and Higgs similar degeneracy is obtained for both wormhole throats.

3. Since induced gamma matrices and modified gamma matrices are mixtures of M^4 and CP_2 gamma matrices right handed neutrino is mixed with the left handed neutrino meaning breaking of R-parity. The simplest decays of sparticles are of form $P \rightarrow P + \nu$ and can be said to be gravitationally induced since the mixing of gamma matrices is indeed a characteristic phenomenon of induced spinor structure. Also more complex decays with neutrino replaced with charge lepton are possible. The basic signature is lonely lepton not possible in decays of weak bosons.

4. The basic outcome of SUSY QFT limit of TGD [11] is that wormhole throat can carry only spin 0,1/2,1 corresponding to fermion and fermion pair if one wants to obtain standard propagator: otherwise one obtains $1/p^n$, $n > 2$ and this is not an ordinary particle pole. The reason is that one cannot assign to fermionic oscillator operators independent momenta but only common momentum so they propagate effectively collinearly.

   One can criticize this argument as being inconsistent with the twistorial approach combined with zero energy ontology implying that wormhole throats are massless even for on mass shell states. In this approach one in principle avoids completely the use of propagators which would of course diverge for on shell wormhole throats. Also for twistor diagrams the counterparts of virtual particles are massless and off shell. The so called region momentum replaces momentum in Grassmannian twistor approach and has a direct counterpart as eigenvalue of the modified Dirac operator so that the analog of propagator exists in TGD framework. Since QFT limit must be a reasonable approximation to the full theory, one might hope that the QFT based argument makes sense when one replaces momentum with region momentum (or pseudo momentum as I have called it in TGD framework).

5. Should one allow both $\nu_R$and its antiparticle as SUSY generators? This would mean more states as in standard SUSY for which only $\overline{\nu}_R$ would be allowed for fermion. This would assign to a given wormhole throat with fermion number 1 spin 1 and spin 0 super partner and companion of fermion containing $\nu_R - \overline{\nu}_R$ pair. For this state however propagator would behave like $1/p^3$ should that again strong SUSY breaking would occur for this extended SUSY. Only one half of SUSY would be broken weakly by the mixing of M^4 and CP_2 gamma matrices appearing in modified gamma matrices: the mixing would not involve weak or color interactions but could be said to be gravitational but not in the sense of abstract for geometry but induced geometry.

   The breaking of symmetries by this mechanicsm would be a beautiful demonstration that it is sub-manifold geometry rather than abstract manifold geometry that matters. Again string theorists managed to miss the point by effectively elimating induced geometry from the original string model by inducing the metric of space-time sheet as an independent variable. The motivation was that it became easy to calculate! The price paid was symmetry breaking mechanisms involving hundreds of three parameters.

6. Single wormhole contact could carry spin J=2 and give rise to graviton like state. If one constructs from this gravitino by adding right-handed neutrinos, and if SUSY QFT limit makes sense, one obtains particle with propagator decreasing faster at either throat so that gravitino in standard sense would not exist. This would represent strong SUSY breaking in gravitational sector. These results are of utmost importance since the basic argument in favor dimension D=10 or D=11 for the target space of superstring models is that higher dimensions would give fundamental massless particles with higher spin.

   Note that the replament of wormhole throats by flux tubes having neutrino pair at the second end of the flux tube complicates the situation since one can add right handed neutrino also to the neutrino end. The SUSY QFT criterion would however suggest that these states are not particle like.

4.1.5 Super-symplectic bosons

TGD predicts also exotic bosons which are analogous to fermion in the sense that they correspond to single wormhole throat associated with CP_2 type vacuum extremal whereas ordinary gauge bosons corresponds to a pair of wormhole contacts assignable to wormhole contact connecting positive and negative energy space-time sheets. These bosons have super-conformal partners with quantum numbers of right handed neutrino and thus having no electro-weak couplings. The bosons are created by the purely bosonic part of super-symplectic algebra [2, 1] , whose generators belong to the representations of the color group and 3-D rotation group but have vanishing electro-weak quantum numbers. Their spin is analogous to orbital angular momentum whereas the spin of ordinary gauge bosons reduces to fermionic spin. Recall that super-symplectic algebra is crucial for the construction of configuration space Kähler geometry. If one assumes that super-symplectic gluons suffer topological mixing identical with that suffered by say U type quarks, the conformal weights would be (5,6,58) for the three lowest generations. The application of super-symplectic bosons in TGD based model of hadron masses is discussed in [19] and here only a brief summary is given.

As explained in [19] , the assignment of these bosons to hadronic space-time sheet is an attractive idea.

1. Quarks explain only a small fraction of the baryon mass and that there is an additional contribution which in a good approximation does not depend on baryon. This contribution should correspond to the non-perturbative aspects of QCD. A possible identification of this contribution is in terms of super-symplectic gluons. Baryonic space-time sheet with $k = 107$ would contain a many-particle

state of super-symplectic gluons with net conformal weight of 16 units. This leads to a model of baryons masses in which masses are predicted with an accuracy better than 1 per cent.

2. Hadronic string model provides a phenomenological description of non-perturbative aspects of QCD and a connection with the hadronic string model indeed emerges. Hadronic string tension is predicted correctly from the additivity of mass squared for $J = 2$ bound states of super-symplectic quanta. If the topological mixing for super-symplectic bosons is equal to that for U type quarks then a 3-particle state formed by 2 super-symplectic quanta from the first generation and 1 quantum from the second generation would define baryonic ground state with 16 units of conformal weight. A very precise prediction for hadron masses results by assuming that the spin of hadron correlates with its super-symplectic particle content.

3. Also the baryonic spin puzzle caused by the fact that quarks give only a small contribution to the spin of baryons, could find a natural solution since these bosons could give to the spin of baryon an angular momentum like contribution having nothing to do with the angular momentum of quarks.

4. Super-symplectic bosons suggest a solution to several other anomalies related to hadron physics. The events observed for a couple of years ago in RHIC [83] suggest a creation of a black-hole like state in the collision of heavy nuclei and inspire the notion of color glass condensate of gluons, whose natural identification in TGD framework would be in terms of a fusion of hadronic space-time sheets containing super-symplectic matter materialized also from the collision energy. In the collision, valence quarks connected together by color bonds to form separate units would evaporate from their hadronic space-time sheets in the collision, and would define TGD counterpart of Pomeron, which experienced a reincarnation for few years ago [84]. The strange features of the events related to the collisions of high energy cosmic rays with hadrons of atmosphere (the particles in question are hadron like but the penetration length is anomalously long and the rate for the production of hadrons increases as one approaches surface of Earth) could be also understood in terms of the same general mechanism.

4.2 Experimental situation

The experimental situation in the case of SUSY is still open but it there are excellent hopes that the results from LHC will determine the fate of the MSSM SUSY and also constraint more general scenarios. Unfortunately, the research concentrates to the signatures of MSSM and its variants quite different from those of TGD SUSY so that it might happen that TGD SUSY will be discovered accidentally if its there: say by the decays of spartner to partner and neutrino. Already from the recent results it is clear that the allowed parameter space for MSSM SUSY is very small and that superpartners of quarks and also weak gauge bosons must be very heavy if MSSM SUSY is realized. This leads to difficulties with the only known evidence for SUSY coming from the g-2 anomaly of muon. TGD based SUSY allows light masses and also SUSY explanation of g-2 anomaly if sneutrino masses are light.

The representation involves a lot of references to blog postings and this might irritate so called serious scientists. I however feel that since blogs provide my only contact to the particle physics it is only fair to make clear that this communication tool is absolutely essential for a scientist working as out-of-law in academic community. Blogs could indeed bring democracy to science and mean end of the era of secrecy and censorship by the referee system.

4.2.1 Goodbye large extra dimensions and MSSM

New results giving strong constraints on large extra dimensions and on the parameters of minimally supersymmetric standard model (MSSM) have come from LHC and one might say that both larger extra dimensions and MSSM are experimentally excluded.

1. The problems of MSSM

According to the article The fine-tuning price of the early LHC by A. Strumia [48] the results from LHC reduce the parameter space of MSSM dramatically. Recall that the king idea of MSSM is that the presence of super partners tends to cancel the loop corrections from ordinary particles giving to Higgs mass much larger correction that the mass itself. Note that the essential assumption is that R-parity is an exact symmetry so that the lightest superpartner is stable. The signature of SUSY is indeed missing energy resulting in the decay chain beginning with the decay of gluino to chargino and quark pair followed by the decay of chargino to W boson and neutralino representing missing energy.

The article Search for supersymmetry using final states with onelepton, jets, and missing transverse momentum with the ATLAS detector in $s^{1/2} = 7$ TeV pp collisions [45] by ATLAS collaboration at LHC poses strong limits on the parameters of MSSM implying that the mass of gluino is above 700 GeV in the case that gluino mass is same as that of squark. In Europhysics 1011 meeting the lower bounds for squark and gluino masses were raised to about 1 TeV. The experimental lower bounds on masses of superpartners are so high and the upper bound on Higgs mass so low that the superpartners cannot give rise to large enough compensating corrections to stabilize Higgs. This requires fine-tuning even in MSSM known as little hierarchy problem.

In typical models this also means that the bounds on slepton masses are too high to be able to explain the muonic g-2 anomaly, which was one of the original experimental motivations for MSSM. Therefore the simplest candidates for supersymmetric unifications are lost. This strengthens the suspicion that something is badly wrong with the standard view about SUSY forcing among other things to assume unstability of proton due to non-conservation of baryon and lepton numbers separately.

2. The difficulties of large extra dimensions

The results from LHC do not leave much about the dream of solving hierarchy problem using SUSY. One must try something else. One example of this something else are large extra dimensions implying massive graviton, which could provide a new mechanism for massivation based on the idea that massive particle in Minkowski space are massless particles in higher dimensional space (also essential element of TGD). This could perhaps the little hierachy problem if the mass of Kaluza-Klein graviton is in TeV range.

The article LHC bounds on large extra dimensions by A. Strumia and collaborators [40] poses very strong constraints on large extra dimensions and mass and effective coupling constant parameter of massive graviton. Kaluza-Klein graviton would appear in exchange diagrams and loop diagrams for 2-jet production and could become visible in higher energy proton-proton collisions at LHC. KK graviton would be also produced as invisible KK-graviton energy in proton-proton collisions. The general conclusion from data gathered hitherto shrinks dramatically the allowed parameter space for the KK-graviton. Does this mean that we are left with the antrophic option?

3. Also M-theorists admit that there are reasons for the skepticism

Michael Dine admits in the article Supersymmetry From the Top Down [46] that there are strong reasons for skepticism. Dine emphasizes that the hierarchy problem related to the in-stablity of Higgs mass due to the radiative corrections is the main experimental motivation for SUSY but that little hierarchy problem remains the greatest challenge of the approach. As noticed, in TGD this problem is absent. The same basic vision based on zero energy ontology and twistors predicts among other things

- the cancellation of UV and IR infinities in generalized Feynman (or more like twistor-) diagrammatics,
- predicts that in the electroweak scale the stringy character of particles identifiable as magnetically charged wormhole flux tubes should begin to make itself manifest,
- particles regarded usually as massless eat all Higgs like particles accompanying them (here "predict" is perhaps too strong a statement),
- also pseudo-scalar counterparts of Higgs-like particles which avoid the fate of their scalar variants (there already exist indications for pseudo-scalar gluons).

Combined with the powerful predictions of p-adic thermodynamics for particle masses these qualitative successes make TGD a respectable candidate for the follower of string theory.

4.3 Could TGD approach save super-symmetry?

In TGD framework the situation is not at all so desolate. Due to the differences between the induced spinor structure and ordinary spinors, Higgs corresponds to SU(2) triplet and singlet in TGD framework rather than complex doublet. The recent view about particles as bound states of massless wormhole throats forced by twistorial considerations and emergence of physical particles as bound states of wormhole contacts carrying fermion number and vibrational degrees of freedom strongly suggests- I do not quite dare to say "implies"- that also photon and gluons become massive and eat their Higgs partners to get longitudinal polarization they need. No Higgs- no fine tuning of Higgs mass- no hierarchy problems.

Note that super-symmetry is not given up in TGD but differs in many essential respects from that of MSSM. In particular, super-symmetry breaking and breaking of R-parity are automatically present from the beginning and relate very closely to the massivation.

1. If the gamma matrices were induced gamma matrices, the mixing would be large by the light-likeness of wormhole throats carrying the quantum numbers. Induced gamma matrices are however excluded by internal consistency requiring modified gamma matrices obtained as contractions of canonical momentum densities with imbedding space gamma matrices. Induced gamma matrices would require the replacement of Kähler action with 4-volume and this is unphysical option.

2. In the interior Kähler action defines the canonical momentum densities and near wormhole throats the mixing is large: one should note that the condition that the modified gamma matrices multiplied by square root of metric determinant must be finite. One should show that the weak form of electric-magnetic duality guarantees this: it could even imply the vanishing of the limiting values of these quantities with the interpretation that the space-time surfaces becomes the analog of Abelian instanton with Minkowski signature having vanishing energy momentum tensor near the wormhole throats. If this is the case, Euclidian and Minkowskian regions of space-time surface could provide dual descriptions of physics in terms of generalized Feynman diagrams and fields.

3. At wormhole throats Abelian Chern-Simons-Kähler action with the constraint term guaranteeing the weak form of electric-magnetic duality defines the modified gamma matrices. Without the constraint term Chern-Simons gammas would involve only CP_2 gamma matrices and no mixing of M^4 chiralities would occur. The constraint term transforming TGD from topological QFT to almost topological QFT by bringing in M^4 part to the modified gamma matrices however induces a mixing proportional to Lagrange multiplier. It is difficult to say anything precise about the strength of the constraint force density but one expect that the mixing is large since it is also large in the nearby interior.

If the mixing of the modified gamma matrices is indeed large, the transformation of the right-handed neutrino to its left handed companion should take place rapidly. If this is the case, the decay signatures of spartners are dramatically changed as will be found and the bounds on the masses of squarks and gluinos derived for MSSM do not apply in TGD framework.

4.3.1 Proposal for the mass spectrum of sfermions

In TGD framework p-adic length scale hypothesis (stating that preferred p-adic primes come as $p \simeq 2^k$, k integer) allows to predict the masses of sleptons and squarks modulo scaling by a powers $\sqrt{2}$ determined by the p-adic length scale by using information coming from CKM mixing induced by topological mixing of particle families in TGD framework. Also natural guesses for the mass scales of ew gauginos and gluinos are obtained.

1. If one assumes that the mass scale of SUSY corresponds to Mersenne prime M_{89} assigned with intermediate gauge bosons one obtains unique predictions for the various masses apart from uncertainties due to the mixing of quarks and neutrinos [14].

2. In first order the p-adic mass formulas for fermions read as

$$\begin{aligned} m_F &= \sqrt{\frac{n_F}{5}} \times 2^{(127-k_F)/2} \times m_e \ , \\ n_L &= (5,14,65) \ , \ n_\nu = (4,24,64) \ , \ n_U = (5,6,58) \ , \ n_D = (4,6,59) \ . \end{aligned} \tag{4.3}$$

Here k_F is the integer characterizing p-adic mass scale of fermion via $p \simeq 2^{k_F}$. The values of k_F are not listed here since they are not needed now. Note that electroweak symmetry breaking distinguish U and D type fermions is very small when one uses p-adic length scale as unit.

By taking $k_F = 89$ for super-partners as a reference mass scale, one obtains in good approximation (the first calculation contained erranous scaling factor)

$$\begin{aligned} \frac{m_{\tilde{L}}}{GeV} &= 2^{(89-k_F)/2}(262,439,945) \ , \\ \frac{m_{\tilde{\nu}}}{GeV} &= 2^{(89-k_F)/2}(235,423,938) \ , \\ \frac{m_{\tilde{U}}}{GeV} &= 2^{(89-k_F)/2}(262,287,893) \ , \\ \frac{m_{\tilde{D}}}{GeV} &= 2^{(89-k_F)/2}(235,287,900) \ . \end{aligned} \tag{4.4}$$

Charged leptons correspond to subsequent Mersennes or Gaussian Mersennes. The first guess is that this holds true also for charged sleptons. This would give $k_F(\tilde{e}) = 89$, $k_F(\tilde{\mu}) = 79$, and $k_F(\tilde{\tau}) = 61$. For quarks one has $k_F(q) \geq 113$ ($k = 113$ corresponds to Gaussian Mersenne). If one generalizes this to $k_F(\tilde{q}) \leq 79$, all sfermion masses expect those of selectron and sneutrinos are above 13 TeV. This option might well be consistent with the recent experimental data require that squark masses are above 1 TeV. The possible problem is selectron mass 262 GeV.

3. The simplest possibility is that ew gauginos are characterized by $k = 89$ and have same masses as W and Z in good approximation. Therefore $\tilde{W}$ could be the lightest super-symmetric particle and could be observed directly if the neutrino mixing is not too fast and allowing the decay $\tilde{W} + \nu$. Also gluinos could be characterized by M_{89} and have mass of order intermediate gauge boson mass. For this option to be discussed below the decay scenario of MSSM changes considerably.

4. It should be noticed that the single strange event reported 1995 [96] discussed in [?] gives for the mass of selectron the estimate 131 GeV, which corresponds to M_{91} instead of M_{89} and is thus one half of the selectron mass for Mersenne option. This event allowed also to estimate the masses of Zino and corresponding Higgsino. The results are summarized by the following table:

$$m(\tilde{e}) = 131 \ GeV \ , \ m(\tilde{Z}^0) = 91.2 \ GeV \ , \ m(\tilde{h}) = 45.6 \ GeV \ . \tag{4.5}$$

If one takes these results at face value one must conclude either that M_{89} hypothesis is too strong or M_{SUSY} corresponds to M_{91} or that M_{89} is correct identification but also sfermions can appear in several p-adic mass scales.

The decay cascades searched for in LHC are initiated by the decay $q \rightarrow \tilde{q} + \tilde{g}$ and $g \rightarrow \tilde{q} + \tilde{q}_c$. Consider first R-parity conserving decays. Gluino could decay in R-parity conserving manner via $\tilde{g} \rightarrow \tilde{q} + q$. Squark in turn could decay via $\tilde{q} \rightarrow q_1 + \tilde{W}$ or via $\tilde{q} \rightarrow q + \tilde{Z}^0$. For the proposed first guess about masses the decay $\tilde{W} \rightarrow \nu_e + \tilde{e}$ or $\tilde{Z}^0 \rightarrow \nu_e + \tilde{\nu}_e$ would not be possible on mass shell.

If the mixing of right-handed and left-handed neutrinos is fast enough, R-parity is not conserved and the decays $\tilde{g} \rightarrow g + \nu$ and $\tilde{q} \rightarrow q + \nu$ could take place by the mixing $\nu_R \rightarrow \nu_L$ following by electroweak interaction between ν_L quark or antiquark appearing as composite of gluon. The decay signature in this case would be pair of jets (quark and antiquark or gluon gluon jet both containing a lonely neutrino not accompanied by a charged lepton required by electroweak decays. Also the decays of electroweak gauginos and sleptons could produce similar lonely neutrinos.

The lower bound to quark masses from LHC is about 600 GeV and 800 GeV for gluon masses assuming light neutralino is slightly above the proposed masses of lightest squarks [37]. In Europhysics 2011 lower bounds were raised to 1 TeV for both gluino and squark masses. These bounds are consistent with the above speculative picture. These masses are allowed for R-parity conserving option if the decay rate producing chargino is reduced by the large mass of chargino the bounds become weaker. If the decay via R-parity breaking is fast enough no bounds on masses of squarks and gluinos are obtained in TGD framework but jets with neutrino unbalanced by a charged lepton should be observed.

4.3.2 How to relate MSSM picture to TGD picture?

In order to utilize MSSM calculation in TGD framework one must relate MSSM picture to TGD picture. The basic constraint is that Higgs is absent. This could apply also to Higgsino. This certainly simplifies the formulas. A further conditionis that superpartners obey the same mass formulas as partners for same pa-dic length scale.

It has been proposed that the loops involving superpartners could explain the anomaly [99]. In one-loop order one would have the processes $\mu \rightarrow \tilde{\mu} + \tilde{Z}^0$ and $\mu \rightarrow \tilde{\nu}_\mu + \tilde{W}^0$. The situation is complicated by the possible mixing of the gauginos and Higgsinos and in MSSM this mixing is described by the mixing matrices called X and Y. The general conclusion is however clear: if muonic sneutrino is light, it is possible to have sizable contribution to the g-2 anomaly.

1. Magnetic moment operator mixes different M^4 chiralities. For simplest one-loop diagrams this corresponds in TGD framework to coupling in the modified Dirac equation mixing different chiralities describable as an effective mass term. The couplings between right and left handed sfermions also contributes to the magnetic moment and these couplings reduce to those of sfermions being basically induced by the fermionic chirality mixing which reduces to the fact that modified gamma matrices are superpositions of M^4 and CP_2 gamma matrices.

2. The basic outcome in the standard SUSY approach is that the mixing is proportional to the factor m_μ^2/m_{SUSY}^2. One expects that in the recent situation $m_{SUSY} = m_W$ is a reasonable first guess so that the mixing is large and could explain the anomaly. Second guess is as M_{89} p-adic mass scale.

3. MSSM calculations for anomalous g-2 involve the mixing of both $\tilde{f}_L$ and $\tilde{f}_R$ and of gauginos and Higgsinos. In MSSM the mixing matrices involve the parameter $tan(\beta)$ where the angle β characterizes the ratio of mass scales of U and D type fermions fixed by the ratio of Higgs expectations for the two complex Higgs doublets [99]. $tan(\beta)$ also characterizes in MSSM the ratio of vacuum expectation values of two Higgses assignable to U and D type quarks and cannot be fixed from this criterion since in TGD framework one has one scalar Higgs and pseudo-scalar Higgs decomposing to triplet and singlet under SU(2) and the mass ratio is fixed by p-adic mass calculations.

The question is what happens if Higgs and Higgsino are absent and what one can conclude about the value of β in TGD framework where p-adic mass calculations give the dominant contribution to fermion masses and the mass formulas for particles and sparticles should be identical for a fixed p-adic prime.

1. Mixing of charged gauginos and Higgsinos

Consider first the mixing between charged gauginos and Higgsinos. The angle β characterizes also the mixing of $\tilde{W}$ and charged Higgsino parametrized by the mass matrix

$$X = \begin{pmatrix} M_2 & M_W\sqrt{2}sin(\beta) \\ M_W\sqrt{2}cos(\beta) & \mu \end{pmatrix} . \tag{4.6}$$

The $tan(\beta)$ gives the ratio of mass scales of U and D type quarks in MSSM. In MSSM $tan(\beta)$ reduces to the ratio of Higgs vacuum expectations and it would be better to get rid of the entire parameter in

TGD framework. The maximally symmetric situation corresponds to the same mass scale for U and D type quarks and this suggests that one has $sin(\beta) = cos(\beta) = 1/\sqrt{2}$ implying $tan(\beta) = 1$. In MSSM $tan(\beta) > 2$ is required and this is due to the large value of the m_{SUSY}.

Whether this parameterization makes sense in TGD framework depends on whether one allows Higgsino.

1. If also Higgsino is absent the formula does not make sense. A natural condition is that the value of $tan(\beta)$ does not appear at all in the limiting formulas for the anomalous g-2. Note that in p-adic mass calculations do not contain this kind of a priori continuous parameter. There the simplest TGD based option is that the Higgsino is just absent and the mass matrix reduces 1×1 matrix M_2 giving wino mass. The idea that particle and sparticles have identical masses for the same p-adic mass scale would give $M_2 = M_W$. One must however remember that in TGD framework mass operator acts like a preferred combination of gamma matrices in CP_2 degrees of freedom mixing M^3 chiralities.

2. If one allows Higgsinos, the simplest guess is that apart from p-adic mass scale same has $M_2 = -\mu = m$: this guarantees identical masses for the mixed states in accordance with the ideas that different masses for particles and sparticles result from the different p-adic length scale. For $cos(\beta) = 1/\sqrt{2}$ this would give mass matrix with eigen values $(M, -M)$, $M = \sqrt{m^2 + m_W^2}$ so that mass squared values of of the mixed states would be identical and above m_W mass for $p = M_{89}$. Symmetry breaking by an increase of the p-adic length scale could however reduce the mass of other state by a power of $\sqrt{2}$.

   If also winos and zinos eat the higgsinos, one can argue that the determinant of X must vanish so that the eigenstate with vanishing eigen value would correspond to unphysical statete meaning the elimination of second state from the spetrum. This would require $M_2\mu - M_W^2 sin(2\beta) = 0$. $sin(\beta) = 1/\sqrt{2}$ and $M_2 = \mu = M_W$ is the simplest solution to the condition. This looks tricky.

2. Mixing of neutral gauginos and Higgsinos

In MSSM 4 × 4 matrix is needed to describe the mixing of neutral gauginos and two kinds of neutral Higgsinos. In TGD framework second Higgs (if it exists at all) is pseudo-scalar and does not contribute and the 2 × 2 matrices describe the mixing also now.

$$X = \begin{pmatrix} \begin{pmatrix} M_1 & 0 \\ 0 & M_2 \end{pmatrix} & M_Z \begin{pmatrix} s_W cos(\beta) & s_W sin(\beta) \\ c_W cos(\beta) & c_W sin(\beta) \end{pmatrix} \\ M_Z \begin{pmatrix} s_W cos(\beta) & s_W sin(\beta) \\ c_W cos(\beta) & c_W sin(\beta) \end{pmatrix} & -\mu \begin{pmatrix} 0 & 1 \\ 1 & 0 \end{pmatrix} \end{pmatrix} . \quad (4.7)$$

For $sin(\beta) = cos(\beta) = 0$ the non-diagonal part of the mass matrix is degenerate.

Again there are two options depending on whether Higgsinos are present and if they are absent the dependence on the angle β vanishes. Indeed, if Higgsinos are absent the matrix reduces to a diagonal 2×2 mass matrix for U(1) gaugino $\tilde{B}$ and neutral SU(2) gaugino $\tilde{W}^3$. If one takes seriously MSSM, there would be no mixing. On the other hand, TGD suggests that neutral gauginos mix in the same manner as neutral gauge bosons so that Weinberg angle would characterize the mixing with photino and zino appearing as mass eigen states. Again for same value of p-adic prime the values of mass squared for gauge bosons and gauginos should be identical.

One can also consider the option with Higgsino.

1. Since Higgs and Higgsino have representation content 3+1 with respect to electroweak SU(2) in TGD framework, one can speak about $\tilde{h}_B$, $B = W, Z, \gamma$. An attractive assumption is that Weinberg angle characterizes also the mixing giving rise to $\tilde{Z}$ and $\tilde{\gamma}$ on one hand and $\tilde{h}_\gamma$ and $\tilde{h}_Z$ on the other hand if these belong to the spectrum. This would reduce the mixing matrix to two 2 × 2 matrices: the first one for $\tilde{\gamma}$ and $\tilde{h}_\gamma$ and the second one for $\tilde{Z}$ and $\tilde{h}_Z$.

2. A further attractive assumption is that the mass matrices describing mixing of gauginos and corresponding Higgsinos are in some sense universal with respect to electroweak interactions. The form of the mixing matrix would be essentially same for all cases. This would suggest that M_W is replaced in the above formula with the mass of Z^0 and photon in these matrices (recall that it is assumed that photon gets small mass by eating the neutral Higgs). Note that for photino and corresponding Higgsino the mixing would be small. The guess is $M_2 = -\mu = m_Z$. For photino one can guess that M_2 corresponds to M_{89} mass scale.

These assumptions of course define only the first maximally symmetric guess and the simplest modification that one can imagine is due to the different p-adic mass scales. If the above discussed values for zino and neutralino masses deduced from the 1995 event [96] are taken at face value, the eigenvalues would be $\pm\sqrt{M_Z^2+m^2}$ with $m = M_2 = -\mu$ for $\tilde{Z} - \tilde{h}_Z$-mixing and the other state would have p-adic length scale $k = 91$ rather than $k = 89$. M and μ would have opposite signs as required by the correct sign for the $g-2$ anomaly for muon assuming that smuons correspond to $p = M_{89}$ as will be found.

3. The relationship between masses of charged sleptons and sneutrinos

In MSSM approach one has also the formula relating the masses of sneutrinos and charged sleptons [99]:

$$m_{\tilde{\nu}}^2 = m_{\tilde{L}}^2 + \frac{1}{2}M_Z^2 cos(2\beta) \ . \quad (4.8)$$

For $\beta = \pm\pi/4$ one would have $tan(\beta) = 1$ and

$$m_{\tilde{\nu}}^2 = m_{\tilde{L}}^2 \ .$$

In p-adic mass calculations this kind of formula is highly questionable and could make sense only if thesp particles involved correspond to same value of p-adic prime and therefore would not make sense after symmetry breaking.

4.3.3 The anomalous magnetic moment of muon as a constraint on SUSY

The anomalous magnetic moment $a_\mu \equiv (g-2)/2$ of muon has been used as a further constraint on SUSY. The measured value of a_μ is $a_\mu^{exp} = 11659208.0(6.3) \times 10^{10}$. The theoretical prediction decomposes to a sum of reliably calculable contributions and hadronic contribution for which the low energy photon appearing as vertex correction decays to virtual hadrons. This contribution is not easy to calculate since non-perturbative regime of QCD is involved. The deviation between prediction and experimental value is $\Delta a_\mu(exp-SM) = 23.9(9.9) \times 10^{-10}$ giving $\Delta a_\mu(exp-SM/)a_\mu = 2 \times 10^{-6}$. The hadronic contribution is estimated to be 692.3×10^{-10} so that the anomaly is 3 per cent from the hadronic contribution [99]. One can ask whether the uncertainties due to the non-perturbative effects could explain the anomaly.

The following calculation is a poor man's version of MSSM calculation [99]. Also now SUSY requires that the electroweak couplings between particles dictate those between sparticles. Supersymmetry for massivation suggests that in TGD framework higgisinos do not belong to the spectrum. Light sfermions appear as single copy with vanishing fermion number so that various mixing matrices of MSSM reduce to unit matrices. This leads to a rough recipe: take only the one loop contributions to g-2 and assume trivial mixing matrices and drop off summations. At least a good order of magnitude estimate should result in this manner.

1. A rough MSSM inspired estimate g-2 anomaly

Consider now a rough estimate for the g-2 anomaly by using the formulas 56-58 of [99]. One obtains for the charged loop the expression

$$\Delta a_\mu^\pm = -\frac{21g_2^2}{32\pi^2} \times (\frac{m_\mu}{m_W})^2 \times sign(\mu M_2) \ . \quad (4.9)$$

This however involves a formula relating sneutrino and charged slepton masses. There is no reason to expect this formula to hold true in TGD framework.

For neutral contribution the expression is more difficult to deduce. As physical intuition suggests, the expression inversely proportional to $1/m_W^2$ since m_W corresponds now m_{SUSY} although this is not obvious on the basis of the general formulas suggesting the proportionality toi $1/m_{\tilde{\nu}_\mu}^2$. The p-adic mass scale corresponding to M_{89} is the natural guess for M_{SUSY} and would give $M_{SUSY} = 104.9$ GeV. The fact that the correction has positive sign requires that μ and M_2 have opposite signs unlike in MSSM. The sign factor is opposite to that in MSSM because sfermion mass scales are assumed to be much higher than weak gaugino mass scale.

The ratio of the correction to the lowest QED estimate $a_{\mu,0} = \alpha/2\pi$ can be written as

$$\frac{\Delta a_\mu^+}{a_{\mu,0}} = \frac{21}{4sin^2(\theta_W)} \times (\frac{m_\mu}{m_{SUSY}})^2 \simeq 2.73 \times 10^{-5} \ . \tag{4.10}$$

which is roughly 10 times larger than the observed correction. The contribution Δa_μ^0 could reduce this contribution. At this moment I am however not yet able to transform the formula for it to TGD context. Also the scaling up of the m_{SUSY} by a factor of order $2^{3/2}$ could reduce the correction.

The parameter values $(tan(\beta) = 1, M_{SUSY} = 100$ GeV) corresponds to the boundary of the region allowed by the LHC data and $g-2$ anomaly is marginally consistent with these parameter values (see figure 16 of [99]). The reason is that in the recent case the mass of lightest Higgs particle does not pose any restrictions (the brown region in the figure). Due to the different mixing pattern of gauginos and higgsinos in neutral sector TGD prediction need not be identidal with MSSM prediction.

The contribution from Higgs loop is not present if Higgs is eaten by photon [91]. This contribution by a factor of order $(m_\mu/h_H)^2$ smaller than the estimate for the SUSY contribution so that the dropping of Higgs contribution does not affect considerably the situation.

$$\Delta a_\mu^H = \frac{2}{2.24^2}(\frac{m_\mu}{m_H})^4) \times (log((\frac{m_H}{m_\tau})^2) - \frac{3}{2}) \ . \tag{4.11}$$

The proposed estimate is certainly poor man's estimate since it is not clear how near the proposed twistorial approach relying on zero energy ontology is to QFT approach. It is however encouraging that the simplest possible scenario might work and that this is essentially due to the p-adic length scale hypothesis.

2. An improved estimate for g-2 anomaly

An attractive scenario for sfermion masses marginally consistent with the recent data from LHC generalizes the observation that charged lepton masses correspond to subsequent Mersenne primes of Gaussian Mersennes. The only sfermions lighter than about 13 TeV are selectron with mass 262 GeV ($k =$ 89) and sneutrinos, which can have much smaller masses. $\tilde{W}\tilde{\nu}_\mu$ virtual state would be mostly responsible for the muonic g-2 anomaly since the largest term in the correction is proportional to $m(\mu)m(\tilde{W})/m^2(\tilde{\nu}_\mu)$ and the anomaly might allow to determine $m(\tilde{\nu}_\mu)$. This option should be explain the g-2 anomaly.

The following estimate demonstrates that there are hopes about this. Using the formulas of [99] one can write the one loop contributions to the anomalous contribution $a(\mu)$ as

$$\begin{aligned} a_\mu^{\chi^0} &= \frac{m(\mu)}{16\pi^2}\sum_{i,m} X_{im} \ , \\ X_{im} &= -\frac{m(\mu)}{12m^2(\tilde{\mu}_m)}\left[|n_{im}^L|^2 + |n_{im}^R|^2\right] F_1^N(x_{im}) + \frac{m(\chi_i^0)}{3m^2(\tilde{\mu}_m)} Re\left[n_{im}^L n_{im}^R\right] F_2^N(x_{im}) \ , \end{aligned} \tag{4.12}$$

and

$$\begin{aligned} a_{\mu}^{\chi^{\pm}} &= \frac{m(\mu)}{16\pi^2}\sum_k X_k \ , \\ X_k &= -\frac{m(\mu)}{12m^2(\tilde{\nu}_{\mu})}\left[|c_k^L|^2+|c_k^R|^2\right]F_1^C(x_k)+\frac{2m(\chi_k^{\pm})}{3m^2(\tilde{\nu}_{\mu})}Re\left[c_k^Lc_k^R\right]F_2^C(x_k) \ . \end{aligned} \tag{4.13}$$

Here $i = 1, ..., 4$ denotes neutralino indices which should reduce to two if also Higgsinos disappear from the spectrum. $k = 1,2$ denotes the neutral and charginos indices reducing to single index now. $m = 1,2$ denotes smuon index. Note that TGD suggests strongly that the masses of $\tilde{\mu}_R$ and $\tilde{\mu}_L$ are degenerate. The matrices n_{im}^L, n_{im}^L and c_k^L and c_k^R relate to the mixing of mass eigenstates and are given explicitly in MSSM [99].

The kinematic variables are defined as the mass ratios $x_{im} = m^2(\chi_i^0)/m^2(\tilde{\mu}_m)$ and $x_k = m^2(\chi_k^{\pm})/m^2(\tilde{\nu}_{\mu})$ and the loop functions are given by

$$\begin{aligned} F_1^N(x) &= \frac{2}{(1-x)^4}\left[1-6x+3x^2+2x^3-6x^2log(x)\right] \ , \\ F_2^N(x) &= \frac{3}{(1-x)^3}\left[1-x^2+2xlog(x)\right] \ , \\ F_1^C(x) &= \frac{2}{(1-x)^4}\left[2+3x-6x^2+x^3+6xlog(x)\right] \ , \\ F_2^C(x) &= \frac{3}{(1-x)^3}\left[-3+4x-x^2-2log(x)\right] \ . \end{aligned} \tag{4.14}$$

If one does not assume any relationship betwen sneutrino and charged slepton masses then for $m(\tilde{\nu}_{\mu})/m(\tilde{\mu}) << 1$, $m(\mu)/m(\chi^{\pm}) << 1$, and $m(\chi_k^0)/m(\tilde{\mu}) << 1$ the functions F_1^N and $F_2^N(x)$ are in good approximation constant and the corresponding contributions are neglilible. One has $F_1^C(x) \simeq 1/x$ and $F_2^C(x) \simeq 3/x$. It rurns out that the terms proportional to $F_1^C(x)$ and $F_2^C(x_k)$ are of the same order of magnitude. If Higgsinos do not belong to the spectrum one has $U_{k2} = 0$ giving $V_{k1}U_{k2} = 0$ leaving only the F_1^C contribution.

Consider now the mixing matrices for sfermions.

1. One has

$$\begin{aligned} c_k^L &= -g_2V_{k1} \ , \ c_k^R = y_{\mu}U_{k2} \ , \\ y_{\mu} &= \frac{m(\mu)}{m(W)}\frac{g_2}{\sqrt{2}cos(\beta)} \ , g_2 = \frac{e}{sin(\theta_W)} \ . \end{aligned} \tag{4.15}$$

   Here the index k refers to the mixed states of L and R type sfermions. Since they are formed from fermion and right-handed neutrino, one expects that at higher energies the mixing is negligible. Mixing is however present and induced by the mixing of right and left handed fermion so that the mixing matrices are non-trivial at low energies and give relate closely to the massivation of sfermions and fermions.

2. One obtains

$$\begin{aligned} c_k^Lc_k^R &= -g_2^2\frac{m(\mu)}{m(W)}\frac{1}{\sqrt{2}cos(\beta)}V_{k1}U_{k2} = -\frac{m(\mu)}{m(W)}\times\frac{4\pi\alpha}{sin^2(\theta_W)}\times\frac{1}{\sqrt{2}cos(\beta)}V_{k1}U_{k2} \ , \\ |c_k^L|^2+|c_k^R|^2 &= g_2^2\left[|V_{k1}|^2+\frac{m^2(\mu)}{m^2(W)}\frac{1}{2cos^2(\beta)}|U_{k2}|^2\right] \ . \end{aligned} \tag{4.16}$$

Using these results one obtains explicit expressions for the two terms in a_μ.

1. The expressions for the term resulting from mixing of right and left handed sfermions is given by

$$\begin{aligned} a_\mu^{mix,k} &= \frac{m(\mu)}{8\pi^2 m(\chi_k^\pm)} \sum_k Re[c_k^L c_k^R] \\ &= \frac{1}{8\pi^2} \frac{4\pi\alpha}{sin^2(\theta_W)\sqrt{2}cos(\beta)} \frac{m^2(\mu)}{m(W)m(\chi_k^\pm)} Re[V_{k1}U_{k2}] \ . \end{aligned} \quad (4.17)$$

2. Second term is diagonal and non-vanishing also when Higgsino is absent from the spectrum.

$$a_\mu^{diag,k} = \frac{1}{8\pi^2} \frac{m^2(\mu)}{m^2(\chi^\pm)} \left[|c_k^L|^2 + |c_k^R|^2 \right] \ . \quad (4.18)$$

Note that $|c_k^R << c_k^L$ holds true unless $cos(\beta)$ is very small.

3. The ratio of the contributions is

$$|\frac{a_\mu^{diag,k}}{a_\mu^{mix,k}}| = \frac{m(W)}{m(\chi^\pm)_k}\sqrt{2}cos(\beta) \times |\frac{V_{k1}}{U_{k_2}}| \ . \quad (4.19)$$

For $c_k^R = 0$ (no Higgsino) one has

$$a_\mu \simeq a_\mu^{diag,k} = \frac{1}{8\pi^2} \frac{m^2(\mu)}{m^2(\chi^\pm)} \sqrt{2}cos(\beta) \frac{4\pi\alpha}{sin^2(\theta_W)} |V_{k1}|^2 \ . \quad (4.20)$$

The dependence on the mass of muonic sneutrino disappears so that one cannot conclude anything about its value in this approximation. a_μ is determined by the mass scale of $\tilde{W}$, which should be of the same order of magnitude as W boson mass. The sign of the diagonal term is positive so that this contribution gives to g-2 a contribution which is of correct sign. This encourages to consider the option for which Higgsinos disappear from the spectrum.

The experimental value of the anomaly is equal to $\Delta a_\mu \simeq 23.9 \times 10^{-10}$. The order of magnitude estimate obtained by assuming $(cos(\beta) = 1/\sqrt{2}, V_{k1} = 1, U_{k2} = 0)$ one obtains $a_\mu = 82.7 \times 10^{-10} \times (m(W)/m(\chi^\pm)^2$, which for $m(W)/m(\chi^\pm) = 1$ is roughly 3.46 times larger than the anomaly. The p-adic scaling $k(\tilde{W}) = 89 \rightarrow k(\tilde{W}) - 2 = 87$ would give a value of a_μ near to the observed one. The mass of $\tilde{W}$ would be 160.8 GeV. Clearly the TGD inspired view about SUSY leads to a remarkably simple picture explaining the g-2 anomaly.

4.3.4 Basic differences between MSSM and TGD

The basic differences between TGD and MSSM [35] and related approaches deserve to be noticed (see also the article about the experimental side [102]). If Higgses and Higgsinos are absent from the spectrum, SUSY in TGD sense does not introduce flavor non-conserving currents (FNCC problem plaguing MSSM type approaches). In MSSM approach the mass spectrum of superpartners can be only guessed using various constraints and in a typical scenario masses of sfermions are assumed to be same in GUT unification scales so that at long length scales the mass spectrum for sfermions is inverted from that for fermions with stop and stau being the lightest superpartners. In TGD framework p-adic thermodynamics and the topological explanation of family replication phenomenon changes the situation completely and the spectrum of sfermions is very naturally qualitatively similar to that of fermions (genus generation correspondence is the SUSY invariant answer to the famous question of Rabi "Who ordered them?" !). This is essential for the explanation of g-2 anomaly for instance. Note that the experimental searches concentrating on finding the production of stop or stau pairs are bound to fail in TGD Universe.

Another key difference is that in TGD the huge number of parameters of MSSM is replaced with a single parameter- the universal coupling characterizing the decay

sparticle $\rightarrow$ particle+right handed neutrino,

which by its universality is very "gravitational". The gravitational character suggests that it is small so that SUSY would not be badly broken meaning for instance that sparticles are rather longlived and R-parity is a rather good symmetry.

One can try to fix the coupling by requiring that the decay rate of sfermion is proportional to gravitational constant G or equivalently, to the square of CP_2radius

$$R \simeq 10^{7+1/2}(\frac{G}{\hbar_0})^{1/2} .$$

Sfermion-fermion-neutrino vertex coupling to each other same fermion M^4 chiralities involves the gradient of the sfermion field. Yukawa coupling - call it L - would have dimension of length. For massive fermions in M^4 it would reduce to dimensionless coupling g different M^4 chiralities. In equal mass case g would be proportional to $L(m_1+m_2)/\hbar$, where m_i are the masses of fermions.

1. For the simplest option L is expressible in terms of CP_2 geometry alone and corresponds to

$$L = kR .$$

k is a numerical constant of order unity. $\hbar_0$denotes the standard value of Planck constant, whose multiple the efffective value of Planck constant is in TGD Universe in dark matter sectors. The decay rate of sfermion would be proportional to

$$k^2R^2(\frac{M}{hbar})^3 \simeq k^2 \times 10^7 \times \frac{G}{\hbar_0} \times (\frac{M}{\hbar})^3 ,$$

where M is the mass scale characterizing the phase space volume for the decays of sfermion and is given by the mass of sfermion multiplied by a dimensionless factor depending on mass ratios. The decay rate is extremely low so that R-parity conservation would be an excellent approximate symmetry. In cosmology this could mean that zinos and photinos would decay by an exchange of sfermions rather than directly and could give rise to dark matter like phase as in MSSM.

2. Second option carries also information about Kähler action one would have apart from a numerical constant of order unity $k = \alpha_K$. The Kähler coupling strength

$$\alpha_K = \frac{g_K^2}{4\pi \times \hbar_0} \simeq 1/137$$

is the fundamental dimensionless coupling of TGD analogous to critical temperature.

3. For the option which "knows" nothing about CP_2 geometry the length scale would be proportional to the Schwartchild radius

$$L = kGM .$$

In this case the decay rate would be proportional to $k^2G^2M^2(M/\hbar)^3$ and extremely low.

4. The purely kinematic option which one cannot call "gravitational" "knows" only about sfermion mass and f Planck constant, and one would have

$$L = k \times \frac{\hbar}{M} .$$

The decay rate would be proportional to the naive order of magnitude guess $k^2(M/\hbar)$ and fast unlike in all "gravitational cases". R-parity would be badly broken. Again$k \propto \alpha_K$ option can be considered.

Note that also in mSUGRA gravitatational sector in short length scales determines MSSM parameters via flavor blind interactions and also breaking of SUSY via breaking of local SUSY in short scales.

5 Dark matter in TGD Universe

TGD based explanation of dark matter means one of the strongest departures of TGD from the more standard approaches. In standard approaches dark matter corresponds to some very weakly interaction exotic particles contributing to the mass density of the Universe a fraction considerably larger than the contributions of "visible" matter. In TGD Universe dark matter corresponds to phases with non-standard value of Planck constant and also ordinary particles could be in dark phase.

5.1 Dark matter and energy in TGD Universe

In TGD framework the identification of dark matter comes from targuments which could start from the strange finding that ELF em fields in frequency range of EEG have quantal effects on vertebrate brain [6]. This is impossible in standard physics since the energies of photons many orders of magnitude below the thermal energy.

The proposal is that Planck constant is dynamical having a discrete integer valued spectrum so that for given frequency the energy of photon can be above thermal energy for sufficiently large value of Planck constant. Large values of Planck constant make possible macroscopic quantum coherence so that the hypothesis would explain how living matter manages to be quantum system in macroscopic scales. Particles characterized by different values of Planck constant cannot appear in same interaction vertices so that in this sense particles with different values of $\hbar$ are dark relative to each other. This however allows interactions by particle exchange involving phase transition changing the value of Planck constant and also the interaction via classical fields.

The observation of Nottale [103] that planetary orbits could be understood as Bohr orbits with a gigantic value of gravitational Planck constant leads also to the same idea [21, 20]. The expression $\hbar_{gr} = GMm/v_0$, where v_0 has dimensions of velocity, forces to identify the Planck constant as a characterizer of the space-time sheets mediating the gravitational interaction between Sun and planet. Quite generally, there is a strong temptation to assign dark matter with the field bodies (or magnetic bodies) of physical systems and this assumption is made in the model of living matter based on the notion of the magnetic body.

One must be cautious with the identification of galactic dark matter in terms of phases with large value of Planck constant. One explanation for the galactic dark matter would be in terms of string like objects containing galaxies like pearls in the necklace [4]. The Newtonian gravitational potential of the long galactic string would give rise to constant velocity spectrum. It could of course be that dark matter in TGD sense resides as particles at the long strings which could also carry antimatter. At least part of dark matter could be in this form.

What can one the concldue about dark energy in this framework?

1. Dark energy might allow interpretation as dark matter at the space-time sheets mediating gravitational interaction and macroscopically quantum coherent in cosmological scales. The enormous Compton wave lengths would imply that the density of dark energy would be constant as required by the interpretation in terms of cosmological constant.

2. This is however not the only possible interpretation. The magnetic tension of the magnetic flux tubes gives rise to the negative "pressure" inducing the accelerated expansion of the Universe serving as the basic motivation for the dark energy [22].

3. The Robertson-Walker cosmologies with critical or over-critical mass density imbeddable to the imbedding space are characterized by their necessarily finite duration and possess a negative pressure. The interpretation as a constraint force due to the imbeddability to $M^2 \times CP_2$ might explain dark energy [22].

4. The GRT limit of TGD based on Einstein-Maxwell system with cosmological constant assigned with Eudlidian regions of space-time allowing to get CP_2 as a special solution of field equation suggests that cosmological constant equals to the cosmological constant of CP_2 multiplied by the fraction of 3-volume with Euclidian signature of metric and representing generalized Feynman graphs [27].

Whether these explanations represent different manners to say one and the same thing is not clear.

One could add the hierarchy of Planck constants as a separate postulate to TGD but it has turned out that the vacuum degeneracy characterizing TGD could imply this hierarchy as an effective hierarchy so that at the fundamental level one would have just the standard value of Planck constant [9]. For both options the geometric realization for the hierarchy of Planck constants comes in terms of local covering spaces of imbedding space.

1. If the hierarchy is postulated rather than derived, the coverings in questions would be those of the causal diamond $CD \times CP_2$ such that the number of sheets of the covering equals to the value of Planck constant. The coverings of both CD and CP_2 are possible so that Planck constant is product of integers.

2. The hierarchy of local coverings would follow from the fact that time derivatives of imbedding space coordinates are in general many-valued functions of canonical momentum densities by the vacuum degeneracy of Kähler action. In this case the covering would be covering of H assignable to a regions of space-time sheet. Note that, for the vacuum extremals for which induced Kähler gauge field is pure gauge and CP_2 projection any 2-D Lagrangian of CP_2, an infinite number of branches of the covering co-incide. The situation can be characterized in terms of a generalization of catastrophe theory [31] to infinite-D context.

An open question is whether dark matter phases can/must correspond to same p-adic length scale and therefore same mass. Dark matter would correspond to particles with non-standard values of Planck constant and also ordinary particles with standard values of masses could appear in dark phase and is assumed in TGD inspired quantum biology. Even quarks with Compton lengths scaled up to cell length scale appear in the model of DNA as topological quantum computer [8]. The model of leptopions [26] in terms of colored excitations of leptons would suggests that colored excitations of leptons have same mass as leptons or possibibly p-adically scaled octave of it in the case of colored ta lepton. The colored excitation of lepton with ordinary value of Planck constant must have mass larger than one half of intermediate gauge boson mass scale. Same applies to possible colored excitations of quarks.

This picture modifies profoundly the ideas about how to detect dark matter.

1. For instance, it might be possible to photograph dark matter and it might be that Peter Gariaev and his group have actually achieved this. What they observe are strange flux tube like structures associated with DNA sample [105]: a TGD based model for the findings is developed in [15]. If dark matter is what TGD claims it to be, the experimental methods used to detect dark matter might be on wrong track.

2. One should try to find a situation in which the particles must be created in dark phase and in this respect colored excitations of leptons are a good candidate since the decay widths of intermediate gauge boson do not allow new light fermions so that if these excitations exist they must have non-standard value of Planck constant.

3. The recent results of DAMA and Cogent suggesting the existence of dark matter particles with mass around 7 GeV are in conflict with the findings of CDMS and Xenon100 experiments. It is encouraging that this conflict could be explained by using the fact that the detection criteria in these experiments are different and by assuming that the dark matter particles involved are tau-pions formed as bound states of colored excitations of tau-leptons.

5.2 Shy positrons

The latest weird looking effect in atomic physics is the observation that positrium atoms consisting of positron and electron scatter particles almost as if they were lonely electrons [94, 86]. The effect has been christened cloaking effect for positron.

The following arguments represent the first attempts to understand the cloaking of positron in terms of these notions.

1. Let us start with the erratic argument since it comes first in mind. If positron and electron correspond to different space-time sheets and if the scattered particles are at the space-time sheet of electron then they do not see positron's Coulombic field at all. The objection is obvious. If positron interacts with the electron with its full electromagnetic charge to form a bound state, the corresponding electric flux at electron's space-time sheet is expected to combine with the electric flux of electron so that positronium would look like neutral particle after all. Does the electric flux of positron return back to the space-time sheet of positronium at some distance larger than the radius of atom? Why should it do this? No obvious answer.

2. Assume that positron dark but still interacts classically with electron via Coulomb potential. In TGD Universe darkness means that positron has large $\hbar$ and Compton size much larger than positronic wormhole throat (actually wormhole contact but this is a minor complication) would have more or less constant wave function in the volume of this larger space-time sheet characterized by zoomed up Compton length of electron. The scattering particle would see pointlike electron plus background charge diffused in a much larger volume. If the value of $\hbar$ is large enough, the effect of this constant charge density to the scattering is small and only electron would be seen.

3. As a matter fact, I have proposed this kind of mechanism to explain how the Coulomb wall, which is the basic argument against cold fusion could be overcome by the incoming deuteron nucleus [29] ,[29]. Some fraction of deuteron nuclei in the palladium target would be dark and have large size just as positron in the above example. It is also possible that only the protons of these nuclei are dark. I have also proposed that dark protons explain the effective chemical formula $H_{1.5}O$ of water in scattering by neutrons and electrons in attosecond time scale [29] ,[29]. The connection with cloaked positrons is highly suggestive.

4. Also one of TGD inspired proposals for the absence of antimatter is that antiparticles reside at different space-time sheets as dark matter and are apparently absent [22]. Also the modified Dirac equation with measurement interaction term suggests that fermions and antifermions reside at different space-time sheets, in particulart that bosons correspond to wormhole contacts [10]. Cloaking positrons (shy as also their discoverer Dirac!) might provide an experimental supports for these ideas.

The recent view about the detailed structure of elementary particles forces to consider the above proposal in more detail.

1. According to this view all particles are weak string like objects having wormhole contacts at its ends and magnetically charged wormhole throats (four altogether) at the ends of the string like objects with length given by the weak length cale connected by a magnetic flux tube at both space-time sheets. Topological condensation means that these structures in turn are glued to larger space-time sheets and this generates one or more wormhole contacts for which also particle interpretation is highly suggestive and could serve as space-time correlate for interactons described in terms of particle exchanges. As far electrodynamics is considered, the second ends of weak strings containing neutrino pairs are effectively non-existing. In the case of fermions also only the second wormhole throat carrying the fermion number is effectively present so that for practical purposes weak string is only responsible for the massivation of the fermions. In the case of photons both wormhole throats carry fermion number.

2. An interesting question is whether the formation of bound states of two charged particles at the same space-time sheet could involve magnetic flux tubes connecting magnetically charged wormhole

throats associated with the two particles. If so, Kähler magnetic monopoles would be part of even atomic and molecular physics. I have proposed already earlier that gravitational interaction in astrophysical scales involves magnetic flux tubes. These flux tubes would have o interpretation as analogs of say photons responsible for bound state energy. In principle it is indeed possible that the energies of the two wormhole throats are of opposite sign for topological sum contact so that the net energy of the wormhole contact pair responsible for the interaction could be negative.

3. Also the interaction of positron and electron would be based on topological condensation at the same space-time sheet and the formation of wormhole contacts mediating the interaction. Also now bound states could be glued together by magnetically charged wormhole contacts. In the case of dark positron, the details of the interaction are rather intricate since dark positron would correspond to a multi-sheeted structure analogous to Riemann surface with different sheets identified in terms of the roots of the equation relating generalized velocities defined by the time derivatives of the imbedding space coordinates to corresponding canonical momentum densities.

5.3 Dark matter puzzle

Sean Carroll has explained in Cosmic Variance the latest rather puzzling situation in dark matter searches. Some experiments support the existence of dark matter particles with mass of about 7 GeV, some experiments exclude them. The following arguments show that TGD based explanation allows to understand the discrepancy.

5.3.1 How to detect dark matter and what's the problem?

Consider first the general idea behind the attempts to detect dark matter particles and how one ends up with the puzzling situation.

1. Galactic nucleus serves as a source of dark matter particles and these one should be able to detect. There is an intense cosmic ray flux of ordinary particles from galactic center which must be eliminated so that only dark matter particles interacting very weakly with matter remain in the flux. The elimination is achieved by going sufficiently deep underground so that ordinary cosmic rays are shielded but extremely weakly interacting dark matter particles remain in the flux. After this one can in the ideal situation record only the events in which dark matter particles scatter from nuclei provided one eliminates events such as neutrino scattering.

2. DAMA experiment does not detect dark matter events as such but annual variations in the rate of events which can include besides dark matter events and other kind of events. DAMA finds annual variation interpreted as dark matter signal since other sources of events are not expected to have this kind of variation [73]. Also CoGENT has reported the annual variation with 2.8 sigma confidence level [90]. The mass of the dark matter particle should be around 7 GeV rather than hundreds of GeVs as required by many models. An unidentified noise with annual variation having nothing to do with dark matter could of course be present and this is the weakness of this approach.

3. For a few weeks ago we learned that XENON100 experiment detects no dark matter [74]. Also CDMS has reported a negative result [65]. According to Sean Carroll, the detection strategy used by XENON100 is different from that of DAMA: individual dark matter scatterings on nuclei are detected. This is a very significant difference which might explain the discrepancy since the theory laden prejudices about what dark matter particle scattering can look like, could eliminate the particles causing the annual variations. For instance, these prejudices are quite different for the habitants of the main stream Universe and TGD Universe.

5.3.2 TGD based explanation of the DAMA events and related anomalies

I have commented earlier the possible interpretation of DAMA events in terms of tau-pions. The spirit is highly speculative.

1. Tau-pions would be identifiable as the particles claimed by Fermi Gamma Ray telescope with mass around 7 GeV and decaying into tau pairs so that one could cope with several independent observations instead of only single one.

2. Recall that the CDF anomaly gave for two and half years ago support for tau-pions whereas earlier anomalies dating back to seventies give support for electro-pions and mu-pions. The existence of these particles is purely TGD based phenomenon and due to the different view about the origin of color quantum numbers. In TGD colored states would be partial waves in CP_2 and spin like quantum numbers in standard theories so that leptons would not have colored excitations.

3. Tau-pions are of course highly unstable and would not come from the galactic center. Instead, they would be created in cosmic ray events at the surface of Earth and if they can penetrate the shielding eliminating ordinary cosmic rays they could produce events responsible for the annual variation caused by that for the cosmic ray flux from galactic center.

Can one regard tau-pion as dark matter in some sense? Or must one do so? The answer is affirmative to both questions on both theoretical and experimental grounds.

1. The existence of colored variants of leptons is excluded in standard physics by intermediate gauge boson decay widths. They could however appear as states with non-standard value of Planck constant and therefore not appearing in same vertices with ordinary gauge bosons so that they would not contribute to the decay widths of weak bosons. In this minimal sense they would be dark and this is what is required in order to understand what we know about dark matter.

   Of course, all particles can in principle appear in states with non-standard value of Planck constant so that tau-pion would be one special instance of dark matter. For instance, in living matter the role of dark variants of electrons and possibly also other stable particles would be decisive. To put it bluntly: in mainstream approach dark matter is identified as some exotic particle with ad hoc properties whereas in TGD framework dark matter is outcome of a generalization of quantum theory itself.

2. DAMA experiment requires that the tau-pions behave like dark matter: otherwise they would never reach the strongly shielded detector. The interaction with the nuclei of detector would be preceded by a transformation to a particle-tau-pion or something else- with ordinary value of Planck constant.

5.3.3 TGD based explanation for the dark matter puzzle

The criteria used in experiments to eliminate events which definitely are not dark matter events - according to the prevailing wisdom of course - dictates to high degree what interactions of tau pions with solid matter detector are used as a signature of dark matter event. It could well be that the criteria used in XENON100 do not allow the scatterings of tau-pions with nuclei. This is indeed the case. The clue comes from the comments of Jester in Resonaances. From a comment of Jester one learns that CoGENT - and also DAMA utilizing the same detections strategy - "does not cut on ionization fraction". Therefore, if dark matter mimics electron recoils (as Jester says) or if dark matter produced in the collisions of cosmic rays with the nuclei of the atmosphere decays to charged particles one can understand the discrepancy.

The TGD based model [26] explaining the more than two years old CDF anomaly [61, 95] indeed explains also the discrepancy between XENON100 and CDMS on one hand and DAMA and CoGENT on the other hand. The TGD based model for the CDF anomaly can be found here. See also blog postings such as this and also two and half year old What's News at my homepage.

1. To explain the observations of CDF [61, 95] one had to assume that tau-pions and therefore also color excited tau-leptons inside them appear as several p-adically scaled up variants so that one would have several octaves of the ground state of tau-pion with masses in good approximation equal to 3.6 GeV (two times the tau-lepton mass), 7.2 GeV, 14.4 GeV. The 14.4 GeV tau-pion was assumed to decay in a cascade like manner via lepto-strong interactions to lighter tau-pions- both charged and neutral- which eventually decayed to ordinary charged leptons and neutrinos.

2. Also other decay modes -say the decay of neutral tau-pions to gamma pair and to a pair of ordinary leptons- are possible but the corresponding rates are much slower than the decay rates for cascade like decay via multi-tau-pion states proceeding via lepto-strong interactions.

3. Just this cascade would take place also now after the collision of the incoming cosmic ray with the nucleus of atmosphere. The mechanism producing the neutral tau-pions -perhaps a coherent state of them- would degenerate in the collision of charged cosmic ray with nucleus generating strong non-orthogonal electric and magnetic fields and the production amplitude would be essentially the Fourier transform of the "instanton density" $E \cdot B$. The decays of 14 GeV neutral tau-pions would produce 7 GeV charged tau-pions, which would scatter from the protons of nuclei and generate the events excluded by XENON100 but not by DAMA and Cogent.

4. In principle the model predicts to a high degree quantitatively the rate of the events. The scattering rates are proportional to an unknown parameter characterizing the transformation probability of tau-pion to a particle with ordinary value of Planck constant and this allows to perform some parameter tuning. This parameter would correspond to a mass insertion in the tau-pion line changing the value of Planck constant and have dimensions of mass squared.

The overall conclusion is that the discrepany between DAMA and XENON100 might be interpreted as favoring TGD view about dark matter and it is fascinating to see how the situation develops. This confusion is not the only confusion in recent day particle physics. All believed-to-be almost-certainties are challenged.

6 Scaled variants of quarks and leptons

6.1 Fractally scaled up versions of quarks

The strange anomalies of neutrino oscillations [75] suggesting that neutrino mass scale depends on environment can be understood if neutrinos can suffer topological condensation in several p-adic length scales [14]. The obvious question whether this could occur also in the case of quarks led to a very fruitful developments leading to the understanding of hadronic mass spectrum in terms of scaled up variants of quarks. Also the mass distribution of top quark candidate exhibits structure which could be interpreted in terms of heavy variants of light quarks. The ALEPH anomaly [101], which I first erratically explained in terms of a light top quark has a nice explanation in terms of b quark condensed at $k = 97$ level and having mass ~ 55 GeV. These points are discussed in detail in [19].

The emergence of ALEPH results [101] meant a an important twist in the development of ideas related to the identification of top quark. In the LEP 1.5 run with $E_{cm} = 130 - 140\ GeV$, ALEPH found 14 e^+e^- annihilation events, which pass their 4-jet criteria whereas 7.1 events are expected from standard model physics. Pairs of dijets with vanishing mass difference are in question and dijets could result from the decay of a new particle with mass about 55 GeV.

The data do not allow to conclude whether the new particle candidate is a fermion or boson. Top quark pairs produced in e^+e^- annihilation could produce 4-jets via gluon emission but this mechanism does not lead to an enhancement of 4-jet fraction. No $b\bar{b}b\bar{b}$ jets have been observed and only one event containing b has been identified so that the interpretation in terms of top quark is not possible unless there exists some new decay channel, which dominates in decays and leads to hadronic jets not initiated by b quarks. For option 2), which seems to be the only sensible option, this kind of decay channels are absent.

Super symmetrized standard model suggests the interpretation in terms of super partners of quarks or/and gauge bosons [89]. It seems now safe to conclude that TGD does not predict sparticles. If the exotic particles are gluons their presence does not affect Z^0 and W decay widths. If the condensation level of gluons is $k = 97$ and mixing is absent the gluon masses are given by $m_g(0) = 0$, $m_g(1) = 19.2\ GeV$ and $m_g(2) = 49.5\ GeV$ for option 1) and assuming $k = 97$ and hadronic mass renormalization. It is however very difficult to understand how a pair of $g = 2$ gluons could be created in e^+e^- annihilation. Moreover, for option 2), which seems to be the only sensible option, the gluon masses are $m_g(0) = 0$,

$m_g(1) = m_g(2) = 30.6$ GeV for $k = 97$. In this case also other values of k are possible since strong decays of quarks are not possible.

The strong variations in the order of magnitude of mass squared differences between neutrino families [75] can be understood if they can suffer a topological condensation in several p-adic length scales. One can ask whether also t and b quark could do the same. In absence of mixing effects the masses of $k = 97$ t and b quarks would be given by $m_t \simeq 48.7$ GeV and $m_b \simeq 52.3$ GeV taking into account the hadronic mass renormalization. Topological mixing reduces the masses somewhat. The fact that b quarks are not observed in the final state leaves only $b(97)$ as a realistic option. Since Z^0 boson mass is ~ 94 GeV, $b(97)$ does not appreciably affect Z^0 boson decay width. The observed anomalies concentrate at cm energy about 105 GeV. This energy is 15 percent smaller than the total mass of top pair. The discrepancy could be understood as resulting from the binding energy of the $b(97)\bar{b}(97)$ bound states. Binding energy should be a fraction of order $\alpha_s \simeq .1$ of the total energy and about ten per cent so that consistency is achieved.

The following arguments suggest that p-adically scaled up variants of quarks might appear not only at very high energies but even in low energy hadron physics.

6.1.1 Aleph anomaly and scaled up copy of b quark

The prediction for the b quark mass is consistent with the explanation of the long since forgotten Aleph anomaly [101] suggesting the exietence of a particle with 55 GeV mass which might represent something real. If b quark condenses at $k(b) = 97$ level, the predicted mass is $m(b, 97) = 52.3$ GeV for $n_b = 59$ for the maximal CP_2 mass consistent with η' mass and interpretation as Aleph particle. If the mass of the particle candidate is defined experimentally as one half of the mass of resonance, b quark mass is actually by a factor $\sqrt{2}$ higher and scaled up b corresponds to $k(b) = 96 = 2^5 \times 3$. The prediction is consistent with the estimate 55 GeV for the mass of the Aleph particle and gives additional support for the model of topological mixing. Also the decay characteristics of Aleph particle are consistent with the interpretation as a scaled up b quark.

6.1.2 Could top quark have scaled variants?

Tony Smith has emphasized the fact that the distribution for the mass of the top quark candidate has a clear structure suggesting the existence of several states, which he interprets as excited states of top quark [97]. According to the figures 6.1.2 and 6.1.2 representing published FermiLab data, this structure is indeed clearly visible.

There is evidence for a sharp peak in the mass distribution of the top quark in 140-150 GeV range (Fig. 6.1.2). There is also a peak slightly below 120 GeV, which could correspond to a p-adically scaled down variant t quark with $k = 95$ having mass 119.6 GeV for ($Y_e = 0, Y_t = 1$) There is also a small peak also around 265 GeV which could relate to $m(t(93)) = 240.4$ GeV. There top could appear at least for the p-adic scales $k = 93, 94, 95$ as also u and d quarks seem to appear as current quarks.

6.1.3 Scaled up variants of d, s, u, c in top quark mass scale

The fact that all neutrinos seem to appear as scaled up versions in several scales, encourages to look whether also u, d, s, and c could appear as scaled up variants transforming to the more stable variants by a stepwise increase of the size scale involving the emission of electro-weak gauge bosons. In the following the scenario in which t and b quarks mix minimally is considered.

q	$m(92)/GeV$	$m(91)/GeV$	$m(90)/GeV$
u	134	189	267
d	152	216	304
c	140	198	280
s	152	216	304

Table 1. The masses of $k = 92, 91$ and $k = 90$ scaled up variants of u,d,c,s quarks assuming same integers n_{q_i} as for ordinary quarks in the scenario $(n_d, n_s, n_b) = (5, 5, 59)$ and $(n_u, n_c, n_t) = (5, 6, 58)$ and

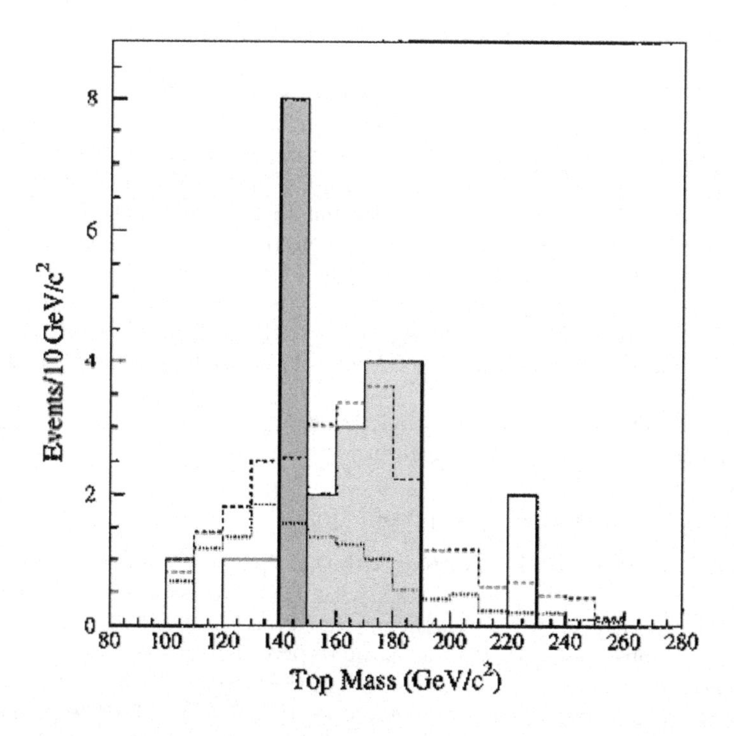

Figure 1: Fermilab semileptonic histogram for the distribution of the mass of top quark candidate (FERMILAB-PUB-94/097-E).

maximal CP_2 mass consistent with the η' mass.

1. For $k = 92$, the masses would be $m(q, 92)$ =134,140,152,152 GeV in the order q= u,c,d,s so that all these quarks might appear in the critical region where the top quark mass has been wandering.

2. For $k = 91$ copies would have masses $m(q, 91)$ =189, 198, 256, 256 GeV in the order q= u,c,d,s. The masses of u and c are somewhat above the value of latest estimate 170 GeV for top quark mass [77].

Note that it is possible to distinguish between scaled up quarks of M_{107} hadron physics and the quarks of M_{89} hadron physics since the unique signature of M_{89} hadron physics would be the increase of the scale of color Coulombic and magnetic energies by a factor of 512. As will be found, this allows to estimate the masses of corresponding mesons and baryons by a direct scaling. For instance, M_{89} pion and nucleon would have masses 71.7 GeV and 481 GeV.

It must be added that the detailed identifications are sensitive to the exact value of the CP_2 mass scale. The possibility of at most 2.5 per cent downward scaling of masses occurs is allowed by the recent value range for top quark mass.

6.1.4 Fractally scaled up copies of light quarks and low mass hadrons?

One can of course ask, whether the fractally scaled up quarks could appear also in low lying hadrons. The arguments to be developed in detail later suggest that u, d, and s quark masses could be dynamical in the sense that several fractally scaled up copies can appear in low mass hadrons and explain the mass differences between hadrons.

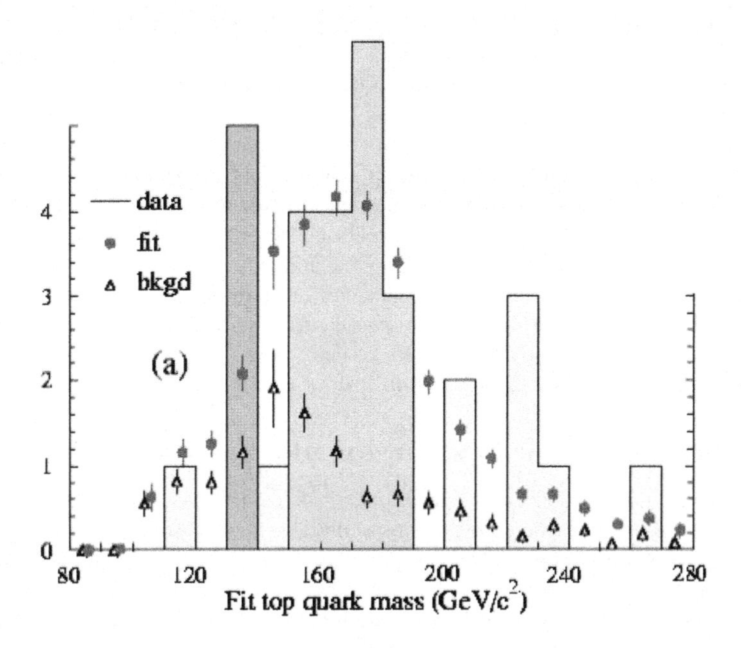

Figure 2: Fermilab D0 semileptonic histogram for the distribution of the mass of top quark candidate (hep-ex/9703008, April 26, 1994

In this picture the mass splittings of low lying hadrons with different flavors would result from fractally scaled up excitations of s and also u and d quarks in case of mesons. This notion would also throw light into the paradoxical presence of two kinds of quark masses: constituent quark masses and current quark masses having much smaller values than constituent quarks masses. That color spin-spin splittings are of same order of magnitude for all mesons supports the view that color gauge fluxes are feeded to $k = 107$ space-time sheet.

The alert reader has probably already asked whether also proton mass could be understood in terms of scaled up copies of u and d quarks. This does not seem to be the case, and an argument predicting with 23 per cent error proton mass scale from $\rho - \pi$ and $\Delta - N$ color magnetic splittings emerges.

To sum up, it seems quite possible that the scaled up quarks predicted by TGD have been observed for decade ago in FermiLab about that the prevailing dogmas has led to their neglect as statistical fluctuations. Even more, scaled up variants of s quarks might have been in front of our eyes for half century! Phenomenon is an existing phenomenon only if it is an understood phenomenon.

6.1.5 The mystery of two Ω_b baryons

Tommaso Dorigo has three interesting postings [78] about the discovery of Ω_b baryon containing two strange quarks and one bottom quark. Ω_b has been discovered -even twice. This is not a problem. The problem is that the masses of these Ω_bs differ quite too much. D0 collaboration discovered Ω_b with a significance of 5.4 sigma and a mass of 6165 $\pm$ 16.4 MeV [68]. Later CDF collaboration announced the discovery of the same particle with a significance of 5.5 sigma and a mass of 6054.4 $\pm$ 6.9 MeV. Both D0 and CDF agree that the particle is there at better than 5 sigma significance and also that the other collaboration is wrong. They cannot both be right. Or could they? In some other Universe that that of standard model and all its standard generalizations, maybe in some less theoretically respected Universe, say TGD Universe?

The mass difference between the two Ω_b candidates is 111 MeV, which represents the mass scale of strange quark. TDG inspired model for quark masses relies on p-adic thermodynamics and predicts that quarks can appear in several p-adic mass scales forming a hierarchy of half octaves - in other words mass scales comes as powers of square root of two. This property is absolutely essential for the TGD based model for masses of even low lying baryons and mesons where strange quarks indeed appear with several different p-adic mass scales. It also explains the large difference of the mass scales assigned to current quarks and constituent quarks. Light variants of quarks appear also in nuclear string model where nucleons are connected by color bonds containing light quark and antiquark at their ends.

Ω_b contains two strange quarks and the mass difference between the two candidates is of order of mass of strange quark. Could it be that both Ω_b s are real and the discrepancy provides additional support for p-adic length scale hypothesis? The prediction of p-adic mass calculations for the mass of s quark is 105 MeV (see Table 1) so that the mass difference can be understood if the second s-quark in Ω_b has mass which is twice the "standard" value. Therefore the strange finding about Ω_b could give additional support for quantum TGD. Before buying a bottle of champaigne, one should however understand why D0 and CDF collaborations only one Ω_b instead of both of them.

6.2 Could neutrinos appear in several p-adic mass scales?

There are some indications that neutrinos can appear in several mass scales from neutrino oscillations [44]. These oscillations can be classified to vacuum oscillations and to solar neutrino oscillations believed to be due to the so called MSW effect in the dense matter of Sun. There are also indications that the mixing is different for neutrinos and antineutrinos [76, 43].

In TGD framework padic length scale hypothesis might explain these findings. The basic vision is that the p-adic length scale of neutrino can vary so that the mass squared scale comes as octaves. Mixing matrices would be universal. The large discrepancy between LSND and MiniBoone results [76] contra solar neutrino results could be understood if electron and muon neutrinos have same p-adic mass scale for solar neutrinos but for LSND and MiniBoone the mass scale of either neutrino type is scaled up. The existence of a sterile neutrino [92] suggested as an explanation of the findings would be replaced by p-adically scaled up variant of ordinary neutrino having standard weak interactions. This scaling up can be different for neutrinos and antineutrinos as suggested by the fact that the anomaly is present only for antineutrinos.

The different values of Δm^2 for neutrinos and antineutrinos in MINOS experiment [43] can be understood if the p-adic mass scale for neutrinos increases by one unit. The breaking of CP and CPT would be spontaneous and realized as a choice of different p-adic mass scales and could be understood in zero energy ontology. Similar mechanism would break supersymmetry and explain large differences between the mass scales of elementary fermions, which for same p-adic prime would have mass scales differing not too much.

6.2.1 Experimental results

There several different type of experimental approaches to study the oscillations. One can study the deficit of electron type solar electron neutrinos (Kamiokande, Super-Kamiokande); one can measure the deficit of muon to electron flux ratio measuring the rate for the transformation of ν_μ to ν_τ (super-Kamiokande); one can study directly the deficit of ν_e ($\overline{\nu}_e$) neutrinos due to transformation to ν_μ $\nu_{\overline{\mu}}$ coming from nuclear reactor with energies in the same range as for solar neutrinos (KamLAND); and one can also study neutrinos from particle accelerators in much higher energy range such as solar neutrino oscillations (K2K,LSND,Miniboone,Minos).

1. Solar neutrino experiments and atmospheric neutrino experiments

The rate of neutrino oscillations is sensitive to the mass squared differences Δm_{12}^2, Δm_{12}^2, Δm_{13}^2 and corresponding mixing angles θ_{12}, θ_{13}, θ_{23} between ν_e, ν_μ, and ν_τ (ordered in obvious manner). Solar neutrino experiments allow to determine $sin^2(2\theta_{12})$ and Δm_{12}^2. The experiments involving atmospheric neutrino oscillations allow to determine $sin^2(2\theta_{23})$ and Δm_{23}^2.

The estimates of the mixing parameters obtained from solar neutrino experiments and atmospheric neutrino experiments are $sin^2(2\theta_{13}) = 0.08$, $sin^2(2\theta_{23}) = 0.95$, and $sin^2(2\theta_{12}) = 0.86$. The mixing between ν_e and ν_τ is very small. The mixing between ν_e and ν_μ, and ν_μ and ν_τ tends is rather near to maximal. The estimates for the mass squared differences are $\Delta m_{12}^2 = 8\times 10^{-5}$ eV^2, $\Delta m_{23}^2 \simeq \Delta m_{13}^2 =$

2.4×10^{-3} eV^2. The mass squared differences have obviously very different scale but this need not means that the same is true for mass squared values.

2. The results of LSND and MiniBoone

LSND experiment measuring the transformation of $\overline{\nu}_\mu$ to $\overline{\nu}_e$ gave a totally different estimate for Δm_{12}^2 than solar neutrino experiments MiniBoone,[92]. If one assumes same value of $sin^2(\theta_{12})^2 \simeq .86$ one obtains $\Delta m_{23}^2 \sim .1$ eV^2 to be compared with $\Delta m_{12}^2 = 8 \times 10^{-5}$ eV^2. This result is known as LSND anomaly and led to the hypothesis that there exists a sterile neutrino having no weak interactions and mixing with the ordinary electron neutrino and inducing a rapid mixing caused by the large value of Δm^2 The purpose of MiniBoone experiment [76] was to test LSND anomaly.

1. It was found that the two-neutrino fit for the oscillations for $\nu_\mu \to \nu_e$ is not consistent with LSND results. There is an unexplained 3σ electron excess for $E < 475$ MeV. For $E > 475$ MeV the two-neutrino fit is not consistent with LSND fit. The estimate for Δm^2 is in the range $.1 - 1$ eV^2 and differs dramatically from the solar neutrino data.

2. For antineutrinos there is a small 1.3σ electron excess for $E < 475$ MeV. For $E > 475$ MeV the excess is 3 per cent consistent with null. Two-neutrino oscillation fits are consistent with LSND. The best fit gives $(\Delta m_{12}^2, sin^2(2\theta_{12}) = (0.064\ eV^2, 0.96)$. The value of Δm_{12}^2 is by a factor 800 larger than that estimated from solar neutrino experiments.

All other experiments (see the table of the summary of [92] about sterile neutrino hypothesis) are consistent with the absence of $\nu_\mu \to n_e$ and $\overline{\nu}_\mu \to \overline{\nu}_e$ mixing and only LSND and MiniBoone report an indication for a signal. If one however takes these findings seriously they suggest that neutrinos and antineutrinos behave differently in the experimental situations considered. Two-neutrino scenarios for the mixing (no sterile neutrinos) are consistent with data for either neutrinos or antineutrinos but not both [92].

3. The results of MINOS group

The MINOS group at Fermi National Accelerator Laboratory has reported evidence that the mass squared differences between neutrinos are not same for neutrinos and antineutrinos [43]. In this case one measures the disappearance of ν_μ and $\overline{\nu}_\mu$ neutrinos from high energy beam beam in the range .5-1 GeV and the dominating contribution comes from the transformation to τ neutrinos. Δm_{23}^2 is reported to be about 40 percent larger for antineutrinos than for neutrinos. There is 5 percent probability that the mass squared differences are same. The best fits for the basic parameters are $(\Delta m_{23}^2 = 2.35 \times 10^{-3}, sin^2(2\theta_{23} = 1)$ for neutrinos with error margin for Δm^2 being about 5 per cent and $(\Delta m_{23}^2 = 3.36 \times 10^{-3}, sin^2(2\theta_{23}) = .86)$ for antineutrinos with errors margin around 10 per cent. The ratio of mass squared differences is $r \equiv \Delta m^2(\overline{\nu})/\Delta m^2(\nu) = 1.42$. If one assumes $sin^2(2\theta_{23}) = 1$ in both cases the ratio comes as $r = 1.3$.

6.2.2 Explanation of findings in terms of p-adic length scale hypothesis

p-Adic length scale hypothesis predicts that fermions can correspond to several values of p-adic prime meaning that the mass squared comes as octaves (powers of two). The simplest model for the neutrino mixing assumes universal topological mixing matrices and therefore for CKM matrices so that the results should be understood in terms of different p-adic mass scales. Even CP breaking and CPT breaking at fundamental level is un-necessary although it would occur spontaneously in the experimental situation selecting different p-adic mass scales for neutrinos and antineutrinos. The expression for the mixing probability a function of neutrino energy in two-neutrino model for the mixing is of form

$$P(E) = sin^2(2\theta) sin^2(X) \ , \ X = k \times \Delta m^2 \times \frac{L}{E} \ .$$

Here k is a numerical constant, L is the length travelled, and E is neutrino energy.

1. LSND and MiniBoone results

LSND and MiniBoone results are inconsistent with solar neutrino data since the value of Δm_{12}^2 is by a factor 800 larger than that estimated from solar neutrino experiments. This could be understood if in solar neutrino experiments ν_μ and ν_w correspond to the same p-adic mass scale $k = k_0$ and have very nearly identical masses so that Δm^2 scale is much smaller than the mass squared scale. If either p-adic scale is changed from k_0 to $k_0 + k$, the mass squared difference increases dramatically. The counterpart of the sterile neutrino would be a p-adically scaled up version of the ordinary neutrino having standard electro-weak interactions. The p-adic mass scale would correspond to the mass scale defined by Δm^2 in LSND and MiniBoone experiments and therefore a mass scale in the range .3-1 eV. The p-adic length scale assignable to eV mass scale could correspond to $k = 167$, which corresponds to cell length scale of 2.5 μm. $k = 167$ defines one of the Gaussian Mersennes $M_{G,k} = (1+i)^k - 1$ $k = 151, 157, 163, 167$ varying in the range 10 nm (celle membrane thickness) and 2.5 μm defining the size of cell nucleus proposed to be fundamental for the understanding of living matter [6].

2. MINOS results

One must assume also now that the p-adic mass scales for ν_τ and $\overline{\nu}_\tau$ are near to each other in the "normal" experimental situation. Assuming that the mass squared scales of ν_μ or $\overline{\nu}_\mu$ come as 2^{-k} powers of $m^2_{\nu_\mu} = m^2_{\nu_\tau} + \Delta m^2$, one obtains

$$m^2_{\nu_\tau}(k_0) - m^2_{\overline{\nu}_\mu}(k_0 + k) = (1 - 2^{-k})m^2_{\nu_\tau} - 2^{-k}\Delta m_0^2 \ .$$

For $k = 1$ this gives

$$r = \frac{\Delta m^2(k=2)}{\Delta m^2(k=1)} = \frac{\frac{3}{2} - \frac{2r}{3}}{1-r} \ , \ r = \frac{\Delta m_0^2}{m^2_{\nu_\tau}} \ . \qquad (6.1)$$

One has $r \geq 3/2$ for $r > 0$ if one has $m_{\nu_\tau} > m_{\nu_\mu}$ for the same p-adic length scale. The experimental ratio $r \simeq 1.3$ could be understood for $r \simeq -.31$. The experimental uncertainties certainly allow the value $r = 1.5$ for $k(\overline{\nu}_\mu) = 1$ and $k(\nu_\mu) = 2$.

This result implies that the mass scale of ν_μ and ν_τ differ by a factor 1/2 in the "normal" situation so that mass squared scale of ν_τ would be of order 5×10^{-3} eV^2. The mass scales for $\overline{\nu}_\tau$ and ν_τ would about .07 eV and .05 eV. In the LSND and MiniBoone experiments the p-adic mass scale of other neutrino would be around .1-1 eV so that different p-adic mass scale large by a factor $2^{k/2}$, $2 \leq 2 \leq 7$ would be in question. The different resuts from various experiments could be perhaps understood in terms of the sensitivity of the p-adic mass scale to the experimental situation. Neutrino energy could serve as a control parameter.

6.2.3 CP and CPT breaking

Different values of Δm^2_{ij} for neutrinos and antineutrinos would require in standard QFT framework not only the violation of CP but also CP [32] which is the cherished symmetry of quantum field theories. CPT symmetry states that when one reverses time's arrow, reverses the signs of momenta and replaces particles with their antiparticles, the resulting Universe obeys the same laws as the original one. CPT invariance follows from Lorentz invariance, Lorentz invariance of vacuum state, and from the assumption that energy is bounded from below. On the other hand, CPT violation requires the breaking of Lorentz invariance.

In TGD framework this kind of violation does not seem to be necessary at fundamental level since p-adic scale hypothesis allowing neutrinos and also other fermions to have several mass scales coming as half-octaves of a basic mass scale for given quantum numbers. In fact, even in TGD inspired low enery hadron physics quarks appear in several mass scales. One could explain the different choice of the p-adic mass scales as being due to the experimental arrangement which selects different p-adic length scales for neutrinos and antineutrinos so that one could speak about spontaneous breaking of CP and possibly CPT. The CP breaking at the fundamental level which is however expected to be small in the case considered. The basic prediction of TGD and relates to the CP breaking of Chern-Simons action inducing CP breaking in the modified Dirac action defining the fermionic propagator [30].

One can indeed consider the possibility of a spontaneous breaking of CPT symmetry in TGD framework since for a given CD (causal diamond defined as the intersection of future and past directed light-cones whose size scales are assumed to come as octaves) the Lorentz invariance is broken due to the preferred time direction (rest system) defined by the time-like line connecting the tips of CD. Since the world of classical worlds is union of CDs with all boosts included the Lorentz invariance is not violated at the level of WCW. Spontaneous symmetry breaking would be analogous to that for the solutions of field equations possessing the symmetry themselves. The mechanism of breaking would be same as that for supersymmetry. For same p-adic length scale particles and their super-partners would have same masses and only the selection of the p-adic mass scale would induces the mass splitting.

There is an article about CPT violation [34] of the dynamics defined by what the authors also call Chern-Simons term. This term is not identical with the measurement interaction term introduced in TGD framework. It is however linear in momentum as is also the measurement interaction term added to Chern-Simons Dirac action and this is what is essential from the point of view of CPT. The measurement interaction term has a formal interpretation as $U(1)$ gauge transform but having non-trivial physical effect since it is added only to the Chern-Simons Dirac action term but not to Kähler-Dirac action. The linearity with respect to momentum suggests CPT oddness of the measurement interaction term. In absence of the measurement interaction CPT would be intact but the change of the sign of the measurement interaction term in PT would bring in CPT violation. One must however notice that in TGD framework both imbedding space level and space-time level are involved and this does not allow straightforward application of strandard arguments.

7 Scaled variants of hadron physics

7.1 Leptohadron physics

TGD suggest strongly ('predicts' is perhaps too strong expression) the existence of color excited leptons. The mass calculations based on p-adic thermodynamics and p-adic conformal invariance lead to a rather detailed picture about color excited leptons.

1. The simplest color excited neutrinos and charged leptons belong to the color octets ν_8 and L_{10} and $L_{\overline{10}}$ decuplet representations respectively and lepto-hadrons are formed as the color singlet bound states of these and possible other representations. Electro-weak symmetry suggests strongly that the minimal representation content is octet and decuplets for both neutrinos and charged leptons.

2. The basic mass scale for lepto-hadron physics is completely fixed by p-adic length scale hypothesis. The first guess is that color excited leptons have the levels $k = 127, 113, 107, ...$ ($p \simeq 2^k$, k prime or power of prime) associated with charged leptons as primary condensation levels. p-Adic length scale hypothesis allows however also the level $k = 11^2 = 121$ in case of electronic lepto-hadrons. Thus both $k = 127$ and $k = 121$ must be considered as a candidate for the level associated with the observed lepto-hadrons. If also lepto-hadrons correspond non-perturbatively to exotic Super Virasoro representations, lepto-pion mass relates to pion mass by the scaling factor $L(107)/L(k) = k^{(107-k)/2}$. For $k = 121$ one has $m_{\pi_L} \simeq 1.057$ MeV which compares favorably with the mass $m_{\pi_L} \simeq 1.062$ MeV of the lowest observed state: thus $k = 121$ is the best candidate contrary to the earlier beliefs. The mass spectrum of lepto-hadrons is expected to have same general characteristics as hadronic mass spectrum and a satisfactory description should be based on string tension concept. Regge slope is predicted to be of order $\alpha' \simeq 1.02/MeV^2$ for $k = 121$. The masses of ground state lepto-hadrons are calculable once primary condensation levels for colored leptons and the CKM matrix describing the mixing of color excited lepton families is known.

The strongest counter arguments against color excited leptons are the following ones.

1. The decay widths of Z^0 and W boson allow only $N = 3$ light particles with neutrino quantum numbers. The introduction of new light elementary particles seems to make the decay widths of Z^0 and W intolerably large.

2. Lepto-hadrons should have been seen in e^+e^- scattering at energies above few MeV. In particular, lepto-hadronic counterparts of hadron jets should have been observed.

A possible resolution of these problems is provided by the loss of asymptotic freedom in lepto-hadron physics. Lepto-hadron physics would effectively exist in a rather limited energy range about one MeV.

The development of the ideas about dark matter hierarchy [12, 23, 7, 5] led however to a much more elegant solution of the problem.

1. TGD predicts an infinite hierarchy of various kinds of dark matters which in particular means a hierarchy of color and electro-weak physics with weak mass scales labelled by appropriate p-adic primes different from M_{89}: the simplest option is that also ordinary photons and gluons are labelled by M_{89}.

2. There are number theoretical selection rules telling which particles can interact with each other. The assignment of a collection of primes to elementary particle as characterizer of p-adic primes characterizing the particles coupling directly to it, is inspired by the notion of infinite primes [24] , and discussed in [12]. Only particles characterized by integers having common prime factors can interact by the exchange of elementary bosons: the p-adic length scale of boson corresponds to a common primes.

3. Also the physics characterized by different values of $\hbar$ are dark with respect to each other as far quantum coherent gauge interactions are considered. Laser beams might well correspond to photons characterized by p-adic prime different from M_{89} and de-coherence for the beam would mean decay to ordinary photons. De-coherence interaction involves scaling down of the Compton length characterizing the size of the space-time of particle implying that particles do not anymore overlap so that macroscopic quantum coherence is lost.

4. Those dark physics which are dark relative to each other can interact only via graviton exchange. If lepto-hadrons correspond to a physics for which weak bosons correspond to a p-adic prime different from M_{89}, intermediate gauge bosons cannot have direct decays to colored excitations of leptons irrespective of whether the QCD in question is asymptotically free or not. Neither are there direct interactions between the QED:s and QCD:s in question if M_{89} characterizes also ordinary photons and gluons. These ideas are discussed and applied in detail in [12, 23, 7].

Skeptic reader might stop the reading after these counter arguments unless there were definite experimental evidence supporting the lepto-hadron hypothesis.

1. The production of anomalous e^+e^- pairs in heavy ion collisions (energies just above the Coulomb barrier) suggests the existence of pseudoscalar particles decaying to e^+e^- pairs. A natural identification is as lepto-pions that is bound states of color octet excitations of e^+ and e^-.

2. The second puzzle, Karmen anomaly, is quite recent [93]. It has been found that in charge pion decay the distribution for the number of neutrinos accompanying muon in decay $\pi \to \mu + \nu_\mu$ as a function of time seems to have a small shoulder at $t_0 \sim ms$. A possible explanation is the decay of charged pion to muon plus some new weakly interacting particle with mass of order 30 MeV [100] : the production and decay of this particle would proceed via mixing with muon neutrino. TGD suggests the identification of this state as color singlet leptobaryon of, say type $L_B = f_{abc}L_8^a L_8^b \bar{L}_8^c$, having electro-weak quantum numbers of neutrino.

3. The third puzzle is the anomalously high decay rate of orto-positronium. [53]. e^+e^- annihilation to virtual photon followed by the decay to real photon plus virtual lepto-pion followed by the decay of the virtual lepto-pion to real photon pair, $\pi_L\gamma\gamma$ coupling being determined by axial anomaly, provides a possible explanation of the puzzle.

4. There exists also evidence for anomalously large production of low energy e^+e^- pairs [87, 82, 85, 52] in hadronic collisions, which might be basically due to the production of lepto-hadrons via the decay of virtual photons to colored leptons.

In the following a revised form of lepto-hadron hypothesis is described.

1. Sigma model realization of PCAC hypothesis allows to determine the decay widths of lepto-pion and lepto-sigma to photon pairs and e^+e^- pairs. Ortopositronium anomaly determines the value of $f(\pi_L)$ and therefore the value of lepto-pion-lepto-nucleon coupling and the decay rate of lepto-pion to two photons. Various decay widths are in accordance with the experimental data and corrections to electro-weak decay rates of neutron and muon are small.

2. One can consider several alternative interpretations for the resonances.

   Option 1: For the minimal color representation content, three lepto-pions are predicted corresponding to $8, 10, \overline{10}$ representations of the color group. If the lightest lepto-nucleons e_{ex} have masses only slightly larger than electron mass, the anomalous e^+e^- could be actually $e^+_{ex} + e^-_{ex}$ pairs produced in the decays of lepto-pions. One could identify 1.062, 1.63 and 1.77 MeV states as the three lepto-pions corresponding to $8, 10, \overline{10}$ representations and also understand why the latter two resonances have nearly degenerate masses. Since d and s quarks have same primary condensation level and same weak quantum numbers as colored e and μ, one might argue that also colored e and μ correspond to $k = 121$. From the mass ratio of the colored e and μ, as predicted by TGD, the mass of the muonic lepto-pion should be about 1.8 MeV in the absence of topological mixing. This suggests that 1.83 MeV state corresponds to the lightest $g = 1$ lepto-pion.

   Option 2: If one believes sigma model (in ordinary hadron physics the existence of sigma meson is not established and its width is certainly very large if it exists), then lepto-pions are accompanied by sigma scalars. If lepto-sigmas decay dominantly to e^+e^- pairs (this might be forced by kinematics) then one could adopt the previous sceneario and could identify 1.062 state as lepto-pion and 1.63, 1.77 and 1.83 MeV states as lepto-sigmas rather than lepto-pions. The fact that muonic lepto-pion should have mass about 1.8 MeV in the absence of topological mixing, suggests that the masses of lepto-sigma and lepto-pion should be rather close to each other.

   Option 3: One could also interpret the resonances as string model 'satellite states' having interpretation as radial excitations of the ground state lepto-pion and lepto-sigma. This identification is not however so plausible as the genuinely TGD based identification and will not be discussed in the sequel.

3. PCAC hypothesis and sigma model leads to a general model for lepto-hadron production in the electromagnetic fields of the colliding nuclei and production rates for lepto-pion and other lepto-hadrons are closely related to the Fourier transform of the instanton density $\bar{E} \cdot \bar{B}$ of the electromagnetic field created by nuclei. The first source of anomalous e^+e^- pairs is the production of $\sigma_L\pi_L$ pairs from vacuum followed by $\sigma_L \rightarrow e^+e^-$ decay. If $e^+_{ex}e^-_{ex}$ pairs rather than genuine e^+e^- pairs are in question, the production is production of lepto-pions from vacuum followed by lepto-pion decay to lepto-nucleon pair.

   Option 1: For the production of lepto-nucleon pairs the cross section is only slightly below the experimental upper bound for the production of the anomalous e^+e^- pairs and the decay rate of lepto-pion to lepto-nucleon pair is of correct order of magnitude.

   Option 2: The rough order of magnitude estimate for the production cross section of anomalous e^+e^- pairs via $\sigma_l\pi_l$ pair creation followed by $\sigma_L \rightarrow e^+e^-$ decay, is by a factor of order $1/\sum N_c^2$ (N_c is the total number of states for a given colour representation and sum over the representations contributing to the ortopositronium anomaly appears) smaller than the reported cross section in case of 1.8 MeV resonance. The discrepancy could be due to the neglect of the large radiative corrections (the coupling $g(\pi_L\pi_L\sigma_L) = g(\sigma_L\sigma_L\sigma_L)$ is very large) and also due to the uncertainties in the value of the measured cross section.

   Given the unclear status of sigma in hadron physics, one has a temptation to conclude that anomalous e^+e^- pairs actually correspond to lepto-nucleon pairs.

4. The vision about dark matter suggests that direct couplings between leptons and lepto-hadrons are absent in which case no new effects in the direct interactions of ordinary leptons are predicted. If colored leptons couple directly to ordinary leptons, several new physics effects such as resonances in photon-photon scattering at cm energy equal to lepto-pion masses and the production of $e_{ex}\bar{e}_{ex}$

(e_{ex} is leptobaryon with quantum numbers of electron) and $e_{ex}\bar{e}$ pairs in heavy ion collisions, are possible. Lepto-pion exchange would give dominating contribution to $\nu - e$ and $\bar{\nu} - e$ scattering at low energies. Lepto-hadron jets should be observed in e^+e^- annihilation at energies above few MeV:s unless the loss of asymptotic freedom restricts lepto-hadronic physics to a very narrow energy range and perhaps to entirely non-perturbative regime of lepto-hadronic QCD.

During 18 years after the first published version of the model also evidence for colored μ has emerged. Towards the end of 2008 CDF anomaly gave a strong support for the colored excitation of τ. The lifetime of the light long lived state identified as a charged τ-pion comes out correctly and the identification of the reported 3 new particles as p-adically scaled up variants of neutral τ-pion predicts their masses correctly. The observed muon jets can be understood in terms of the special reaction kinematics for the decays of neutral τ-pion to 3 τ-pions with mass scale smaller by a factor 1/2 and therefore almost at rest. A spectrum of new particles is predicted. The discussion of CDF anomaly led to a modification and generalization of the original model for lepto-pion production and the predicted production cross section is consistent with the experimental estimate.

7.2 First evidence for M_{89} hadron physics?

The first evidence -or should we say indication- for the existence of M_{89} hadron physics has emerged from CDF which for two and half years ago provided evidence also for the colored excitations of tau lepton and for leptohadron physics.

7.2.1 Has CDF discovered a new boson with mass around 145 GeV?

The story began when The eprint of CDF collaboration [38] reported evidence for a new resonance like state, presumably a boson decaying to a dijet (jj) with mass around 145 GeV. The dijet is produced in association with W boson. The interpretation as Higgs is definitely excluded.

Bloggers reacted intensively to the possibility of a new particle. Tommaso Dorigo gave a nice detailed analysis about the intricacies of the analysis of the data leading to the identification of the bump. Also Lubos and Resonaances commented the new particle. Probably the existence of the bump had been known for months in physics circles. The flow of eprints to arXiv explaining the new particle begun immediately.

One should not forget that 3 sigma observation was in question and that 5 sigma is required for discovery. It is quite possible that the particle is just a statistical fluke due to an erratic estimation of the background as Tommaso Dorigo emphasizes. Despite this anyone who has a theory able to predict something is extremely keen to see whether the possibly existing new particle has a natural explanation. This also provides the opportunity for dilettantes like me to develop the theoretical framework in more detail. We also know from general consistency conditions that New Physics must emerge in TeV scale: what we do not know what this New Physics is. Therefore all indications for it must be taken seriously.

CDF bump did not disappear and the most recent analysis assigns 4.1 sigma signicance to it. The mass of the bump was reported to be at 147 ± 5 GeV. Also some evidence that the entire Wjj system results in a decay of a resonance with mass slightly below 300 GeV has emerged. D0 was however not able to confirm the existence of the bump and the latest reincarnation of the bump is as 2.8 sigma evidence for Higgs candidate in the range 140-150 GeV range and one can of ask whether this is actually evidence for the familiar 145 GeV boson which cannot be Higgs. The story involves many twists and turns and teaches how cautiously theoretician should take also the claims of experimentalists. In the following I pretend that the 145 GeV bump is real but this should not confuse the reader to believe that this is really the case.

7.2.2 Why an exotic weak boson a la TGD cannot be in question?

For the inhabitant of the TGD Universe the most obvious identification of the new particle would be as an exotic weak boson. The TGD based explanation of family replication phenomenon predicts that gauge bosons come in singlets and octets of a dynamical SU(3) symmetry associated with three fermion generations (fermion families correspond to topologies of partonic wormhole throats characterized by the number of handles attached to sphere). Exotic Z or W boson could be in question.

If the symmetry breaking between octet and singlet is due to different value of p-adic prime alone then the mass would come as an power of half-octave of the mass of Z or W. For W boson one would obtain 160 GeV only marginally consistent with 145 GeV. Z would give 180 GeV mass which is certainly too high. The Weinberg angle could be however different for the singlet and octet so that the naive p-adic scaling need not hold true exactly.

Note that the strange forward backward asymmetry in the production of top quark pairs [62, 88] might be understood in terms of exotic gluon octet whose existence means neutral flavor changing currents as discussed in this article.

The *extremely* important data bit is that the decays to two jets favor quark pairs over lepton pairs. A model assuming exotic Z -called $Z^{'}$- produced together with W and decaying preferentially to quark pairs has been proposed as an explanation [41]. Neither ordinary nor the exotic weak gauge bosons of TGD Universe have this kind of preference to decay to quark pairs so that my first guess was wrong.

7.2.3 Is a scaled up copy of hadron physics in question?

The natural explanation for the preference of quark pairs over lepton pairs would be that strong interactions are somehow involved. This suggests a state analogous to a charged pion decaying to W boson two gluons annihilating to the quark pair (box diagram). This kind of proposal is indeed made in Technicolor at the Tevatron [47]: the problem is also now why the decays to quarks are favored. Techicolor has as its rough analog second fundamental prediction of TGD that p-adically scaled up variants of hadron physics should exist and one of them is waiting to be discovered in TeV region. This prediction emerged already for about 15 years ago as I carried out p-adic mass calculations and discovered that Mersenne primes define fundamental mass scales.

Also colored excitations of leptons and therefore leptohadron physics are predicted [26]. What is amusing that CDF discovered towards the end of 2008 what became known as CDF anomaly giving support for tau-pions. The evidence for electro-pions and mu-pions had emerged already earlier (for references see [26]). All these facts have been buried underground because they simply do not fit to the standard model wisdom. TGD based view about dark matter is indeed needed to circumvent the fact that the lifetimes of weak bosons do not allow new light particles. There is also a long series of blog postings in my blog summarizing development of the TGD based model for CDF anomaly.

As should have become already clear, TGD indeed predicts p-adically scaled up copy of hadron physics in TeV region and the lightest hadron of this physics is a pion like state produced abundantly in the hadronic reactions. Ordinary hadron physics corresponds to Mersenne prime $M_{107} = 2^{107} - 1$ whereas the scaled up copy would correspond to M_{89}. The mass scale would be 512 times the mass scale 1 GeV of ordinary hadron physics so that the mass of M_{89} proton should be about 512 GeV. The mass of the M_{89} pion would be by a naive scaling 71.7 GeV and about two times smaller than the observed mass in the range 120-160 GeV and with the most probable value around 145 GeV as Lubos reports in his blog. $2 \times 71.7 GeV = 143.4$ GeV would be the guess of the believer in the p-adic scaling hypothesis and the assumption that pion mass is solely due to quarks. It is important to notice that this scaling works precisely only if CKM mixing matrix is same for the scaled up quarks and if charged pion consisting of u-d quark pair is in question. The well-known current algebra hypothesis that pion is massless in the first approximation would mean that pion mass is solely due to the quark masses whereas proton mass is dominated by other contributions if one assumes that also valence quarks are current quarks with rather small masses. The alternative which also works is that valence quarks are constituent quarks with much higher mass scale.

According to p-adic mass calculations the mass of pion is just the sum of mass squared for the quarks composing. If one assumes that u and d quarks of M_{89} hadron physics correspond to $k = 93$ (top corresponds to $k = 94$, the mass of these quarks is predicted to be 102 GeV whereas the pion mass is predicted to be 144.3 GeV (the argument will be discussed in detail later). My guess based on deep ignorance about the experimental side is that this signature should be easily testable: one should try to detect mono-chromatic gamma pairs with gamma ray energy around 72.2 GeV.

7.2.4 The simplest identification of the 145 GeV resonance

The picture about CDF resonance has become (see the postings Theorists vs. the CDF bump and More details about the CDF bump by Jester [51]. One of the results is that leptophobic Z' can explain only 60 per cent of the production rate. There is also evidence that Wjj corresponds to a resonance with mass slightly below 300 GeV as naturally predicted by technicolor models [80].

The simplest TGD based model indeed relies on the assumption that the entire Wjj corresponds to a resonance with mass slightly below 300 GeV for which there is some evidence as noticed. If one assume that only *neutral pions* are produced in strong non-orthogonal electric and magnetic fields of colliding proton and antiproton, the mother particle must be actually second octave of 147 GeV pion and have mass somewhat below 600 GeV producing in its possibly allowed strong decays pions which are almost at rest for kinematic reasons. Therefore the production mechanism could be exactly the same as proposed for two and one half year old CDF anomaly and for the explanation of DAMA events and DAMA-Xenon100 discrepancy,

1. This suggests that the mass of the mother resonance is in a good accuracy two times the mass of 145 GeV bump for which best estimate is 147 ± 5 GeV. This brings in mind the explanation for the two and half year old CDF anomaly in which tau-pions with masses coming as octaves of basic tau-pion played a key role (masses were in good approximation $2^k \times m(\pi_\tau)$, $m(\pi_\tau) \simeq 2m_\tau$, $k = 1, 2$. The same mechanism would explain the discrepancy between the DAMA and Xenon100 experiments.

2. If this mechanism is at work now, the mass of the lowest M_{89} pion should be around 73 GeV as the naivest scaling estimate gives. One can however consider first the option for which lightest M_{89} has mass around 147 GeV so that the 300 GeV resonance could correspond to its first p-adic octave. This pion would decay to W and neutral M_{89} pion with mass around 147 GeV in turn decaying to two jets. At quark level the simplest diagram would involve the emission of W and exchange of gluon of M_{89} hadron physics. Also the decay to Z and charged pion is possible but in this case the decay of the final state could not take place via annihilation to gluon so that jet pair need not be produced.

3. One could also imagine the mother particle to be ρ meson of M_{89} hadron physics with mass in a good approximation equal to pion mass. At the level of mathematics this option is very similar to the technicolor model of CDF bump based also on the decay of ρ meson discussed in [80]. In this model the decays of π to heavy quarks have been assumed to dominate. In TGD framework the situation is different. If π consists of scaled up u and d quarks, the decays mediated by boson exchanges would produce light quarks. In the annihilation to quark pair by a box diagram involving two gluons and two quarks at edges the information about the quark content of pion is lost. The decays involving emission of Z boson the resulting pion would be charged and its decays by annihilation to gluon would be forbidden so that Wjj final states would dominate over Zjj final states as observed.

4. The strong decay of scaled up pion to charged and neutral pion are forbidden by parity conservation. The decay can however proceed by via the exchange of intermediate gauge boson as a virtual particle. The first quark would emit virtual W/Z boson and second quark the gluon of the hadron physics. Gluon would decay to a quark pair and second quark would absorb the virtual W boson so that a two-pion final state would be produced. The process would involve same vertices as the decay of ρ meson to W boson and pion. The proposed model of the two and one half year old CDF anomaly and the explanation of DAMA and Xenon100 experiments assumes cascade like decay of pion at given level of hierarchy to two pions at lower level of hierarchy and the mechanism of decay should be this.

Consider next the masses of the M_{89} mesons. Naive scaling of the mass of ordinary pion gives mass about 71 GeV for M_{89} pion. One can however argue that color magnetic spin-spin splitting need not obey scaling formula and that it becomes small because if is proportional to eB/m where B denotes typical value of color magnetic field and m quark mass scale which is now large. The mass of pion at the limit of vanishing color magnetic splitting given by m_0 could however obey the naive scaling.

1. For (ρ,π) system the QCD estimate for the color magnetic spin-spin splitting would be

$$(m(\rho),m(\pi)) = (m_0 + 3\Delta/4, m_0 - \Delta/4) \ .$$

p-Adic mass calculations are for mass squared rather than mass and the calculations for the mass splittings of mesons [19] force to replace this formula with

$$(m^2(\rho),m^2(\pi)) = (m_0^2 + 3\Delta^2/4, m_0^2 - \Delta^2/4) \ . \tag{7.1}$$

The masses of ρ and ω are very near to each other: $(m(\rho),m(\omega) = (.770,.782)$ GeV and obey the same mass formula in good approximation. The same is expected to hold true also for M_{89}.

2. One obtains for the parameters Δ and m_0 the formulas

$$\Delta = [m^n(\rho) - m^n(\pi)]^{1/n} \ , \ m_0 = [(m^2(\rho) + 3m(\pi)^2)/4]^{1/n} \ . \tag{7.2}$$

Here $n = 1$ corresponds to ordinary QCD and $n = 2$ to p-adic mass calculations.

3. Assuming that m_0 experiences an exact scaling by a factor 512, one can deduce the value of the parameter Δ from the mass 147 GeV of M_{89} pion and therefore predict the mass of ρ_{89}. The results are following

$$m_0 = 152.3\ GeV \ , \ \Delta = 21.3\ GeV \ , \ m(\rho_{89}) = 168.28\ GeV \tag{7.3}$$

for QCD model for spin-spin splitting and

$$m_0 = 206.7\ GeV \ , \ \Delta = 290.5\ GeV \ , \ m(\rho_{89}) = 325.6\ GeV \ . \tag{7.4}$$

for TGD model for spin-spin splitting.

4. Rather remarkably, there are indications from D0 [49] for charged and from CDF [49, 50] for neutral resonances with masses around 325 GeV such that the neutral one is split by .2 GeV: the splitting could correspond to $\rho - \omega$ mass splitting. Hence one obtains support for both M_{89} hadron physics and p-adic formulas for color magnetic spin-spin splitting. Note that the result excludes also the interpretation of the nearly 300 GeV resonance as ρ_{89} in TGD framework.

5. This scenario allows to make estimates also for the masses other resonances and naive scaling argument is expected to improve as the mass increases. For (K_{89}, K_{89}^*) system this would predict mass $m(K_{89}) > 256$ GeV and $m(K_{89}^*) < 456.7$ GeV.

The nasty question is why the octaves of pion are not realized as a resonances in ordinary hadron physics. If they were there, their decays to ordinary pion pairs by this mechanism would very slow.

1. Could it be that also ordinary pion has these octaves but are not produced by ordinary strong interactions in nucleon collisions since the nucleons do not contain the p-adically scaled up quarks fusing to form the higher octave of the pion. Also the fusion rate for two pions to higher octave of pion would be rather small by parity breaking requiring weak interactions.

2. The production mechanism for the octaves of ordinary pions, for M_{89} pions in the collisions of ordinary nucleons, and for leptohadrons would be universal, namely the collision of charged particles with cm kinetic energy above the octave of pion. The presence of strong non-orthogonal electric and magnetic fields varying considerably in the time scale defined by the Compton time of the pion is necessary since the interaction Lagrangian density is essentially the product of the abelian instanton density and pion field. In fact, in [80] it is mentioned that 300 GeV particle candidate is indeed created at rest in Tevatron lab -in other words in the cm system of colliding proton and antiproton beams.

3. The question is whether the production of the octaves of scaled up pions could have been missed in proton-proton and proton antiproton collisions due to the very peculiar kinematics: pions would be created almost at rest in cm system [26]. Whether or not this is the case should be easy to test. For a theorists this kind of scenario does not look impossible but at the era of LHC it would require a diplomatic genius and authority of Witten to persuade experimentalists to check whether low energy collisions of protons produce octaves of pions!

There is also the question about the general production mechanisms for M_{89} hadrons.

1. Besides the production of scalar mesons in strong non-orthogonal magnetic and electric fields also the production via annihilation of quark pairs to photon and weak bosons in turn decaying to the quarks of M_{89} hadron physics serves as a possible production mechanism. These production mechanisms do not give much hopes about the production of nucleons of M_{89} physics.

2. If ordinary gluons couple to M_{89} quarks, also the production via fusion to gluons is possible. If the transition from M_{107} hadron physics corresponds to a phase transition transforming M_{107} hadronic space-time sheets/gluons to M_{89} space-time sheets/gluons, M_{107} gluons do not couple directly to M_{89} gluons. In this case however color spin glass phase for M_{107} gluons could decay to M_{89} gluons in turn producing also M_{89} nucleons. Recall that naive scalings for M_{89} nucleon the mass 481 GeV. The actual mass is expected to be higher but below the scaled up Δ resonance mass predicted to be below 631 GeV.

7.2.5 How could one understand CDF-D0 discrepancy concerning 145 GeV resonance?

The situation concerning 145 GeV bump has become rather paradoxical. CDF claims that 145 GeV resonance is there at 4.3 sigma level. The new results from D0 however fail to support CDF bump [72] (see Lubos, Jester, and Tommaso).

This shows only that either CDF or D0 is wrong, not that CDF is wrong as some of us suddenly want to believe. My own tentative interpretation -not a belief- relies on bigger picture provided by TGD and is that both 145 GeV, 300 GeV, and 325 GeV resonances are there and have interpretations in terms of π and its p-adic octave, ρ, and ω of M_{89} hadron physics. I could of course be wrong. LHC will be the ultimate jury.

In any case, neither CDF and D0 are cheating and one should explain the discrepancy rationally. Resonaances mentions different estimates for QCD background as a possible explanation. What one could say about this in TGD framework allowing some brain storming?

1. There is long history of this kind of forgotten discoveries having same interpretation in TGD framework. Always pionlike states-possibly coherent state of them- would have been produced in strong non-orthogonal magnetic and electric fields of the colliding charges and most pion-like states predicted to be almost at rest in cm frame.

   Electro-pions were observed already at seventies in the collisions of heavy nuclei at energies near Coulomb wall, resonances having interpretation as mu-pions about three years ago, tau-pions detected by CDF for two and half years ago with refutation coming from D0, now DAMA and Cogent observed dark matter candidate having explanation in terms of tau-pion in TGD framework but Xenon100 found nothing (in this case on can understand the discrepancy in TGD framework). The octaves of M_{89} pions would represent the last episode of this strange history. In the previous posting universality of the production mechanism forced to made the proposal that also the collisions of ordinary nuclei could generate octaves of ordinary pions. They have not been observed and as I proposed this might due to the peculiarity of the production mechanism.

What could be a common denominator for this strange sequence of almost discoveries? Light colored excitations of leptons can be of course be argued to be non-existent because intermediate boson decay widths do not allow them but it is difficult to believe that his would have been the sole reason for not taking leptopions seriously.

2. Could the generation of a pionic coherent state as a critical phenomenon very sensitive to the detailed values of the dynamical parameters, say the precise cm energies of the colliding beams? For leptopions a phase transition generating dark colored variants of leptons (dark in the sense having non-standard value of Planck constant) would indeed take place so that criticality might make sense. Could also M_{89} quarks be dark or colored excitations of ordinary quarks which are dark? Could the $M_{107} \rightarrow M_{89}$ phase transition take place only near criticality? This alone does not seem to be enough however.

3. The peculiarity of the production mechanism is that the pion like states are produced mostly at rest in cm frame of the colliding charges. Suppose that the cm frame for the colliding charged particles is not quite the lab frame in D0. Since most dark pions are produced nearly at rest in the cm frame, they could in this kind of situation leave the detector before decaying to ordinary particles: they would behave just like dark matter is expected to behave and would not be detected. The only signature would be missing energy. This would also predict that dark octaves of ordinary pions would not be detected in experiments using target which is at rest in lab frame.

4. This mechanism is actually quite general. Dark matter particles decaying to ordinary matter and having long lifetime remain undetected if they move with high enough velocity with respect to laboratory. Long lifetime would be partially due to the large value of $\hbar$ and relativistic with respect to laboratory velocities also time dilation would increases the lifetime. Dark matter particles could be detected only as a missing energy not identifiable in terms of neutrinos. A special attention should be directed to state candidates which are nearly at rest in laboratory.

An example from ordinary hadron physics is the production of pions and their octaves in the strong electric and magnetic field of nuclei colliding with a target at rest in lab. The lifetime of neutral pion is about 10^{-8} seconds and scaled up for large $\hbar$ and by time dilation when the colliding nucleons have relativistic energies. Therefore the dark pion might leave the measurement volume before decay to two gammas when the the target is at rest in laboratory. It is not even clear whether the gammas need to have standard value of Planck constant.

For the second octave of M_{89} pion the lifetime would be scaled down by the ratio of masses giving a factor 2^{11} and lifetime of order $.5 \times 10^{-11}$ seconds. Large $\hbar$ would scale up the lifetime. For non-relativistic relativistic velocities the distance travelled before the decay to gamma pair would $L = (\hbar/\hbar_0) \times (v/c) \times 1.1$ mm.

If also the gamma pair is dark, the detection would require even larger volume. TGD suggests strongly that also photons have a small mass which they obtain by eating the remaining component of Higgs a la TGD (transforming like 1+3 under vectorial weak SU(2)). If photon mass defines the upper bound for the rate for the transformation to ordinary photons, dark photons would remain undetected.

7.2.6 Higgs or a pion of M_{89} hadron physics?

D0 refuted the 145 GeV bump and after this it was more or less forgotten in blogs, which demonstrates how regrettably short the memory span of blog physicists is. CDF reported it in Europhysics 2011 and it seems that the groups are considering seriously possible explanations for the discrepancy. To my opinion the clarification of his issue is of extreme importance.

The situation changed at the third day of conference (Saturday) when ATLAS reported about average 2.5 sigma evidence for what might be Higgs in the mass range 140-150 GeV. The candidate revealed itself via decays to WW in turn decaying to lepton pairs. Also D0 and CDF told suddenly that they have observed similar evidence although the press release had informed that Higgs had been located to the mass range 120-137 GeV. There is of course no reason to exclude the possibility that the decays of

145 GeV resonance are in question and in this case the interpretation as standard model Higgs would be definitely excluded. If the pion of M_{89} physics is in question it would decay to WW pair instead of quark pair producing two jets. Since weak decay is in question one an expect that the decay rate is small.

If this line of reasoning is correct, standard model Higgs is absent. TGD indeed predicts that the components of TGD Higgs become longitudinal components of gauge bosons since also photon and graviton gain a small mass. This however leaves the two Higgses predicted by MSSM under consideration. The stringent lower bounds for the masses of squarks and gluinos of standard SUSY were tightened in the conference and are now about 1 TeV and this means that the the basic argument justifying MSSM (stability of Higgs mass against radiative corrections) is lost.

The absence of Higgs forces a thorough re-consideration of the fundamental ideas about particle massivation. p-Adic thermodynamics combined with zero energy ontology and the identification of massive particles as bound states of massless fermions is the vision provided by TGD.

7.2.7 Short digression to TGD SUSY

Although the question about TGD variant of SUSY is slightly off-topic, its importance justifies a short discussion. Although SUSY is not needed to stabilize Higgs mass, the anomaly of muonic g-2 suggests TGD SUSY and the question is whether TGD SUSY could explain it.

1. Leptons are characterized by Mersennes or Gaussian Mersennes: $(M_{127}, M_{G,113}, M_{107})$ for (e, μ, τ). If also sleptons correspond to Mersennes of Gaussian Mersennes, then (selectron, smuon, stau) should correspond to $(M_{89}, M_{G,79}, M_{61})$ is one assumes that selectron corresponds to M_{89}. Selectron mass would be 262 GeV and smuon mass 13.9 TeV. g-2 anomaly for muon [16] suggests that the mass of selectron should not be much above .1 TeV and M_{89} fits the bill. Valence quarks correspond to the Gaussian Mersenne $k \leq 113$, which suggests that squarks have $k \geq 79$ so that squark masses should be above 13 TeV. If sneutrinos correspond to Gaussian Mersenne $k = 167$ then sneutrinos could have mass below electron mass scale. Selectron would remain the only experiment signature of TGD SUSY at this moment.

2. One decay channel for selectron would be to electron+ sZ or neutrino+ sW. sZ/sW would eventually decay to possibly virtual Z+ neutrino/W+neutrino: that is weak gauge boson plus missing energy. Neutralino and chargino need not decay in the detection volume. The lower bound for neutralino mass is 46 GeV from intermediate gauge boson decay widths. Hence this option is not excluded by experimental facts.

3. If the sfermions decay rapidly enough to fermion plus neutrino, the signature of TGD SUSY would be excess of events of type lepton+ missing energy or jet+ missing energy. For instance, lepton+missing jet could be mis-identified as decay products of possibly exotic counterpart of weak gauge boson. The decays of 262 GeV selectron would give rise to decays which might be erratically interpreted as decays of W' to electron plus missing energy. The study of CDF at $\sqrt{s}$= 1.96 TeV in p-pbar collisions excludes heavy W′ with mass below 1.12 TeV [59]. The decay rate to electron plus neutrino must therefore be slow.

   There are indications for a tiny excess of muon + missing energy events in the decays of what has been tentatively identified as a heavy W boson W^{prime} (see Figure 1 of [58]). The excess is regarded as insignificant by experimenters. W^{prime} candidate is assumed to have mass 1.0 TeV or 1.4 TeV. If smuon is in question, one must give up the Mersenne hypothesis.

7.2.8 The mass of u and d quarks of M_{89} physics

In the last updating of the chapter about the p-adic model for hadronic masses [19] I found besides some silly numerical errors also a gem that I had forgotten. For pion the contributions to mass squared from color-magnetic spin-spin interaction and color Coulombic interaction and super-symplectic gluons cancel and the mass is in excellent approximation given by the $m^2(\pi) = 2m^2(u)$ with $m(u) = m(d) = 0.1$ GeV in good approximation. That only quarks contribute is the TGD counterpart for the almost Goldstone boson character of pion meaning that its mass is only due to the massivation of quarks. The value of the p-adic prime is $p \simeq 2^k$, with $k(u) = k(d) = 113$ and the mass of charged pion is predicted with error of .2 per cent.

If the reduction of pion mass to mere quark mass holds true for all scaled variants of ordinary hadron physics, one can deduce the value of u and d quark masses from the mass of the pion of M_{89} hadron physics and vice versa. The mass estimate is 145 GeV if one identifies the bump claimed by CDF [64] as M_{89} pion. Recall that D0 did not detect the CDF bump [72] (I have discussed possible reasons for the discrepancy in terms of the hypothesis that dark quarks are in question). From this one can deduce that the p-adic prime $p \simeq 2^k$ for the u and d quarks of M_{89} physics is $k = 93$ using $m(u,93) = 2^{(113-93)/2}m(u,113)$, $m(u,113) \simeq .1$ GeV. For top quark one has $k = 94$ so that a very natural transition takes place to a new hadron physics. The predicted mass of $\pi(89)$ is 144.8 GeV and consistent with the value claimed by CDF. What makes the prediction non-trivial is that possible quark masses comes as as half-octaves meaning exponential sensitivity with respect to the p-adic length scale.

The common mass of $u(89)$ and $d(89)$ quarks is 102 GeV in a good approximation and quark jets with mass peaked around 100 GeV should serve as a signature for them. The direct decays of the $\pi(89)$ to M_{89} quarks are of course non-allowed kinematically.

7.2.9 A connection with the top pair backward-forward asymmetry in the production of top quark pars?

One cannot exclude the possibility that the predicted exotic octet of gluons proposed as an explanation of the anomalous backward-forward asymmetry in top pair production correspond sto the gluons of the scaled up variant of hadron physics. M_{107} hadron physics would correspond to ordinary gluons only and M_{89} only to the exotic octet of gluons only so that a strict scaled up copy would not be in question. Could it be that given Mersenne prime tolerates only single hadron physics or leptohadron physics?

In any case, this would give a connection with the TGD based explanation of the backward-forward asymmetry in the production of top pairs also discussed in this article. In the collision incoming quark of proton and antiquark of antiproton would topologically condense at M_{89} hadronic space-time sheet and scatter by the exchange of exotic octet of gluons: the exchange between quark and antiquark would not destroy the information about directions of incoming and outgoing beams as s-channel annihilation would do and one would obtain the large asymmetry. The TGD based generalized Feynman diagram would involve an exchange of a gluon represented by a wormhole contact. The first wormhole throat would have genus two as also top quark and second throat genus zero. One can imagine that the top quark comes from future and then travels along space-like direction together with antiquark wormhole throat of genus zero a and then turns back to the future. Incoming quark and antiquark perform similar turn around [16].

This asymmetry observed found a further confirmation in Europhysics 2011 conference [67]. The obvious question is whether this asymmetry could be reduced to that in collisions of quarks and antiquarks. Tommaso Dorigo tells that CMS has found that this is not the case, which suggests that the phenomenon might be assignable to valence quarks only.

7.3 Other indications for M_{89} hadron physics

Also other indications for M_{89} hadron physics have emerged during this year and although the fate of these signals is probably the usual one, they deserve to be discussed briefly.

7.3.1 Bumps also at CDF and D0?

It seems that experimentalists have gone totally crazy. Maybe new physics is indeed emerging from LHC and they want to publish every data bit in the hope of getting paid visit to Stockholm. *CDF* and ATLAS have told about bumps and now Lubos [49] tells about a new 3 sigma bump reported by D0 collaboration at mass 325 GeV producing muon in its decay producing W boson plus jets [71]. The proposed identification of bump is in terms of decay of t′ quark producing W boson.

Lubos mentions also second mysterious bump at 324.8 GeV or 325.0 GeV reported by *CDF* collaboration [63] and discussed by Tommaso Dorigo [50] towards the end of the last year. The decays of these particles produce 4 muons through the decays of two Z bosons to two muons. What is peculiar is that two mass values differing by .2 GeV are reported. The proposed explanation is in terms of Higgs decaying to two Z bosons. TGD based view about new physics suggests strongly that the three of four particles forming a multiplet is in question.

One can consider several explanations in TGD framework without forgetting that these bumps very probably disappear. Consider first the D0 anomaly alone.

1. TGD predicts also higher generations but there is a nice argument based on conformal invariance and saying that higher particle families are heavy. What "heavy" means is not clear. It could of mean heavier that intermediate gauge boson mass scale. This explanation does not look convincing to me.

2. Another interpretation would be in terms of scaled up variant of top quark. The mass of top is around 170 GeV and p-adic length scale hypothesis would predict that the mass should equal to a multiple of half octave of top quark mass. Single octave would give mass of 340 GeV. The deviation from predicted mass would be 5 per cent.

3. The third interpretation is in terms of ρ and ω mesons of M_{89}. By assuming that the masses of M_{89} π and ρ in absence of color magnetic spin-spin splitting scale naively in the transition from M_{107} to M_{89} physics and by determining the parameter characterizing color magnetic spin-spin splitting from the condition that M_{89} pion has 157 GeV mass, one predicts that M_{89} ρ and ω have same mass 325.6 GeV in good approximation The .2 GeV mass difference would have interpretation as $\rho - \omega$ mass difference. In TGD framework this explanation is unique.

7.3.2 Indications for M_{89} charmonium from ATLAS

Lubos commented last ATLAS release about dijet production. There is something which one might interpret as the presence of resonances above 3.3 TeV [see Fig. 2) of the article] [56]. Of course, just a slight indication is in question, so that it is perhaps too early to pay attention to the ATLAS release. I am however advocating a new hadron physics and it is perhaps forgivable that I am alert for even tiniest signals of new physics.

In a very optimistic mood I could believe that a new hadron physics is being discovered (145 GeV boson could be identified as charged pion and 325 GeV bumps could allow interpretation as kaons). With this almost killer dose of optimism the natural question is whether this extremely slight indication about new physics might have interpretation as a scaled up J/Ψ and various other charmonium states above it giving rise to what is not single very wide bump to a family of several resonances in the range 3-4 TeV by scaling the 3-4 GeV range for charmonium resonances. For instance, J/Ψ decay width is very small, about .1 MeV, which is about $.3 \times 10^{-4}$ of the mass of J/Ψ. In the recent case direct scaling would give decay of about 300 MeV for the counterpart of J/Ψ if the decay is also now slow for kinematic reasons. For other charmonium resonances the widths are measurement in per cents meaning in the recent case width of order of magnitude 30 GeV: this estimate looks more reasonable as the first estimate.

One can also now perform naive scalings. J/Ψ has mass of about 3 GeV. If the scaling of ordinary pion mass from .14 GeV indeed gives something like 145 GeV then one can be very naive and apply the same scaling factor of about 1030 to get the scaled up J/Ψ; with mass of order 3.1 TeV. The better way to understand the situation is to assume that color-magnetic spin spin splitting is small also for M_{89} charmonium states and apply naive scaling to the mass of J/Ψ; to get a lower bound for the mass of its M_{89} counterpart. This would give mass of 1.55 TeV which is by a factor 1/2 too small. p-Adic mass calculations lead to the conclusion that c quark is characterized by $p \simeq 2^k$, $k = 104$. Naive scaling would give $k = 104 - 18 = 86$ and 1.55 TeV mass for J/Ψ. Nothing however excludes $k = 84$ and the lower bound 3.1 TGD for the mass of J/Ψ. Since color magnetic spin-spin splitting is smaller for M_{89} pion, same is expected to be true also for charmonium states so that the mass might well be around 3.3 TeV.

7.3.3 Blackholes at LHC: or just bottonium of M_{89} hadron physics?

The latest Tommaso Dorigo's posting has a rather provocative title: The Plot Of The Week - A Black Hole Candidate. Some theories inspired by string theories predict micro black holes at LHC. Micro blackholes have been proposed as explanation for certain exotic cosmic ray events such as Centauros, which however seem to have standard physics explanation.

Without being a specialist one could expect that evaporating black hole would be in many respects analogous to quark gluon plasma phase decaying to elementary particles producing jets. Or any particle like system, which has forgot all information about colliding particles which created it- say the information about the scattering plane of partons leading to the jets as a final state and reflecting itself as the coplanarity of the jets. If the information about the initial state is lost, one would expect more or less spherical jet distribution. The variable used as in the study is sum of transverse energies for jets emerging from same point and having at least 50 GeV transverse energy. QCD predicts that this kind of events should be rather scarce and if they are present, one can seriously consider the possibility of new physics.

The LHC document containing the sensational proposal is titled Search for Black Holes in pp collisions at $\sqrt{s} = 7$ TeV [55] and has the following abstract:

An update on a search for microscopic black hole production in pp collisions at a center-of-mass energy of 7 TeV by the CMS experiment at the LHC is presented using a 2011 data sample corresponding to an integrated luminosity of 190 pb1. This corresponds to a six-fold increase in statistics compared to the original search based on 2010 data. Events with large total transverse energy have been analyzed for the presence of multiple energetic jets, leptons, and photons, typical of a signal from an evaporating black hole. A good agreement with the expected standard model backgrounds, dominated by QCD multijet production, has been observed for various multiplicities of the final state. Stringent model-independent limits on new physics production in high-multiplicity energetic final states have been set, along with model-specific limits on semi-classical black hole masses in the 4-5 TeV range for a variety of model parameters. This update extends substantially the sensitivity of the 2010 analysis.

The abstract would suggest that nothing special has been found but in sharp contrast with this the article mentions black hole candidate decaying to 10 jets with total transverse energy S_T. The event is illustrated in the figure 3 of the article. The large number of jets emanating from single point would suggest a single object decaying producing the jets.

Personally I cannot take black holes as an explanation of the event seriously. What can I offer instead? p-Adic mass calculations rely on p-adic thermodynamics and this inspires obvious questions. What p-adic cooling and heating processes could mean? Can one speak about p-adic hot spots? What p-adic overheating and over-cooling could mean? Could the octaves of pions and possibly other mesons explaining several anomalous findings including CDF bump correspond to unstable over-heated hadrons for which the p-adic prime near power of two is smaller than normally and p-adic mass scale is correspondingly scaled up by a power of two?

The best manner to learn is by excluding various alternative explanations for the 10 jet event.

1. M_{89} variants of QCD jets are excluded both because their production requires higher energies and because their number would be small. The first QCD three-jets were observed around 1979 [98]. $q-\overline{q}-g$ three-jet was in question and it was detected in e^+e^- collision with cm energy about 7 GeV. The naive scaling by factor 512 would suggest that something like 5.6 TeV cm energy is needed to observed M_{89} parton jets. The recent energy is 7 TeV so that there are hopes of observing M_{89} three- jets in decays of heavy M_{89}. For instance, the decays of charmonium and bottonium of M_{89} physics to three gluons or two-gluons and photon would create three-jets.

2. Ordinary quark gluon plasma is excluded since in a sufficiently large volume of quark gluon plasma so called jet quenching [39] occurs so that jets have small transverse energies. This would be due to the dissipation of energy in the dense quark gluon plasma. Also ordinary QCD jets are predicted to be rare at these transverse energies: this is of course the very idea of how black hole evaporation might be observed. Creation of quark gluon plasma of M_{89} hadron physics cannot be in question since ordinary quark gluon plasma was created in p-anti-p collision with cm energy of few TeV so that something like 512 TeV of cm energy might be needed!

3. Could the decay correspond to a decay of a blob of M_{89} hadronic phase to M_{107} hadrons? How this process could take place? I proposed for about 15 years ago [16] that the transition from M_{89} hadron physics to M_{107} hadron physics might take place as a p-adic cooling via a cascade like process via highly unstable intermediate hadron physics. The p-adic temperature is quantized and

given by $T_p = n/log(p) \simeq n/klog(2)$ for $p \simeq 2^k$ and p-adic cooling process would proceed in a step-wise manner as $k \rightarrow k+2 \rightarrow k+4+ ...$ Also $k \rightarrow k+1 \rightarrow k+2+..$ with mass scale reduced in powers of $\sqrt{2}$ can be considered. If only octaves are allowed, the p-adic prime characterizing the hadronic space-time sheets and quark mass scale could decrease in nine steps from M_{89} mass scale proportional to $2^{-89/2}$ octave by octave down to the hadronic mass scale proportional $2^{-107/2}$ as $k = 89 \rightarrow 91 \rightarrow 93... \rightarrow 107$. At each step the mass in the propagator of the particle would be changed. In particular on mass shell particles would become off mass shell particles which could decay.

At quark level the cooling process would naturally stop when the value of k corresponds to that characterizing the quark. For instance b quark one has $k(b) = 103$ so that 7 steps would be involved. This would mean the decay of M_{89} hadrons to highly unstable intermediate states corresponding to $k = 91, 93, ..., 107$. At every step states almost at rest could be produced and the final decay would produce large number of jets and the outcome would resemble the spectrum blackhole evaporation. Note that for u, d, s quarks one has $k = 113$ characterizing also nuclei and muon which would mean that valence quark space-time sheets of lightest hadrons would be cooler than hadronic space-time sheet, which could be heated by sea partons. Note also that quantum superposition of phases with several p-adic temperatures can be considered in zero energy ontology.

This is of course just a proposal and might not be the real mechanism. If M_{89} hadrons are dark in TGD sense as the TGD based explanation of CDF-D0 discrepancy suggests, also the transformation changing the value of Planck constant is involved.

4. This picture does not make sense in the TGD inspired model explaining DAMA observations and DAMA-Xenon100 anomaly, CDF bump discussed in this article and two and half year old CDF anomalyd discussed in [26]. The model involves creation of second octave of M_{89} pions decaying in stepwise manner. A natural interpretation of p-adic octaves of pions is in terms of a creation of over-heated unstable hadronic space-time sheet having $k = 85$ instead of $k = 89$ and p-adically cooling down to relatively thermally stable M_{89} sheet and containing light mesons and electroweak bosons. If so then the production of CDF bump would correspond to a creation of hadronic space-time sheet with p-adic temperature corresponding to $k = 85$ cooling by the decay to $k = 87$ pions in turn decaying to $k = 89$. After this the decay to M_{107} hadrons and other particles would take place.

Consider now whether the 10 jet event could be understood as a creation of a p-adic hot spot perhaps assignable to some heavy meson of M_{89} physics. The table below is from [14, 18] and gives the p-adic primes assigned with constituent quarks identified as valence quarks. For current quarks the p-adic primes can be much large so that in the case of u and d quark the masses can be in 10 MeV range (which together with detailed model for light hadrons supports the view that quarks can appear at several p-adic temperatures).

1. According to p-adic mass calculations [18] ordinary charmed quark corresponds to $k = 104 = 107-3$ and that of bottom quark to $k = 103 = 107-4$, which is prime and correspond to the second octave of M_{107} mass scale assignable to the highest state of pion cascade. By naive scaling M_{89} charmonium states (Ψ would correspond to $k = 89-3 = 86$ with mass of about 1.55 TeV by direct scaling. $k = 89-4 = 85$ would give mass about 3.1 GeV and there is slight evidence for a resonance around 3.3 TeV perhaps identifiable as charmonium. Υ (bottonium) consisting of $b\bar{b}$ pair correspond to $k = 89-4 = 85$ just like the second octave of M_{89} pion. The mass of M_{89} Υ meson would be about 4.8 TeV for $k = 85$. $k = 83$ one obtains 9.6 TeV, which exceeds the total cm energy 7 TeV.

2. Intriguingly, $k = 85$ for the bottom quark and for first octave of charmonium would correspond to the second octave of M_{89} pion. Could it be that the hadronic space-time sheet of Υ is heated to the p-adic temperature of the bottom quark and then cools down in a stepwise manner? If so, the decay of Υ could proceed by the decay to higher octaves of light M_{89} mesons in a process involving two steps and could produce a large number jets.

3. For the decay of ordinary Υ meson 81.7 per cent of the decays take place via ggg state. In the recent case they would create three M_{89} parton jets producing relativistic M_{89} hadrons. 2.2 per cent of

decays take place via γgg state producing virtual photon plus M_{89} hadrons. The total energies of the three jets would be about 1.6 TeV each and much higher than the energies of QCD jets so that this kind of jets would serve as a clearcut signature of M_{89} hadron physics and its bottom quark. Note that there already exists slight evidence for charmonium state. Recall that the total transverse energy of the 10 jet event was about 1 TeV.

Also direct decays to M_{89} hadrons take place. η' +anything- presumably favored by the large contribution of $b\bar{b}$ state in η'- corresponds to 2.9 per cent branching ratio for ordinary hadrons. If second octaves of η' and other hadrons appear in the hadron state, the decay product could be nearly at rest and large number of M_{89} would result in the p-adic cooling process (the naive scaling of η' mass gives .5 TeV and second octave would correspond to 2 TeV.

4. If two octave p-adic over-heating is dynamically favored, one must also consider the first octave of of scaled variant of J/Ψ state with mass around 3.1 GeV scaled up to 3.1 TeV for the first octave. The dominating hadronic final state in the decay of J/Ψ is $\rho^{\pm}\pi^{\mp}$ with branching ratio of 1.7 per cent. The branching fractions of $\omega\pi^+\pi^+\pi^-\pi^-$, $\omega\pi^+\pi^-\pi^0$, and $\omega\pi^+\pi^+pi^-$ are 8.5×10^{-3} 4.0×10^{-3}, and 8.6×10^{-3} respectively. The second octaves for the masses of ρ and π would be 1.3 TeV and .6 TeV giving net mass of 1.9 TeV so that these mesons would be relativistic if charmonium state with mass around 3.3 TeV is in question. If the two mesons decay by cooling, one would obtain two jets decaying two jets. Since the original mesons are relativistic one would probably obtain two wide jets decomposing to sub-jets. This would not give the desired fireball like outcome.

   The decays $\omega\pi^+\pi^+\pi^-\pi^-$ (see Particle Data Tables would produce five mesons, which are second octaves of M_{89} mesons. The rest masses of M_{89} mesons would in this case give total rest mass of 3.5 TeV. In this kind of decay -if kinematically possible- the hadrons would be nearly at rest. They would decay further to lower octaves almost at rest. These states in turn would decay to ordinary quark pairs and electroweak bosons producing a large number of jets and black hole like signatures might be obtained. If the process proceeds more slowly from M_{89} level, the visible jets would correspond to M_{89} hadrons decaying to ordinary hadrons. Their transverse energies would be very high.

q	d	u	s	c	b	t
n_q	4	5	6	6	59	58
s_q	12	10	14	11	67	63
$k(q)$	113	113	113	104	103	94
$m(q)/GeV$	.105	.092	.105	2.191	7.647	167.8

Constituent quark masses predicted for diagonal mesons assuming $(n_d, n_s, n_b) = (5, 5, 59)$ and $(n_u, n_c, n_t) = (5, 6, 58)$, maximal CP_2 mass scale($Y_e = 0$), and vanishing of second order contributions.

To sum up, the most natural interpretation for the 10-jet event in TGD framework would be as p-adic hot spots produced in collision.

7.3.4 Has CMS detected λ baryon of M_{89} hadron physics?

In his recent posting Lubos tells about a near 3-sigma excess of 390 GeV 3-jet RPV-gluino-like signal reported by CMS collaboration in article Search for Three-Jet Resonances in p-p collisions at $\sqrt{s} = 7$ TeV [57]. This represents one of the long waited results from LHC and there are good reason to consider it at least half-seriously.

Gluinos are produced in pairs and in the model based on standard super-symmetry decay to three quarks. The observed 3-jets in question would correspond to a decay to uds quark triplet. The decay would be R-parity breaking. The production rate would however too high for standard SUSY so that something else is involved if the 3 sigma excess is real.

1. Signatures for standard gluinos correspond to signatures for M_{89} baryons in TGD framework

In TGD Universe gluinos would decay to ordinary gluons and right-handed neutrino mixing with the left handed one so that gluino in TGD sense is excluded as an explanation of the 3-jets. In TGD framework the gluino candidate would be naturally replaced with $k = 89$ variant of strange baryon λ decaying to uds quark triplet. Also the 3-jets resulting from the decays of proton and neutron and Δ resonances are predicted. The mass of ordinary λ is $m(\lambda, 107) = 1.115$ GeV. The naive scaling by a factor 512 would give mass $m(\lambda, 107) = 571$ GeV, which is considerably higher than 390 GeV. Naive scaling would predict the scaled up copies of the ordinary light hadrons so that the model is testable.

It is quite possible that the bump is a statistical fluctuation. One can however reconsider the situation to see whether a less naive scaling could allow the interpretation of 3-jets as decay products of M_{89} λ-baryon.

2. Massivation of hadrons in TGD framework

Let us first look the model for the masses of nucleons in p-adic thermodynamics [19].

1. The basic model for baryon masses assumes that mass squared -rather than energy as in QCD and mass in naive quark model- is additive at space-time sheet corresponding to given p-adic prime whereas masses are additive if they correspond to different p-adic primes. Mass contains besides quark contributions also "gluonic contribution" which dominates in the case of baryons. The additivity of mass squared follows naturally from string mass formula and distinguishes dramatically between TGD and QCD. The value of the p-adic prime $p \simeq 2^k$ characterizing quark depends on hadron: this explains the mass differences between baryons and mesons. In QCD approach the contribution of quark masses to nucleon masses is found to be less than 2 per cent from experimental constraints. In TGD framework this applies only to sea quarks for which masses are much lighter whereas the light valence quarks have masses of order 100 MeV.

   For a mass formula for quark contributions additive with respect to quark mass squared quark masses in proton would be around 100 MeV. The masses of u, d, and s quarks are in good approximation 100 MeV if p-adic prime is $k = 113$, which characterizes the nuclear space-time sheet and also the space-time sheet of muon. The contribution to proton mass is therefore about $\sqrt{3} \times 100$ MeV.

   Remark: The masses of u and d sea quarks must be of order 10 MeV to achieve consistency with QCD. In this case p-adic primes characterizing the quarks are considerably larger. Quarks with mass scale of order MeV are important in nuclear string model which is TGD based view about nuclear physics [29].

2. If color magnetic spin-spin splitting is neglected, p-adic mass calculations lead to the following additive formula for mass squared.

$$M(baryon) = M(quarks) + M(gluonic) \ , \quad M^2(gluonic) = nm^2(107) \ . \tag{7.5}$$

   The value of integer n can almost predicted from a model for the TGD counterpart of the gluonic contribution [19] to be $n = 18$. $m^2(107)$ corresponds to p-adic mass squared associated with the Mersenne prime $M_{107} = 2^{107} - 1$ characterizing hadronic space-time sheet responsible for the gluonic contribution to the mass squared. One has $m(107) = 233.55$ MeV from electron mass $m_e \simeq \sqrt{5} \times m(127) \simeq 0.5$ MeV and from $m(107) = 2^{(127-107)/2} \times m(127)$.

3. For proton one has

$$M(quarks) = (\sum_{quarks} m^2(quark))^{1/2} \simeq 3^{1/2} \times \ 100 \ MeV$$

   for $k(u) = k(d) = 113$ [19].

3. Super-symplectic gluons as TGD counterpart for non-perturbative aspects of QCD

A key difference as compared to QCD is that the TGD counterpart for the gluonic contribution would contain also that due to "super-symplectic gluons" besides the possible contribution assignable to ordinary gluons.

1. Super-symplectic gluons do not correspond to pairs of quark and and antiquark at the opposite throats of wormhole contact as ordinary gluons do but to single wormhole throat carrying purely bosonic excitation corresponding to color Hamiltonian for CP_2. They therefore correspond directly to wave functions in WCW ("world of classical worlds") and could therefore be seen as a genuinely non-perturbative objects allowing no description in terms of a quantum field theory in fixed background space-time.

2. The description of the massivation of super-symplectic gluons using p-adic thermodynamics allows to estimate the integer n characterizing the gluonic contribution. Also super-symplectic gluons are characterized by genus g of the partonic 2-surface and in the absence of topological mixing $g = 0$ super-symplectic gluons are massless and do not contribute to the ground state mass squared in p-adic thermodynamics. It turns out that a more elegant model is obtained if the super-symplectic gluons suffer a topological mixing assumed to be same as for U type quarks. Their contributions to the mass squared would be $(5, 6, 58) \times m^2(107)$ with these assumptions.

3. The quark contribution $(M(nucleon) - M(gluonic))/M(nucleon)$ is roughly 82 per cent of proton mass. In QCD approach experimental constraints imply that the sum of quark masses is less that 2 per cent about proton mass. Therefore one has consistency with QCD approach if one assumes that the light quarks correspond to sea quarks.

4. What happens in $M_{107} \rightarrow M_{89}$ transition?

What happens in the transition $M_{107} \rightarrow M_{89}$ depends on how the quark and gluon contributions depend on the Mersenne prime.

1. One can also scale the "gluonic" contribution to baryon mass which should be same for proton and λ. Assuming that the color magnetic spin-spin splitting and color Coulombic conformal weight expressed in terms of conformal weight are same as for the ordinary baryons, the gluonic contribution to the mass of $p(89)$ corresponds to conformal weight $n = 11$ reduced from its maximal value $n = 3 \times 5 = 15$ corresponding to three topologically mixed super-symplectic gluons with conformal weight 5 [19]. The reduction is due to the negative colour Coulombic conformal weight. This is equal to $M_g = \sqrt{11} \times 512 \times m(107)$, $m(107) = 233.6$ MeV, giving $M_g = 396.7$ GeV which happens to be very near to the mass about 390 GeV of CMS bump. The facts that quarks appear already in light hadrons in several p-adic length scales and quark and gluonic contributions to mass are additive, raises the question whether the state in question corresponds to p-adically hot $(1/T_p \propto log(p) \simeq klog(2)$ gluonic/hadronic space-time sheet with $k = 89$ containing ordinary quarks giving a small contribution to the mass squared. Kind of overheating of hadronic space-time sheet would be in question.

2. The option for which quarks have masses of thermally stable M_{89} hadrons with quark masses deduced from the questionable 145 GeV CDF bump identified as the pion of M_{89} physics does not work.

   (a) If both contributions scale up by factor 512, one obtains $m(p, 89) = 482$ GeV and $m(\lambda) = 571$ GeV. The values are too large.

   (b) A more detailed estimate gives the same result. One can deduce the scaling of the quark contribution to the baryon mass by generalizing the condition that the mass of pion is in a good approximation just $m(\pi) = \sqrt{2}m(u, 107)$ (Goldstone property). One obtains that u and d quarks of M_{89} hadron physics correspond to $k = 93$ whereas top quark corresponds to $k = 94$: the transition between hadron physics would be therefore natural. One obtains $m(u, 89) = m(d, 89) = 102$ GeV in good approximation: note that this predicts quark jets with mass around 100 GeV as a signature of M_{89} hadron physics.
   The contribution of quarks to proton mass would be $M_q = \sqrt{3} \times 2^{(113-93)/2}m(u, 107) \simeq 173$ GeV. By adding the quark contribution to gluonic contribution $M_g = 396.7$ GeV, one obtains $m(p, 89) = 469.7$ GeV which is rather near to the naively scaled mass 482 GeV and too large. For $\lambda(89)$ the mass is even larger: if $\lambda(89) - p(89)$ mass difference obeys the naive

scaling one has $m(\lambda,89) - m(p,89) = 512 \times m(\lambda,107) - m(p,107)$. One obtains $m(\lambda,89) = m(p,89)+m(s,89)-m(u,89) = 469.7+89.6$ GeV $= 559.3$ GeV rather near to the naive scaling estimate 571 GeV. This option fails.

Maybe I would be happier if the 390 GeV bump would turn out to be a fluctuation (as it probably does) and were replaced with a bump around 570 GeV plus other bumps corresponding to nucleons and Δ resonances and heavier strange baryons. The essential point is however that the mass scale of the gluino candidate is consistent with the interpretation as λ baryon of M_{89} hadron physics. Quite generally, the signatures of R-parity breaking standard SUSY have interpretation as signatures for M_{89} hadron physics in TGD framework.

7.4 Dark nucleons and genetic code

Water memory is one of the ugly words in the vocabulary of a main stream scientist. The work of pioneers is however now carrying fruit. The group led by Jean-Luc Montagnier, who received Nobel prize for discovering HIV virus, has found strong evidence for water memory and detailed information about the mechanism involved [13, 25] ,[104]. The work leading to the discovery was motivated by the following mysterious finding. When the water solution containing human cells infected by bacteria was filtered in purpose of sterilizing it, it indeed satisfied the criteria for the absence of infected cells immediately after the procedure. When one however adds human cells to the filtrate, infected cells appear within few weeks. If this is really the case and if the filter does what it is believed to do, this raises the question whether there might be a representation of genetic code based on nano-structures able to leak through the filter with pores size below 200 nm.

The question is whether dark nuclear strings might provide a representation of the genetic code. In fact, I posed this question year before the results of the experiment came with motivation coming from attempts to understand water memory. The outcome was a totally unexpected finding: the states of dark nucleons formed from three quarks can be naturally grouped to multiplets in one-one correspondence with 64 DNAs, 64 RNAS, and 20 aminoacids and there is natural mapping of DNA and RNA type states to aminoacid type states such that the numbers of DNAs/RNAs mapped to given aminoacid are same as for the vertebrate genetic code.

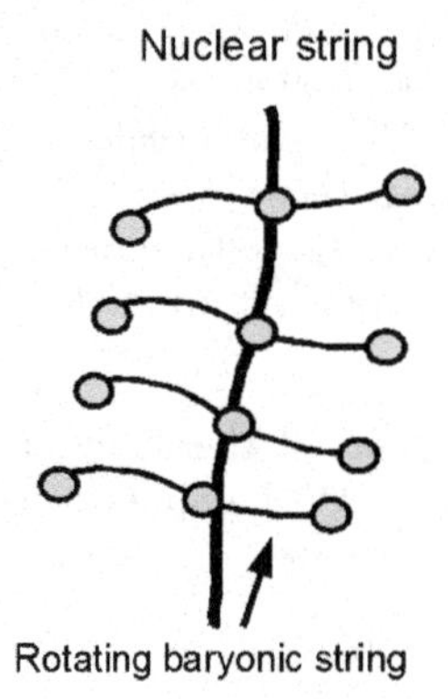

Figure 3: Illustration of a possible vision about dark nucleus as a nuclear string consisting of rotating baryonic strings.

The basic idea is simple. Since baryons consist of 3 quarks just as DNA codons consist of three nucleotides, one might ask whether codons could correspond to baryons obtained as open strings with quarks connected by two color flux tubes. This representation would be based on entanglement rather than letter sequences. The question is therefore whether the dark baryons constructed as string of 3 quarks using color flux tubes could realize 64 codons and whether 20 aminoacids could be identified as equivalence classes of some equivalence relation between 64 fundamental codons in a natural manner.

The following model indeed reproduces the genetic code directly from a model of dark neutral baryons as strings of 3 quarks connected by color flux tubes.

1. Dark nuclear baryons are considered as a fundamental realization of DNA codons and constructed as open strings of 3 dark quarks connected by two colored flux tubes, which can be also charged. The baryonic strings cannot combine to form a strictly linear structure since strict rotational invariance would not allow the quark strings to have angular momentum with respect to the quantization axis defined by the nuclear string. The independent rotation of quark strings and breaking of rotational symmetry from SO(3) to SO(2) induced by the direction of the nuclear string is essential for the model.

   (a) Baryonic strings could form a helical nuclear string (stability might require this) locally parallel to DNA, RNA, or aminoacid) helix with rotations acting either along the axis of the DNA or along the local axis of DNA along helix. The rotation of a flux tube portion around an axis parallel to the local axis along DNA helix requires that magnetic flux tube has a kink in this portion. An interesting question is whether this kink has correlate at the level of DNA too. Notice that color bonds appear in two scales corresponding to these two strings. The model of DNA as topological quantum computer [8] allows a modification in which dark nuclear string of this kind is parallel to DNA and each codon has a flux tube connection to the lipid of cell membrane or possibly to some other bio-molecule.

   (b) The analogs of DNA -, RNA -, and of amino-acid sequences could also correspond to sequences of dark baryons in which baryons would be 3-quark strings in the plane transversal to the dark nuclear string and expected to rotate by stringy boundary conditions. Thus one would have nuclear string consisting of short baryonic strings not connected along their ends (see Fig. 7.4). In this case all baryons would be free to rotate.

2. The new element as compared to the standard quark model is that between both dark quarks and dark baryons can be charged carrying charge $0, \pm 1$. This is assumed also in nuclear string model and there is empirical support for the existence of exotic nuclei containing charged color bonds between nuclei.

3. The net charge of the dark baryons in question is assumed to vanish to minimize Coulomb repulsion:

$$\sum_q Q_{em}(q) = -\sum_{flux\ tubes} Q_{em}(flux\ tube) \ . \tag{7.6}$$

   This kind of selection is natural taking into account the breaking of isospin symmetry. In the recent case the breaking cannot however be as large as for ordinary baryons (implying large mass difference between Δ and nucleon states).

4. One can classify the states of the open 3-quark string by the total charges and spins associated with 3 quarks and to the two color bonds. Total em charges of quarks vary in the range $Z_B \in \{2, 1, 0, -1\}$ and total color bond charges in the range $Z_b \in \{2, 1, 0, -1, -2\}$. Only neutral states are allowed. Total quark spin projection varies in the range $J_B = 3/2, 1/2, -1/2, -3/2$ and the total flux tube spin projection in the range $J_b = 2, 1, -1, -2$. If one takes for a given total charge assumed to be vanishing one representative from each class (J_B, J_b), one obtains $4 \times 5 = 20$ states which is the number of amino-acids. Thus genetic code might be realized at the level of baryons by mapping the neutral states with a given spin projection to single representative state with the same spin projection. The problem is to find whether one can identify the analogs of DNA, RNA and aminoacids as baryon like states.

7.4.1 States in the quark degrees of freedom

One must construct many-particle states both in quark and flux tube degrees of freedom. These states can be constructed as representations of rotation group SU(2) and strong isospin group SU(2) by using the standard tensor product rule $j_1 \times j_2 = j_1 + j_2 \oplus j_1 + j_2 - 1 \oplus ... \oplus |j_1 - j_2|$ for the representation of SU(2) and Fermi statistics and Bose-Einstein statistics are used to deduce correlations between total spin and total isospin (for instance, $J = I$ rule holds true in quark degrees of freedom). Charge neutrality is assumed and the breaking of rotational symmetry in the direction of nuclear string is assumed.

Consider first the states of dark baryons in quark degrees of freedom.

1. The tensor product $2\otimes2\otimes2$ is involved in both cases. Without any additional constraints this tensor product decomposes as $(3 \oplus 1) \otimes 2 = 4 \oplus 2 \oplus 2$: 8 states altogether. This is what one should have for DNA and RNA candidates. If one has only identical quarks uuu or ddd, Pauli exclusion rule allows only the 4-D spin 3/2 representation corresponding to completely symmetric representation -just as in standard quark model. These 4 states correspond to a candidate for amino-acids. Thus RNA and DNA should correspond to states of type uud and ddu and aminoacids to states of type uuu or ddd. What this means physically will be considered later.

2. Due to spin-statistics constraint only the representations with $(J, I) = (3/2, 3/2)$ (Δ resonance) and the second $(J, I) = (1/2, 1/2)$ (proton and neutron) are realized as free baryons. Now of course a dark -possibly p-adically scaled up - variant of QCD is considered so that more general baryonic states are possible. By the way, the spin statistics problem which forced to introduce quark color strongly suggests that the construction of the codons as sequences of 3 nucleons - which one might also consider - is not a good idea.

3. Second nucleon like spin doublet - call it 2_{odd} - has wrong parity in the sense that it would require $L = 1$ ground state for two identical quarks (uu or dd pair). Dropping 2_{odd} and using only $4 \oplus 2$ for the rotation group would give degeneracies $(1, 2, 2, 1)$ and 6 states only. All the representations in $4 \oplus 2 \oplus 2_{odd}$ are needed to get 8 states with a given quark charge and one should transform the wrong parity doublet to positive parity doublet somehow. Since open string geometry breaks rotational symmetry to a subgroup SO(2) of rotations acting along the direction of the string and since the boundary conditions on baryonic strings force their ends to rotate with light velocity, the attractive possibility is to add a baryonic stringy excitation with angular momentum projection $L_z = -1$ to the wrong parity doublet so that the parity comes out correctly. $L_z = -1$ orbital angular momentum for the relative motion of uu or dd quark pair in the open 3-quark string would be in question. The degeneracies for spin projection value $J_z = 3/2, ..., -3/2$ are $(1, 2, 3, 2)$. Genetic code means spin projection mapping the states in $4 \oplus 2 \oplus 2_{odd}$ to 4.

7.4.2 States in the flux tube degrees of freedom

Consider next the states in flux tube degrees of freedom.

1. The situation is analogous to a construction of mesons from quarks and antiquarks and one obtains the analogs of π meson (pion) with spin 0 and ρ meson with spin 1 since spin statistics forces $J = I$ condition also now. States of a given charge for a flux tube correspond to the tensor product $2 \otimes 2 = 3 \oplus 1$ for the rotation group.

2. Without any further constraints the tensor product $3 \otimes 3 = 5 \oplus 3 \oplus 1$ for the flux tubes states gives 8+1 states. By dropping the scalar state this gives 8 states required by DNA and RNA analogs. The degeneracies of the states for DNA/RNA type realization with a given spin projection for $5 \oplus 3$ are $(1, 2, 2, 2, 1)$. $8\times$ 8 states result altogether for both uud and udd for which color bonds have different charges. Also for ddd state with quark charge -1 one obtains $5 \oplus 3$ states giving 40 states altogether.

3. If the charges of the color bonds are identical as the are for uuu type states serving as candidates for the counterparts of aminoacids bosonic statistics allows only 5 states ($J = 2$ state). Hence 20 counterparts of aminoacids are obtained for uuu. Genetic code means the projection of the states of $5 \oplus 3$ to those of 5 with the same spin projection and same total charge.

7.4.3 Analogs of DNA,RNA, aminoacids, and of translation and transcription mechanisms

Consider next the identification of analogs of DNA, RNA and aminoacids and the baryonic realization of the genetic code, translation and transcription.

1. The analogs of DNA and RNA can be identified dark baryons with quark content *uud*, *ddu* with color bonds having different charges. There are 3 color bond pairs corresponding to charge pairs $(q_1, q_2) = (-1, 0), (-1, 1), (0, 1)$ (the order of charges does not matter). The condition that the total charge of dark baryon vanishes allows for *uud* only the bond pair $(-1, 0)$ and for *udd* only the pair $(-1, 1)$. These thus only single neutral dark baryon of type *uud resp. udd*: these would be the analogous of DNA and RNA codons. Amino-acids would correspond to *uuu* states with identical color bonds with charges $(-1, -1)$, $(0, 0)$, or $(1, 1)$. *uuu* with color bond charges (-1,-1) is the only neutral state. Hence only the analogs of DNA, RNA, and aminoacids are obtained, which is rather remarkable result.

2. The basic transcription and translation machinery could be realized as processes in which the analog of DNA can replicate, and can be transcribed to the analog of mRNA in turn translated to the analogs of amino-acids. In terms of flux tube connections the realization of genetic code, transcription, and translation, would mean that only dark baryons with same total quark spin and same total color bond spin can be connected by flux tubes. Charges are of course identical since they vanish.

3. Genetic code maps of $(4 \oplus 2 \oplus 2) \otimes (5 \oplus 3)$ to the states of 4×5. The most natural map takes the states with a given spin to a state with the same spin so that the code is unique. This would give the degeneracies $D(k)$ as products of numbers $D_B \in \{1, 2, 3, 2\}$ and $D_b \in \{1, 2, 2, 2, 1\}$: $D = D_B \times D_b$. Only the observed degeneracies $D = 1, 2, 3, 4, 6$ are predicted. The numbers $N(k)$ of aminoacids coded by D codons would be

$$[N(1), N(2), N(3), N(4), N(6)] = [2, 7, 2, 6, 3] \ .$$

The correct numbers for vertebrate nuclear code are $(N(1), N(2), N(3), N(4), N(6)) = (2, 9, 1, 5, 3)$. Some kind of symmetry breaking must take place and should relate to the emergence of stopping codons. If one codon in second 3-plet becomes stopping codon, the 3-plet becomes doublet. If 2 codons in 4-plet become stopping codons it also becomes doublet and one obtains the correct result $(2, 9, 1, 5, 3)$!

4. Stopping codons would most naturally correspond to the codons, which involve the $L_z = -1$ relative rotational excitation of *uu* or *dd* type quark pair. For the 3-plet the two candidates for the stopping codon state are $|1/2, -1/2\rangle \otimes \{|2, k\rangle\}$, $k = 2, -2$. The total spins are $J_z = 3/2$ and $J_z = -7/2$. The three candidates for the 4-plet from which two states are thrown out are $|1/2, -3/2\rangle \otimes \{|2, k\rangle, |1, k\rangle\}$, $k = 1, 0, -1$. The total spins are now $J_z = -1/2, -3/2, -5/2$. One guess is that the states with smallest value of J_z are dropped which would mean that $J_z = -7/2$ states in 3-plet and $J_z = -5/2$ states 4-plet become stopping codons.

5. One can ask why just vertebrate code? Why not vertebrate mitochondrial code, which has unbroken $A - G$ and $T - C$ symmetries with respect to the third nucleotide. And is it possible to understand the rarely occurring variants of the genetic code in this framework? One explanation is that the baryonic realization is the fundamental one and biochemical realization has gradually evolved from non-faithful realization to a faithful one as kind of emulation of dark nuclear physics. Also the role of tRNA in the realization of the code is crucial and could explain the fact that the code can be context sensitive for some codons.

7.4.4 Understanding the symmetries of the code

Quantum entanglement between quarks and color flux tubes would be essential for the baryonic realization of the genetic code whereas chemical realization could be said to be classical. Quantal aspect means that one cannot decompose to codon to letters anymore. This raises questions concerning the symmetries of the code.

1. What is the counterpart for the conjugation $ZYZ \rightarrow X_cY_cZ_c$ for the codons?

2. The conjugation of the second nucleotide Y having chemical interpretation in terms of hydrophoby-hydrophily dichotomy in biology. In DNA as tqc model it corresponds to matter-antimatter conjugation for quarks associated with flux tubes connecting DNA nucleotides to the lipids of the cell membrane. What is the interpretation in now?

3. The A-G, T-C symmetries with respect to the third nucleotide Z allow an interpretation as weak isospin symmetry in DNA as tqc model. Can one identify counterpart of this symmetry when the decomposition into individual nucleotides does not make sense?

Natural candidates for the building blocks of the analogs of these symmetries are the change of the sign of the spin direction for quarks and for flux tubes.

1. For quarks the spin projections are always non-vanishing so that the map has no fixed points. For flux tube spin the states of spin $S_z = 0$ are fixed points. The change of the sign of quark spin projection must therefore be present for both $XYZ \rightarrow X_cY_cZ_c$ and $Y \rightarrow Y_c$ but also something else might be needed. Note that without the symmetry breaking $(1,3,3,1) \rightarrow (1,2,3,2)$ the code table would be symmetric in the permutation of 2 first and 2 last columns of the code table induced by both full conjugation and conjugation of Y.

2. The analogs of the approximate $A-G$ and $T-C$ symmetries cannot involve the change of spin direction in neither quark nor flux tube sector. These symmetries act inside the A-G and T-C sub-2-columns of the 4-columns defining the rows of the code table. Hence this symmetry must permute the states of same spin inside 5 and 3 for flux tubes and 4 and 2 for quarks but leave 2_{odd} invariant. This guarantees that for the two non-degenerate codons coding for only single amino-acid and one of the codons inside triplet the action is trivial. Hence the baryonic analog of the approximate $A-G$ and $T-C$ symmetry would be exact symmetry and be due to the basic definition of the genetic code as a mapping states of same flux tube spin and quark spin to single representative state. The existence of full 4-columns coding for the same aminoacid would be due to the fact that states with same quark spin inside $(2,3,2)$ code for the same amino-acid.

3. A detailed comparison of the code table with the code table in spin representation should allow to fix their correspondence uniquely apart from permutations of n-plets and thus also the representation of the conjugations. What is clear that Y conjugation must involve the change of quark spin direction whereas Z conjugation which maps typically 2-plets to each other must involve the permutation of states with same J_z for the flux tubes. It is not quite clear what X conjugation correspond to.

7.4.5 Some comments about the physics behind the code

Consider next some particle physicist's objections against this picture.

1. The realization of the code requires the dark scaled variants of spin 3/2 baryons known as Δ resonance and the analogs (and only the analogs) of spin 1 mesons known as ρ mesons. The lifetime of these states is very short in ordinary hadron physics. Now one has a scaled up variant of hadron physics: possibly in both dark and p-adic senses with latter allowing arbitrarily small overall mass scales. Hence the lifetimes of states can be scaled up.

2. Both the absolute and relative mass differences between Δ and N *resp.* ρ and π are large in ordinary hadron physics and this makes the decays of Δ and ρ possible kinematically. This is due to color magnetic spin-spin splitting proportional to the color coupling strength $\alpha_s \sim .1$, which is large. In the recent case α_s could be considerably smaller - say of the same order of magnitude as fine structure constant 1/137 - so that the mass splittings could be so small as to make decays impossible.

3. Dark hadrons could have lower mass scale than the ordinary ones if scaled up variants of quarks in p-adic sense are in question. Note that the model for cold fusion that inspired the idea about genetic code requires that dark nuclear strings have the same mass scale as ordinary baryons. In any case, the most general option inspired by the vision about hierarchy of conscious entities extended to a hierarchy of life forms is that several dark and p-adic scaled up variants of baryons realizing genetic code are possible.

4. The heaviest objection relates to the addition of $L_z = -1$ excitation to $S_z = |1/2, \pm 1/2\rangle_{odd}$ states which transforms the degeneracies of the quark spin states from $(1,3,3,1)$ to $(1,2,3,2)$. The only reasonable answer is that the breaking of the full rotation symmetry reduces $SO(3)$ to $SO(2)$. Also the fact that the states of massless particles are labeled by the representation of $SO(2)$ might be of some relevance. The deeper level explanation in TGD framework might be as follows. The generalized imbedding space is constructed by gluing almost copies of the 8-D imbedding space with different Planck constants together along a 4-D subspace like pages of book along a common back. The construction involves symmetry breaking in both rotational and color degrees of freedom to Cartan sub-group and the interpretation is as a geometric representation for the selection of the quantization axis. Quantum TGD is indeed meant to be a geometrization of the entire quantum physics as a physics of the classical spinor fields in the "world of classical worlds" so that also the choice of measurement axis must have a geometric description.

The conclusion is that genetic code can be understand as a map of stringy baryonic states induced by the projection of all states with same spin projection to a representative state with the same spin projection. Genetic code would be realized at the level of dark nuclear physics and biochemical representation would be only one particular higher level representation of the code. A hierarchy of dark baryon realizations corresponding to p-adic and dark matter hierarchies can be considered. Translation and transcription machinery would be realized by flux tubes connecting only states with same quark spin and flux tube spin. Charge neutrality is essential for having only the analogs of DNA, RNA and aminoacids and would guarantee the em stability of the states.

References

Books related to TGD

[1] M. Pitkänen. Configuration Space Spinor Structure. In *Quantum Physics as Infinite-Dimensional Geometry*. Onlinebook. `http://tgd.wippiespace.com/public_html/tgdgeom/tgdgeom.html#cspin`, 2006.

[2] M. Pitkänen. Construction of Configuration Space Kähler Geometry from Symmetry Principles. In *Quantum Physics as Infinite-Dimensional Geometry*. Onlinebook. `http://tgd.wippiespace.com/public_html/tgdgeom/tgdgeom.html#compl1`, 2006.

[3] M. Pitkänen. Construction of elementary particle vacuum functionals. In *p-Adic length Scale Hypothesis and Dark Matter Hierarchy*. Onlinebook. `http://tgd.wippiespace.com/public_html/paddark/paddark.html#elvafu`, 2006.

[4] M. Pitkänen. Cosmic Strings. In *Physics in Many-Sheeted Space-Time*. Onlinebook. `http://tgd.wippiespace.com/public_html/tgdclass/tgdclass.html#cstrings`, 2006.

[5] M. Pitkänen. Dark Forces and Living Matter. In *p-Adic Length Scale Hypothesis and Dark Matter Hierarchy*. Onlinebook. `http://tgd.wippiespace.com/public_html/paddark/paddark.html#darkforces`, 2006.

[6] M. Pitkänen. Dark Matter Hierarchy and Hierarchy of EEGs. In *TGD and EEG*. Onlinebook. `http://tgd.wippiespace.com/public_html/tgdeeg/tgdeeg.html#eegdark`, 2006.

[7] M. Pitkänen. Dark Nuclear Physics and Condensed Matter. In *p-Adic Length Scale Hypothesis and Dark Matter Hierarchy.* Onlinebook. http://tgd.wippiespace.com/public_html/paddark/paddark.html#exonuclear, 2006.

[8] M. Pitkänen. DNA as Topological Quantum Computer. In *Genes and Memes.* Onlinebook. http://tgd.wippiespace.com/public_html/genememe/genememe.html#dnatqc, 2006.

[9] M. Pitkänen. Does TGD Predict the Spectrum of Planck Constants? In *Towards M-Matrix.* Onlinebook. http://tgd.wippiespace.com/public_html/tgdquant/tgdquant.html#Planck, 2006.

[10] M. Pitkänen. Does the Modified Dirac Equation Define the Fundamental Action Principle? In *Quantum Physics as Infinite-Dimensional Geometry.* Onlinebook. http://tgd.wippiespace.com/public_html/tgdgeom/tgdgeom.html#Dirac, 2006.

[11] M. Pitkänen. Does the QFT Limit of TGD Have Space-Time Super-Symmetry? In *Towards M-Matrix.* Onlinebook. http://tgd.wippiespace.com/public_html/tgdquant/tgdquant.html#susy, 2006.

[12] M. Pitkänen. General Ideas about Many-Sheeted Space-Time: Part I. In *Physics in Many-Sheeted Space-Time.* Onlinebook. http://tgd.wippiespace.com/public_html/tgdclass/tgdclass.html#topcond, 2006.

[13] M. Pitkänen. Homeopathy in Many-Sheeted Space-Time. In *Bio-Systems as Conscious Holograms.* Onlinebook. http://tgd.wippiespace.com/public_html/hologram/hologram.html#homeoc, 2006.

[14] M. Pitkänen. Massless states and particle massivation. In *p-Adic Length Scale Hypothesis and Dark Matter Hierarchy.* Onlinebook. http://tgd.wippiespace.com/public_html/paddark/paddark.html#mless, 2006.

[15] M. Pitkänen. Model for the Findings about Hologram Generating Properties of DNA. In *Genes and Memes.* Onlinebook. http://tgd.wippiespace.com/public_html/genememe/genememe.html#dnahologram, 2006.

[16] M. Pitkänen. New Particle Physics Predicted by TGD: Part I. In *p-Adic Length Scale Hypothesis and Dark Matter Hierarchy.* Onlinebook. http://tgd.wippiespace.com/public_html/paddark/paddark.html#mass4, 2006.

[17] M. Pitkänen. Nuclear String Hypothesis. In *p-Adic Length Scale Hypothesis and Dark Matter Hierarchy.* Onlinebook. http://tgd.wippiespace.com/public_html/paddark/paddark.html#nuclstring, 2006.

[18] M. Pitkänen. p-Adic Particle Massivation: Elementary Particle Masses. In *p-Adic Length Scale Hypothesis and Dark Matter Hierarchy.* Onlinebook. http://tgd.wippiespace.com/public_html/paddark/paddark.html#mass2, 2006.

[19] M. Pitkänen. p-Adic Particle Massivation: Hadron Masses. In *p-Adic Length Scale Hypothesis and Dark Matter Hierarchy.* Onlinebook. http://tgd.wippiespace.com/public_html/paddark/paddark.html#mass3, 2006.

[20] M. Pitkänen. Quantum Astrophysics. In *Physics in Many-Sheeted Space-Time.* Onlinebook. http://tgd.wippiespace.com/public_html/tgdclass/tgdclass.html#qastro, 2006.

[21] M. Pitkänen. TGD and Astrophysics. In *Physics in Many-Sheeted Space-Time.* Onlinebook. http://tgd.wippiespace.com/public_html/tgdclass/tgdclass.html#astro, 2006.

[22] M. Pitkänen. TGD and Cosmology. In *Physics in Many-Sheeted Space-Time.* Onlinebook. http://tgd.wippiespace.com/public_html/tgdclass/tgdclass.html#cosmo, 2006.

[23] M. Pitkänen. TGD and Nuclear Physics. In *p-Adic Length Scale Hypothesis and Dark Matter Hierarchy*. Onlinebook. http://tgd.wippiespace.com/public_html/paddark/paddark.html#padnucl, 2006.

[24] M. Pitkänen. TGD as a Generalized Number Theory: Infinite Primes. In *TGD as a Generalized Number Theory*. Onlinebook. http://tgd.wippiespace.com/public_html/tgdnumber/tgdnumber.html#visionc, 2006.

[25] M. Pitkänen. The Notion of Wave-Genome and DNA as Topological Quantum Computer. In *Genes and Memes*. Onlinebook. http://tgd.wippiespace.com/public_html/genememe/genememe.html#gari, 2006.

[26] M. Pitkänen. The Recent Status of Lepto-hadron Hypothesis. In *p-Adic Length Scale Hypothesis and Dark Matter Hierarchy*. Onlinebook. http://tgd.wippiespace.com/public_html/paddark/paddark.html#leptc, 2006.

[27] M. Pitkänen. The Relationship Between TGD and GRT. In *Physics in Many-Sheeted Space-Time*. Onlinebook. http://tgd.wippiespace.com/public_html/tgdclass/tgdclass.html#tgdgrt, 2006.

[28] M. Pitkänen. Yangian Symmetry, Twistors, and TGD. In *Towards M-Matrix*. Onlinebook. http://tgd.wippiespace.com/public_html/tgdquant/tgdquant.html#Yangian, 2006.

Articles about TGD

[29] M. Pitkänen. Further Progress in Nuclear String Hypothesis. http://tgd.wippiespace.com/public_html/articles/nuclstring.pdf, 2007.

[30] M. Pitkänen. Construction of Configuration Space Spinor Structure. http://tgd.wippiespace.com/public_html/articles/spinstructure.pdf, 2010.

Mathematics

[31] E. C. Zeeman. Catastrophe Theory. Addison-Wessley Publishing Company, 1977.

Theoretical Physics

[32] CPT symmetry. http://en.wikipedia.org/wiki/CPT_symmetry.

[33] The Minimally Super-Symmetric Standard Model. http://en.wikipedia.org/wiki/Minimal_Supersymmetric_Standard_Model.

[34] X. Wang X. Zhang N. Li, Yi-Fu Cai. *CPT* Violating Electrodynamics and Chern-Simons Modified Gravity. http://arxiv.org/abs/0907.5159, 2009.

[35] S. P. Martin. A Supersymmetry Primer).
http://arxiv.org/abs/hep-ph/9709356, 1997.

Particle and Nuclear Physics

[36] Chargino mass and R_b anomaly. http://arxiv.org/pdf/hep-ph/9603310v1.

[37] Implications of Initial LHC Searches for Supersymmetry. http://arxiv.org/abs/1102.4585.

[38] Invariant Mass Distribution of Jet Pairs Produced in Association with a W boson in ppbar Collisions at $\sqrt{s} = 1.96$ TeV. http://arxiv.org/abs/1104.0699.

[39] Jet quenching. http://en.wikipedia.org/wiki/Jet_quenching.

[40] LHC bounds on large extra dimensions. http://arxiv.org/abs/1101.4919.

[41] Light Z' Bosons at the Tevatron. http://arxiv.org/abs/arXiv:1103.6035.

[42] Massive High p_T Jets: Updates from CDF. http://indico.cern.ch/getFile.py/access?contribId=16&resId=0&materialId=slides&confId=113980.

[43] Minos for Scientists. http://www-numi.fnal.gov/PublicInfo/forscientists.html.

[44] Neutrino oscillation, howpublished=http://en.wikipedia.org/wiki/neutrino_oscillation.

[45] Search for supersymmetry using final states with onelepton, jets, and missing transverse momentum with the ATLAS detector in $s^{1/2} = 7$ TeV pp collisions. http://cdsweb.cern.ch/record/1328281/files/susy-1l-arxiv.pdf.

[46] Supersymmetry From the Top Down. http://arxiv.org/abs/1102.3386.

[47] Technicolor at the Tevatron. http://arxiv.org/abs/1104.0976.

[48] The fine-tuning price of the early LHC. http://arxiv.org/abs/1101.2195.

[49] D0: 2.5-sigma evidence for a 325 GeV top prime quark. http://motls.blogspot.com/2011/04/d0-3-sigma-evidence-for-325-gev-top.html, 2011.

[50] If That Were A Higgs At 200 GeV... http://www.science20.com/quantum_diaries_survivor/if_were_higgs_200_gev, 2011.

[51] More details about the CDF bump. http://resonaances.blogspot.com/2011/06/more-details-about-cdf-bump.html, 2011.

[52] S. Barshay. *Mod. Phys. Lett. A*, 7(20):1843, 1992.

[53] R. S. Gidley R. S Conti C. I. Westbrook, D. W Kidley and A. Rich. *Phys. Rev.*, 58:1328, 1987.

[54] D. Choudbury and D. P. Roy. R-parity-breaking SUSY solution to the Rb and ALEPH anomalies. *Phys. Rev. D*, (11), 1996.

[55] CMS colaboration. Search for Black Holes in pp collisions at $\sqrt{s} = 7$ TeV. http://cms-physics.web.cern.ch/cms-physics/public/EXO-11-021-pas.pdf, 2011.

[56] ATLAS collaboration. Update of the Search for New Physics in the Dijet Mass Distribution in 163 pb^{-1} of pp Collisions at $\sqrt{s} = 7$ TeV Measured with the ATLAS Detector. http://cdsweb.cern.ch/record/1355704/files/ATLAS-CONF-2011-081.pdf?version=1.

[57] ATLAS collaboration. Search for Three-Jet Resonances in p-p collisions at $\sqrt{s} = 7$ TeV. http://arxiv.org/abs/1107.3084, 2011.

[58] ATLAS collaboration. Search for W′ boson decaying to a muon and neutrino in pp collisions at $\sqrt{s} = 7$ TeV. 2011.

[59] CDF collaboration. Search for a new heavy gauge boson W with event signature electron+missing transverse energy in ppbar collisions at $\sqrt{s}$=1.96 TeV.

[60] CDF collaboration. High-mass Di-electron Resonance Research in $p\overline{p}$ Collisions at s = 1.96 TeV. http://www-cdf.fnal.gov/physics/exotic/r2a/20080306.dielectron_duke/pub25/cdfnote9160_pub.pdf, 2008.

[61] CDF Collaboration. Study of multi-muon events produced in p-pbar collisions at sqrt(s)=1.96 TeV. http://arxiv.org/PS_cache/arxiv/pdf/0810/0810.0714v1.pdf, 2008.

[62] CDF collaboration. Evidence for a Mass Dependent Forward-Backward Asymmetry in Top Quark Pair Production. http://arxiv.org/abs/1101.0034, 2011.

[63] CDF Collaboration. Observation of Z0Z0 llll at CDF. http://www-cdf.fnal.gov/physics/ewk/2009/ZZllll/ZZWeb/cdf9910_ZZ_4leptons_public.pdf, 2011.

[64] CDF Collaboration. Searches for the Higgs Boson. http://blois.in2p3.fr/2011/transparencies/punzi.pdf, 2011.

[65] CDMS collaboration. Results from the Final Exposure of the CDMS II Experiment. http://arxiv.org/abs/0912.3592, 2009.

[66] CMS collaboration. Measurement of the Charge Asymmetry in Top Quark Pair Production. http://cdsweb.cern.ch/record/1369205/files/TOP-11-014-pas.pdf, 2011.

[67] D0 collaboration. Forward-backward asymmetry in top quark-antiquark production. http://www-d0.fnal.gov/Run2Physics/WWW/results/final/TOP/T110/T110.pdf%.

[68] D0 collaboration. Observation of the doubly strange b baryon Ω_b^-. http://arxiv.org/abs/0808.4142, 2008.

[69] D0 collaboration. Search for high-mass narrow resonances in the di-electron channel at D0. http://www-d0.fnal.gov/Run2Physics/WWW/results/prelim/NP/N66/, 2009.

[70] D0 Collaboration. Evidence for an anomalous like-sign dimuon charge asymmetry. http://arxiv.org/abs/1005.2757, 2010.

[71] D0 Collaboration. Search for a fourth generation t' quark in ppbar collisions at sqrts=1.96 TeV. http://arxiv.org/abs/1104.4522, 2011.

[72] D0 collaboration. Study of the dijet invariant mass distribution in $p\overline{p} \rightarrow W(\rightarrow l\nu) + jj$ final states at $\sqrt{s} = 1.96$ TeV. http://www-d0.fnal.gov/Run2Physics/WWW/results/final/HIGGS/H11B, 2011.

[73] DAMA Collaboration. New results from DAMA/LIBRA. http://arxiv.org/abs/1002.1028, 2010.

[74] Xenon100 collaboration. Dark Matter Results from 100 Live Days of XENON100 Data. http://arxiv.org/abs/1104.2549, 2011.

[75] A. E. Nelson D. B. Kaplan and N. Weiner. Neutrino Oscillations as a Probe of Dark Energy. http://arxiv.org/abs/hep-ph/0401099, 2004.

[76] R. Van de Water. Updated Anti-neutrino Oscillation Results from MiniBooNE. http://indico.cern.ch/getFile.py/access?contribId=208&sessionId=3&resId=0&materialId=slides&confId=73981, 2010.

[77] T. Dorigo. The top quark mass measured from its production rate. http://dorigo.wordpress.com/2007/06/26/a-particle-mass-from-its-production-rate/#more-910, 2007.

[78] T. Dorigo. Nitpicking Ω_b discovery:part I. http://www.scientificblogging.com/quantum_diaries_survivor/nitpicking_omega_b_discovery, 2009.

[79] Tommaso Dorigo. A New Z' Boson at 240 GeV? No, Wait, at 720!? http://www.scientificblogging.com/quantum_diaries_survivor/new_z_boson_240_gev_no_wait_720, 2009.

[80] A. Martin E. J. Eichten, K. Lane. Technicolor at Tevatron. http://arxiv.org/abs/1104.0976, 2011.

[81] A. Alavi-Harati et al. http://xxx.lanl.gov/abs/hep-ex/9905060, 1999.

[82] A.T. Goshaw et al. *Phys. Rev.*, 43, 1979.

[83] B. B. Back et al. *Phys. Rev. Lett.*, 89(22), November 2002.

[84] M. Derrick et al. *Phys. Lett B*, 315:481, 1993.

[85] P.V. Chliapnikov et al. *Phys. Lett. B*, 141, 1984.

[86] S. J. Brawley et al. Electron-Like Scattering of Positronium. *Science*, 330(6005):789, November 2010.

[87] T. Akesson et al. *Phys. Lett. B*, 463:36, 1987.

[88] Y.Takeuchi et al. Measurement of the Forward Backward Asymmetry in Top Pair Production in the Dilepton Decay Channel using 5.1 fb^{-1}. http://www-cdf.fnal.gov/physics/new/top/2011/DilAfb/, 2011.

[89] G. R. Farrar. New Signatures of Squarks. http://arxiv.org/abs/hep-ph/9512306, 1995.

[90] R. Foot. A CoGeNT confirmation of the DAMA signal. http://arxiv.org/abs/1004.1424, 2010.

[91] S. K. Kang and K. Y. Lee. Implications of the muon anomalous magnetic moment and Higgs-mediated flavor changing neutral currents. http://arxiv.org/abs/hep-ph/0103064, 2001.

[92] G. Karagiorgi. Towards Solution of MiniBoone-LSND anomalies. http://indico.cern.ch/contributionDisplay.py?contribId=209&sessionId=3&confId=73981, 2010.

[93] B. Armbruster et al KARMEN Collaboration. *Phys. Lett. B*, 348, 1995.

[94] H. Muir. Cloaking effect in atoms baffles scientists. *New Scientist*, November 2010.

[95] M. J. Kim M. Kruse K. Pitts F. Ptohos S. Torre P. Giromini, F. Happacher. Phenomenological interpretation of the multi-muon events reported by the CDF collaboration. http://arxiv.org/abs/0810.5730, 2008.

[96] Seongwan Park. Search for New Phenomena in CDF-I: Z, W, and leptoquarks. http://lss.fnal.gov/archive/1995/conf/Conf-95-155-E.pdf, 1995.

[97] T. Smith. Truth Quark, Higgs, and Vacua. http://www.innerx.net/personal/tsmith/TQvacua.html, 2003.

[98] P. Söding. On the discovery of the gluon. *Eur. Phys. J. H*, pages 3–28, 2010.

[99] D. Stöckinger. The Muon Magnetic Moment and Supersymmetry. http://arxiv.org/abs/hep-ph/0609168, 2006.

[100] S. Sarkar V. Barger, R. J. N. Phillips. *Phys. Lett. B*, 352:365–371, 1995.

[101] H. Waschmuth. Results from e^+e^- collisions at 130, 136 and 140 GeV center of mass energies in the ALEPH Experiment. http://alephwww.cern.ch/ALPUB/pub/pub_96.html, 1996.

[102] A. C. Kraan. SUSY Searches at LEP. http://arxiv.org/pdf/hep-ex/0505002, 2005.

Cosmology and Astro-Physics

[103] D. Da Roacha and L. Nottale. Gravitational Structure Formation in Scale Relativity. http://arxiv.org/abs/astro-ph/0310036, 2003.

Biology

[104] S. Ferris J.-L. Montagnier L. Montagnier, J. Aissa and C. Lavall'e. Electromagnetic Signals Are Produced by Aqueous Nanostructures Derived from Bacterial DNA Sequences. *Interdiscip. Sci. Comput. Life Sci.*, 2009.

[105] A.V. Tovmash P. P. Gariaev, G. G. Tertishni. Experimental investigation in vitro of holographic mapping and holographic transposition of DNA in conjuction with the information pool encircling DNA. *New Medical Tehcnologies*, 9:42–53, 2007.

Article

New Particle Physics Predicted by TGD: Part II

Matti Pitkänen [1]

Abstract

In this article the focus is on the hadron physics. The applications are to various anomalies discovered during years.

1. Application of the many-sheeted space-time concept in hadron physics

The many-sheeted space-time concept involving also the notion of field body can be applied to hadron physics to explain findings which are difficult to understand in the framework of standard model.

1. The spin puzzle of proton is a two decades old mystery with no satisfactory explanation in QCD framework. The notion of hadronic space-time sheet which could be imagined as string like rotating object suggests a possible approach to the spin puzzle. The entanglement between valence quark spins and the angular momentum states of the rotating hadronic space-time sheet could allow natural explanation for why the average valence quark spin vanishes.
2. The notion of Pomeron was invented during the Bootstrap era preceding QCD to solve difficulties of Regge approach. There are experimental findings suggesting the reincarnation of this concept. The possibility that the newly born concept of Pomeron of Regge theory might be identified as the sea of perturbative QCD in TGD framework is considered. Geometrically Pomeron would correspond to hadronic space-time sheet without valence quarks.
3. The discovery that the charge radius of proton deduced from the muonic version of hydrogen atom is about 4 per cent smaller than from the radius deduced from hydrogen atom is in complete conflict with the cherished belief that atomic physics belongs to the museum of science. The title of the article *Quantum electrodynamics-a chink in the armour?* of the article published in Nature expresses well the possible implications, which might actually go well extend beyond QED. TGD based model for the findings relies on the notion of color magnetic body carrying both electromagnetic and color fields and extends well beyond the size scale of the particle. This gives rather detailed constraints on the model of the magnetic body.
4. The soft photon production rate in hadronic reactions is by an average factor of about four higher than expected. In the article soft photons assignable to the decays of Z^0 to quark-antiquark pairs. This anomaly has not reached the attention of particle physics which seems to be the fate of anomalies quite generally nowadays: large extra dimensions and black-holes at LHC are much more sexy topics of study than the anomalies about which both existing and speculative theories must remain silent. TGD based model is based on the notion of electric flux tube.

2. Quark gluon plasma

QCD predicts that at sufficiently high collision energies de-confinement phase transitions for quarks should take place leading to quark gluon plasma. In heavy ion collisions at RHIC something like this was found to happen. The properties of the quark gluon plasma were however not what was expected. There are long range correlations and the plasma seems to behave like perfect fluid with minimal viscosity/entropy ratio. The lifetime of the plasma phase is longer than expected and its density much higher than QCD would suggest. The experiments at LHC for proton proton collisions suggest also the presence of quark gluon plasma with similar properties.

TGD suggests an interpretation in terms of long color magnetic flux tubes containing the plasma. The confinement to color magnetic flux tubes would force higher density. The preferred extremals of Kähler action have interpretation as defining a flow of perfect incompressible fluid and the perfect fluid property is broken only by the many-sheeted structure of space-time with smaller space-time sheets assignable to sub-CDs representing radiative corrections. The phase in question corresponds to a non-standard value of Planck constant: this could also explain why the lifetime of the phase is longer than expected.

[1] Correspondence: E-mail:matpitka@luukku.com

Contents

1 Introduction

In this article the focus is on the new hadron physics. The applications are to various anomalies discovered during years.

1.1 Application of the many-sheeted space-time concept in hadron physics

The many-sheeted space-time concept involving also the notion of field body can be applied to hadron physics to explain findings which are difficult to understand in the framework of standard model

1. The spin puzzle of proton [32, 34] is at the time of writing a two decades old mystery with no satisfactory explanation in QCD framework. The notion of hadronic space-time sheet which could be imagined as string like rotating object suggests a possible approach to the spin puzzle. The entanglement between valence quark spins and the angular momentum states of the rotating hadronic space-time sheet could allow natural explanation for why the average valence quark spin vanishes.

2. The notion of Pomeron was invented during the Bootstrap era preceding QCD to solve difficulties of Regge approach. There are experimental findings suggesting the reincarnation of this concept [33, 27, 28]. The possibility that the newly born concept of Pomeron of Regge theory might be identified as the sea of perturbative QCD in TGD framework is considered. Geometrically Pomeron would correspond to hadronic space-time sheet without valence quarks.

3. The discovery that the charge radius of proton deduced from the muonic version of hydrogen atom is about 4 per cent smaller than from the radius deduced from hydrogen atom [38, 41] is in complete conflict with the cherished belief that atomic physics belongs to the museum of science. The title of the article *Quantum electrodynamics-a chink in the armour?* of the article published in Nature [35] expresses well the possible implications, which might actually go well extend beyond QED. TGD based model for the findings relies on the notion of color magnetic body carrying both electromagnetic and color fields and extends well beyond the size scale of the particle. This gives rather detailed constraints on the model of the magnetic body.

4. The soft photon production rate in hadronic reactions is by an average factor of about four higher than expected [26]. In the article soft photons assignable to the decays of Z^0 to quark-antiquark pairs. This anomaly has not reached the attention of particle physics which seems to be the fate of anomalies quite generally nowadays: large extra dimensions and black-holes at LHC are much more sexy topics of study than the anomalies about which both existing and speculative theories must remain silent. TGD based model is based on the notion of electric flux tube.

1.2 Quark gluon plasma

QCD predicts that at sufficiently high collision energies de-confinement phase transitions for quarks should take place leading to quark gluon plasma. In heavy ion collisions at RHIC [29] something like this was found to happen. The properties of the quark gluon plasma were however not what was expected. There are long range correlations and the plasma seems to behave like perfect fluid with minimal viscosity/entropy ratio. The lifetime of the plasma phase is longer than expected and its density much higher than QCD would suggest. The experiments at LHC for proton proton collisions suggest also the presence of quark gluon plasma with similar properties.

TGD suggests an interpretation in terms of long color magnetic flux tubes containing the plasma so that additional support for the notion of field would would emerge. The confinement to color magnetic flux tubes would force higher density. The preferred extremals of Kähler action have interpretation as defining a flow of perfect incompressible fluid and the perfect fluid property is broken only by the many-sheeted structure of space-time with smaller space-time sheets assignable to sub-CDs representing radiative corrections. The phase in question corresponds to a non-standard value of Planck constant: this could also explain why the lifetime of the phase is longer than expected.

2 New space-time concept applied to hadrons

2.1 A new twist in the spin puzzle of proton

The so called proton spin crisis or spin puzzle of proton was an outcome of the experimental finding that the quarks contribute only 13-17 per cent of proton spin [32, 34] whereas the simplest valence quark model predicts that quarks contribute about 75 per cent to the spin of proton with the remaining 25 per cent being due to the orbital motion of quarks. Besides the orbital motion of valence quarks also gluons could contribute to the spin of proton. Also polarized sea quarks can be considered as a source of proton spin.

Quite recently, the spin crisis got a new twist [37] . One of the few absolute predictions of perturbative QCD (pQCD) is that at the limit, when the momentum fraction of quark approaches unity, quark spin should be parallel to the proton spin. This is due to the helicity conservation predicted by pQCD in the lowest order. The findings are consistent with this expectation in the case of protonic u quarks but not in the case of protonic d quark. The discovery is of a special interest from the point of view of TGD since it might have an explanation involving the notions of many-sheeted space-time, of color-magnetic flux tubes, the predicted super-symplectic "vacuum" spin, and also the concept of quantum parallel dissipation.

2.1.1 The experimental findings

In the experiment performed in Jefferson Lab [37] neutron spin asymmetries A_1^n and polarized structure functions $g_{1,2}^n$ were deduced for three kinematic configurations in the deep inelastic region from e-^{3}He scattering using 5.7 GeV longitudinally polarized electron beam and a polarized ^3He target. A_1^n and $g_{1,2}^n$ were deduced for $x = .33, .47$, and $.60$ and $Q^2 = 2.7, 3.5$ and 4.8 $(\text{GeV/c})^2$. A_1^n and g_1^n at $x = .33$ are consistent with the world data. At $x = .47$ A_1^n crosses zero and is significantly positive at $x = 0.60$. This finding agrees with the next-to-leading order QCD analysis of previous world data without the helicity conservation constraint. The trend of the data agrees with the predictions of the constituent quark model but disagrees with the leading order pQCD assuming hadron helicity conservation.

By isospin symmetry one can translate the result to the case of proton by the replacement $u \leftrightarrow d$. By using world proton data, the polarized quark distribution functions were deduced for proton using isospin symmetry between neuron and proton. It was found that $\Delta u/u$ agrees with the predictions of various models while $\Delta d/d$ disagrees with the leading-order pQCD.

Let us denote by $q(x) = q^{\uparrow} + q^{\downarrow}(x)$ the spin independent quark distribution function. The difference $\Delta q(x) = q^{\uparrow} - q^{\downarrow}(x)$ measures the contribution of quark q to the spin of hadron. The measurement allowed to deduce estimates for the ratios $(\Delta q(x) + \Delta \overline{q}(x))/(q(x) + \overline{q}(x))$.

The conclusion of [37] is that for proton one has

$$\frac{\Delta u(x) + \Delta \overline{u}(x)}{u(x) + \overline{u}(x)} \simeq .737 \pm .007 \ , \quad \text{for } x = .6 \ .$$

This is consistent with the pQCD prediction. For d quark the experiment gives

$$\frac{\Delta d(x) + \Delta \overline{d}(x)}{d(x) + \overline{d}(x)} \simeq -.324 \pm .083 \ \text{ for } x = .6 \ .$$

The interpretation is that d quark with momentum fraction $x > .6$ in proton spends a considerable fraction of time in a state in which its spin is opposite to the spin of proton so that the helicity conservation predicted by first order pQCD fails. This prediction is of special importance as one of the few absolute predictions of pQCD.

The finding is consistent with the relativistic $SU(6)$ symmetry broken by spin-spin interaction and the QCD based model interpolated from data but giving up helicity conservation [37] . $SU(6)$ is however not a fundamental symmetry so that its success is probably accidental.

It has been also proposed that the spin crisis might be illusory [40] and due to the fact that the vector sum of quark spins is not a Lorentz invariant quantity so that the sum of quark spins in infinite-momentum frame where quark distribution functions are defined is not same as, and could thus be smaller than, the spin sum in the rest frame. The correction due to the transverse momentum of the quark brings in a non-negative numerical correction factor which is in the range $(0, 1)$. The negative sign of $\Delta d/d$ is not consistent with this proposal.

2.1.2 TGD based model for the findings

The TGD based explanation for the finding involves the following elements.

1. TGD predicts the possibility of vacuum spin due to the super-symplectic symmetry. Valence quarks can be modelled as a star like formation of magnetic flux tubes emanating from a vertex with the conservation of color magnetic flux forcing the valence quarks to form a single coherent structure. A good guess is that the super-symplectic spin corresponds classically to the rotation of the the star like structure.

2. By parity conservation only even values of super-symplectic spin J are allowed and the simplest assumption is that the valence quark state is a superposition of ordinary $J = 0$ states predicted by

pQCD and $J = 2$ state in which all quarks have spin which is in a direction opposite to the direction of the proton spin. The state of $J = 1/2$ baryon is thus replaced by a new one:

$$\begin{aligned} |B, \frac{1}{2}, \uparrow\rangle &= a|B, 1/2, \frac{1}{2}\rangle|J = J_z = 0\rangle + b|B, \frac{3}{2}, -\frac{3}{2}\rangle|J = J_z = 2\rangle \ , \\ |B, 1/2, \frac{1}{2}\rangle &= \sum_{q_1,q_2,q_3} c_{q_1,q_2,q_3} q_1^{\uparrow} q_2^{\uparrow} q_3^{\downarrow} \ , \\ |B, \frac{3}{2}, -\frac{3}{2}\rangle &= d_{q_1,q_2,q_3} q_1^{\downarrow} q_2^{\downarrow} q_3^{\downarrow} \ . \end{aligned} \tag{2.1}$$

$|B, 1/2, \frac{1}{2}\rangle$ is in a good approximation the baryon state as predicted by pQCD. The coefficients c_{q_1,q_2,q_3} and d_{q_1,q_2,q_3} depend on momentum fractions of quarks and the states are normalized so that $|a|^2 + |b|^2 = 1$ is satisfied: the notation $p = |a|^2$ will be used in the sequel. The quark parts of $J = 0$ and $J = 2$ have quantum numbers of proton and Δ resonance. $J = 2$ part need not however have the quark distribution functions of Δ.

3. The introduction of $J = 0$ and $J = 2$ ground states with a simultaneous use of quark distribution functions makes sense if one allows quantum parallel dissipation. Although the system is coherent in the super-symplectic degrees of freedom which correspond to the hadron size scale, there is a de-coherence in quark degrees of freedom which correspond to a shorter p-adic length scale and smaller space-time sheets.

4. Consider now the detailed structure of the $J = 2$ state in the case of proton. If the d quark is at the rotation axis, the rotating part of the triangular flux tube structure resembles a string containing u-quarks at its ends and forming a di-quark like structure. Di-quark structure is taken to mean correlations between u-quarks in the sense that they have nearly the same value of x so that $x < 1/2$ holds true for them whereas the d-quark behaving more like a free quark can have $x > 1/2$.

   A stronger assumption is that di-quark behaves like a single colored hadron with a small value of x and only the d-quark behaves as a free quark able to have large values of x. Certainly this would be achieved if u quarks reside at their own string like space-time sheet having $J = 2$.

From these assumptions it follows that if u quark has $x > 1/2$, the state effectively reduces to a state predicted by pQCD and $u(x) \rightarrow 1$ for $x \rightarrow 1$ is predicted. For the d quark the situation is different and introducing distribution functions $q^{J)}(x)$ for $J = 0, 2$ separately, one can write the spin asymmetry at the limit $x \rightarrow 1$ as

$$\begin{aligned} A_d &\equiv \frac{\Delta d(x) + \Delta \overline{d}(x)}{d(x) + \overline{d}(x)} = \frac{p(\Delta d_0 + \Delta \overline{d}_0) + (1-p)(\Delta d_2 + \Delta \overline{d}_2)}{p(d_0 + \overline{d}_0) + (1-p)(d_2 + \overline{d}_2)} \ , \\ p &= |a|^2 \ . \end{aligned} \tag{2.2}$$

Helicity conservation gives $\Delta d_0/d_0 \rightarrow 1$ at the limit $x \rightarrow 1$ and one has trivially $\Delta d_2/d_2 = -1$. Taking the ratio

$$y = \frac{d_2}{d_0}$$

as a parameter, one can write

$$A_d \rightarrow \frac{p - (1-p)y}{p + (1-p)y} \tag{2.3}$$

at the limit $x \rightarrow 1$. This allows to deduce the value of the parameter y once the value of p is known:

$$y = \frac{p}{1-p} \times \frac{1-A_d}{1+A_d} . \quad (2.4)$$

From the requirement that quarks contribute a fraction $\Sigma = \sum_q \Delta q \in (13, 17)$ per cent to proton spin, one can deduce the value of p using

$$\frac{p \times \frac{1}{2} - (1-p) \times \frac{3}{2}}{\frac{1}{2}} = \Sigma \quad (2.5)$$

giving $p = (3 + \Sigma)/4 \simeq .75$.

Eq. 2.4 allows estimate the value of y. In the range $\Sigma \in (.13, .30)$ defined by the lower and upper bounds for the contribution of quarks to the proton spin, $A_d = -.32$ gives $y \in (6.98, 9.15)$. $d_2(x)$ would be more strongly concentrated at high values of x than $d_0(x)$. This conforms with the assumption that u quarks tend to carry a small fraction of proton momentum in $J = 2$ state for which uu can be regarded as a string like di-quark state.

A further input to the model comes from the ratio of neutron and proton F_2 structure functions expressible in terms of quark distribution functions of proton as

$$R^{np} \equiv \frac{F_2^n}{F_2 p} = \frac{u(x) + 4d(x)}{4u(x) + d(x)} . \quad (2.6)$$

According to [37] $R^{np}(x)$ is a straight line starting with $R^{np}(x \to 0) \simeq 1$ and dropping below $1/2$ as $x \to 1$. The behavior for small x can be understood in terms of sea quark dominance. The pQCD prediction for R^{np} is $R^{np} \to 3/7$ for $x \to 1$, which corresponds to $d/u \to z = 1/5$. TGD prediction for R^{np} for $x \to 1$

$$\begin{aligned} R^{np} &\equiv \frac{F_2^n}{F_2^p} = \frac{pu_0 + 4(pd_0 + (1-p)d_2)}{4pu_0 + pd_0 + (1-p)d_2} \\ &= \frac{p + 4z(p + (1-p)y}{4p + z(p + (1-p)y)} . \end{aligned} \quad (2.7)$$

In the range $\Sigma \in (.13, .30)$ which corresponds to $y \in (6.98, 9.15)$ for $A_d = -.32$ $R^{np} = 1/2$ gives $z \simeq .1$, which is 20 per cent of pQCD prediction. 80 percent of d-quarks with large x predicted to be in $J = 0$ state by pQCD would be in $J = 2$ state.

2.2 Topological evaporation and the concept of Pomeron

Topological evaporation provides an explanation for the mysterious concept of Pomeron originally introduced to describe hadronic diffractive scattering as the exchange of Pomeron Regge trajectory [43]. No hadrons belonging to Pomeron trajectory were however found and via the advent of QCD Pomeron was almost forgotten. Pomeron has recently experienced reincarnation [33, 27, 28]. In Hera [33] $e - p$ collisions, where proton scatters essentially elastically whereas jets in the direction of incoming virtual photon emitted by electron are observed. These events can be understood by assuming that proton emits color singlet particle carrying small fraction of proton's momentum. This particle in turn collides with virtual photon (antiproton) whereas proton scatters essentially elastically.

The identification of the color singlet particle as Pomeron looks natural since Pomeron emission describes nicely diffractive scattering of hadrons. Analogous hard diffractive scattering events in pX diffractive scattering with $X = \bar{p}$ [27] or $X = p$ [28] have also been observed. What happens is that proton scatters essentially elastically and emitted Pomeron collides with X and suffers hard scattering so that large rapidity gap jets in the direction of X are observed. These results suggest that Pomeron is real and consists of ordinary partons.

TGD framework leads to two alternative identifications of Pomeron relying on same geometric picture in which Pomeron corresponds to a space-time sheet separating from hadronic space-time sheet and colliding with photon.

2.2.1 Earlier model

The earlier model is based on the assumption that baryonic quarks carry the entire four-momentum of baryon. p-Adic mass calculations have shown that this assumption is wrong. The modification of the model requires however to change only wordings so that I will represent the earlier model first.

The TGD based identification of Pomeron is very economical: Pomeron corresponds to sea partons, when valence quarks are in vapor phase. In TGD inspired phenomenology events involving Pomeron correspond to pX collisions, where incoming X collides with proton, when valence quarks have suffered coherent simultaneous (by color confinement) evaporation into vapor phase. System X sees only the sea left behind in evaporation and scatters from it whereas valence quarks continue without noticing X and condense later to form quasi-elastically scattered proton. If X suffers hard scattering from the sea the peculiar hard diffractive scattering events are observed. The fraction of these events is equal to the fraction f of time spent by valence quarks in vapor phase.

Dimensional argument can be used to derive a rough order of magnitude estimate for f as $f \sim 1/\alpha = 1/137 \sim 10^{-2}$ for f: f is of same order of magnitude as the fraction (about 5 per cent) of peculiar events from all deep inelastic scattering events in Hera. The time spent in condensate is by dimensional arguments of the order of the p-adic legth scale $L(M_{107})$, not far from proton Compton length. Time dilation effects at high collision energies guarantee that valence quarks indeed stay in vapor phase during the collision. The identification of Pomeron as sea explains also why Pomeron Regge trajectory does not correspond to actual on mass shell particles.

The existing detailed knowledge about the properties of sea structure functions provides a stringent test for the TGD scenario. According to [27] Pomeron structure function seems to consist of soft ($(1-x)^5$), hard ($(1-x)$) and super-hard component (delta function like component at $x = 1$). The peculiar super hard component finds explanation in TGD based picture. The structure function $q_P(x,z)$ of parton in Pomeron contains the longitudinal momentum fraction z of the Pomeron as a parameter and $q_P(x,z)$ is obtained by scaling from the sea structure function $q(x)$ for proton $q_P(x,z) = q(zx)$. The value of structure function at $x = 1$ is non-vanishing: $q_P(x = 1, z) = q(z)$ and this explains the necessity to introduce super hard delta function component in the fit of [27].

2.2.2 Updated model

The recent developments in the understanding of hadron mass spectrum involve the realization that hadronic $k = 107$ space-time sheet is a carrier of super-symplectic bosons (and possibly their super-counterparts with quantum numbers of right handed neutrino) [7] . The model leads to amazingly simple and accurate mass formulas for hadrons. Most of the baryonic momentum is carried by super-symplectic quanta: valence quarks correspond in proton to a relatively small fraction of total mass: about 170 MeV. The counterparts of string excitations correspond to super-symplectic many-particle states and the additivity of conformal weight proportional to mass squared implies stringy mass formula and generalization of Regge trajectory picture. Hadronic string tension is predicted correctly. Model also provides a solution to the proton spin puzzle.

In this framework valence quarks would naturally correspond to a color singlet state formed by space-time sheets connected by color flux tubes having no Regge trajectories and carrying a relatively small fraction of baryonic momentum. In the collisions discussed valence quarks would leave the hadronic space-time sheet and suffer a collision with photon. The lightness of Pomeron and and electro-weak neutrality of Pomeron support the view that photon stripes valence quarks from Pomeron, which continues its flight more or less unperturbed. Instead of an actual topological evaporation the bonds connecting valence quarks to the hadronic space-time sheet could be stretched during the collision with photon.

The large value of $\alpha_K = 1/4$ for super-symplectic matter suggests that the criterion for a phase transition increasing the value of Planck constant [2] and leading to a phase, where $\alpha_K \propto 1/hbar$ is reduced, could occur. For α_K to remain invariant, $\hbar_0 \rightarrow 26\hbar_0$ would be required. In this case, the size of hadronic space-time sheet, "color field body of the hadron", would be $26 \times L(107) = 46$ fm, roughly the size of the heaviest nuclei. Hence a natural expectation is that the dark side of nuclei plays a role in the formation of atomic nuclei. Note that the sizes of electromagnetic field bodies of current quarks u and d with masses of order few MeV is not much smaller than the Compton length of electron. This would mean that super-symplectic bosons would represent dark matter in a well-defined sense and Pomeron exchange would represent temporary separation of ordinary and dark matter.

Note however that the fact that super-symplectic bosons have no electro-weak interactions, implies their dark matter character even for the ordinary value of Planck constant: this could be taken as an objection against dark matter hierarchy. My own interpretation is that super-symplectic matter is dark matter in the strongest sense of the world whereas ordinary matter in the large hbar phase is only apparently dark matter because standard interactions do not reveal themselves in the expected manner.

2.2.3 Astrophysical counterpart of Pomeron events

Pomeron events have direct analogy in astrophysical length scales. In the collision of two galaxies dark and visible matter parts of the colliding galaxies have been found to separate by Chandra X-ray Observatory [44] .

Imagine a collision between two galaxies. The ordinary matter in them collides and gets interlocked due to the mutual gravitational attraction. Dark matter, however, just keeps its momentum and keeps going on leaving behind the colliding galaxies. This kind of event has been detected by the Chandra X-Ray Observatory by using an ingenious manner to detect dark matter. Collisions of ordinary matter produces a lot of X-rays and the dark matter outside the galaxies acts as a gravitational lens.

2.3 The incredibly shrinking proton

The discovery that the charge radius of proton deduced from the muonic version of hydrogen atom is about 4 per cent smaller than from the radius deduced from hydrogen atom [38, 41] is in complete conflict with the cherished belief that atomic physics belongs to the museum of science. The title of the article *Quantum electrodynamics-a chink in the armour?* of the article published in Nature [35] expresses well the possible implications, which might actually go well extend beyond QED.

The finding is a problem of QED or to the standard view about what proton is. Lamb shift [19] is the effect distinguishing between the states hydrogen atom having otherwise the same energy but different angular momentum. The effect is due to the quantum fluctuations of the electromagnetic field. The energy shift factorizes to a product of two expressions. The first one describes the effect of these zero point fluctuations on the position of electron or muon and the second one characterizes the average of nuclear charge density as "seen" by electron or muon. The latter one should be same as in the case of ordinary hydrogen atom but it is not. Does this mean that the presence of muon reduces the charge radius of proton as determined from muon wave function? This of course looks implausible since the radius of proton is so small. Note that the compression of the muon's wave function has the same effect.

Before continuing it is good to recall that QED and quantum field theories in general have difficulties with the description of bound states: something which has not received too much attention. For instance, van der Waals force at molecular scales is a problem. A possible TGD based explanation and a possible solution of difficulties proposed for two decades ago is that for bound states the two charged particles (say nucleus and electron or two atoms) correspond to two 3-D surfaces glued by flux tubes rather than being idealized to points of Minkowski space. This would make the non-relativistic description based on Schrödinger amplitude natural and replace the description based on Bethe-Salpeter equation having horrible mathematical properties.

2.3.1 Basic facts and notions

In this section the basic TGD inspired ideas and notions - in particular the notion of field body- are introduced and the general mechanism possibly explaining the reduction of the effective charge radius relying on the leakage of muon wave function to the flux tubes associated with u quarks is introduced. After this the value of leakage probability is estimated from the standard formula for the Lamb shift in the experimental situation considered.

1. Basic notions of TGD which might be relevant for the problem

Can one say anything interesting about the possible mechanism behind the anomaly if one accepts TGD framework? How the presence of muon could reduce the charge radius of proton? Let us first list the basic facts and notions.

1. One can say that the size of muonic hydrogen characterized by Bohr radius is by factor $m_e/m_\mu = 1/211.4 = 4.7 \times 10^{-4}$ smaller than for hydrogen atom and equals to 250 fm. Hydrogen atom Bohr radius is .53 Angstroms.

2. Proton contains 2 quarks with charge 2e/3 and one d quark which charge -e/3. These quarks are light. The last determination of u and d quark masses [31] gives masses, which are $m_u = 2$ MeV and $m_d = 5$ MeV (I leave out the error bars). The standard view is that the contribution of quarks to proton mass is of same order of magnitude. This would mean that quarks are not too relativistic meaning that one can assign to them a size of order Compton wave length of order $4 \times r_e \simeq 600$ fm in the case of u quark (roughly twice the Bohr radius of muonic hydrogen) and $10 \times r_e \simeq 24$ fm in the case of d quark. These wavelengths are much longer than the proton charge radius and for u quark more than twice longer than the Bohr radius of the muonic hydrogen. That parts of proton would be hundreds of times larger than proton itself sounds a rather weird idea. One could of course argue that the scales in question do not correspond to anything geometric. In TGD framework this is not the way out since quantum classical correspondence requires this geometric correlate.

3. There is also the notion of classical radius of electron and quark. It is given by $r = \alpha\hbar/m$ and is in the case of electron this radius is 2.8 fm whereas proton charge radius is .877 fm and smaller. The dependence on Planck constant is only apparent as it should be since classical radius is in question. For u quark the classical radius is .52 fm and smaller than proton charge radius. The constraint that the classical radii of quarks are smaller than proton charge radius gives a lower bound of quark masses: p-adic scaling of u quark mass by $2^{-1/2}$ would give classical radius .73 fm which still satisfies the bound. TGD framework the proper generalization would be $r = \alpha_K\hbar/m$, where α_K is Kähler coupling strength defining the fundamental coupling constant of the theory and quantized from quantum criticality. Its value is very near or equal to fine structure constant in electron length scale.

4. The intuitive picture is that light-like 3-surfaces assignable to quarks describe random motion of partonic 2-surfaces with light-velocity. This is analogous to zitterbewegung assigned classically to the ordinary Dirac equation. The notion of braid emerging from Chern-Simons Dirac equation via periodic boundary conditions means that the orbits of partonic 2-surface effectively reduces to braids carrying fermionic quantum numbers. These braids in turn define higher level braids which would move inside a structure characterizing the particle geometrically. Internal consistency suggests that the classical radius $r = \alpha_K\hbar/m$ characterizes the size scale of the zitterbewegung orbits of quarks.

   I cannot resist the temptation to emphasize the fact that Bohr orbitology is now reasonably well understood. The solutions of field equations with higher than 3-D CP_2 projection describing radiation fields allow only generalizations of plane waves but not their superpositions in accordance with the fact it is these modes that are observed. For massless extremals with 2-D CP_2 projection superposition is possible only for parallel light-like wave vectors. Furthermore, the restriction of the solutions of the Chern-Simons Dirac equation at light-like 3-surfaces to braid strands gives the analogs of Bohr orbits. Wave functions of -say electron in atom- are wave functions for the position of wormhole throat and thus for braid strands so that Bohr's theory becomes part of quantum theory.

5. In TGD framework quantum classical correspondence requires -or at least strongly suggests- that also the p-adic length scales assignable to u and d quarks have geometrical correlates. That quarks would have sizes much larger than proton itself how sounds rather paradoxical and could be used as an objection against p-adic length scale hypothesis. Topological field quantization however leads to the notion of field body as a structure consisting of flux tubes and and the identification of this geometric correlate would be in terms of Kähler (or color-, or electro-) magnetic body of proton consisting of color flux tubes beginning from space-time sheets of valence quarks and having length scale of order Compton wavelength much longer than the size of proton itself. Magnetic loops and electric flux tubes would be in question. Also secondary p-adic length scale characterizes field body.

For instance, in the case of electron the causal diamond assigned to electron would correspond to the time scale of .1 seconds defining an important bio-rhythm.

2. Could the notion of field body explain the anomaly?

The large Compton radii of quarks and the notion of field body encourage the attempt to imagine a mechanism affecting the charge radius of proton as determined from electron's or muon's wave function.

1. Muon's wave function is compressed to a volume which is about 8 million times smaller than the corresponding volume in the case of electron. The Compton radius of u quark more that twice larger than the Bohr radius of muonic hydrogen so that muon should interact directly with the field body of u quark. The field body of d quark would have size 24 fm which is about ten times smaller than the Bohr radius so that one can say that the volume in which muons sees the field body of d quark is only one thousandth of the total volume. The main effect would be therefore due to the two u quarks having total charge of 4e/3.

   One can say that muon begins to "see" the field bodies of u quarks and interacts directly with u quarks rather than with proton via its elecromagnetic field body. With d quarks it would still interact via protons field body to which d quark should feed its electromagnetic flux. This could be quite enough to explain why the charge radius of proton determined from the expectation value defined by its wave function wave function is smaller than for electron. One must of course notice that this brings in also direct magnetic interactions with u quarks.

2. What could be the basic mechanism for the reduction of charge radius? Could it be that the electron is caught with some probability into the flux tubes of u quarks and that Schrödinger amplitude for this kind state vanishes near the origin? If so, this portion of state would not contribute to the charge radius and the since the portion ordinary state would smaller, this would imply an effective reduction of the charge radius determined from experimental data using the standard theory since the reduction of the norm of the standard part of the state would be erratically interpreted as a reduction of the charge radius.

3. This effect would be of course present also in the case of electron but in this case the u quarks correspond to a volume which million times smaller than the volume defined by Bohr radius so that electron does not in practice "see" the quark sub-structure of proton. The probability P for getting caught would be in a good approximation proportional to the value of $|\Psi(r_u)|^2$ and in the first approximation one would have

$$\frac{P_e}{P_\mu} \sim (a_\mu/a_e)^3 = (m_e/m_\mu)^3 \sim 10^{-7} \ .$$

   from the proportionality $\Psi_i \propto 1/a_i^{3/2}$, i=e,$\mu$.

3. A general formula for Lamb shift in terms of proton charge radius

The charge radius of proton is determined from the Lamb shift between 2S- and 2P states of muonic hydrogen. Without this effect resulting from vacuum polarization of photon Dirac equation for hydrogoen would predict identical energies for these states. The calculation reduces to the calculation of vacuum polarization of photon inducing to the Coulomb potential and an additional vacuum polarization term. Besides this effect one must also take into account the finite size of the proton which can be coded in terms of the form factor deducible from scattering data. It is just this correction which makes it possible to determine the charge radius of proton from the Lamb shift.

1. In the article [21] the basic theoretical results related to the Lamb shift in terms of the vacuum polarization of photon are discussed. Proton's charge density is in this representation is expressed in terms of proton form factor in principle deducible from the scattering data. Two special cases can be distinguished corresponding to the point like proton for which Lamb shift is non-vanishing

only for S wave states and non-point like proton for which energy shift is present also for other states. The theoretical expression for the Lamb shift involves very refined calculations. Between 2P and 2S states the expression for the Lamb shift is of form

$$\Delta E(2P_{3/2}^{F=2}2S_{1/2}^{F=1} = a - br_p^2 + cr_p^3 = 209.968(5)5.2248 \times r_p^2 + 0.0347 \times r_p^3 \ meV \ . \quad (2.8)$$

where the charge radius $r_p = .8750$ is expressed in femtometers and energy in meVs.

2. The general expression of Lamb shift is given in terms of the form factor by

$$E(2P - 2S) = \int \frac{d^3q}{2\pi)^3} \times (-4\pi\alpha)\frac{F(q^2)}{q^2}\frac{\Pi(q^2)}{q^2} \times \int (|\Psi_{2P}(r)|^2 - |\Psi_{2S}(r)|^2)exp(iq\cdot r)dV \ . \quad (2.9)$$

Here Π is is a scalar representing vacuum polarization due to decay of photon to virtual pairs.

The model to be discussed predicts that the effect is due to a leakage from "standard" state to what I call flux tube state. This means a multiplication of $|\Psi_{2P}|^2$ with the normalization factor $1/N$ of the standard state orthogonalized with respect to flux tube state. It is essential that $1/N$ is larger than unity so that the effect is a genuine quantum effect not understandable in terms of classical probability.

The modification of the formula is due to the normalization of the 2P and 2S states. These are in general different. The normalization factor $1/N$ is same for all terms in the expression of Lamb shift for a given state but in general different for 2S and 2P states. Since the lowest order term dominates by a factor of ~ 40 over the second one, one one can conclude that the modification should affect the lowest order term by about 4 per cent. Since the second term is negative and the modification of the first term is interpreted as a modification of the second term when r_p is estimated from the standard formula, the first term must increase by about 4 per cent. This is achieved if this state is orthogonalized with respect to the flux tube state. For states Ψ_0 and Ψ_{tube} with unit norm this means the modification

$$\begin{aligned} \Psi_0 &\rightarrow \frac{1}{1-|C|^2} \times (\Psi_i - C\Psi_{tube}) \ , \\ C &= \langle \Psi_{tube}|\Psi_0\rangle \ . \end{aligned} \quad (2.10)$$

In the lowest order approximation one obtains

$$a - br_p^2 + cr_p^3 \rightarrow (1+|C|^2)a - br_p^2 + cr_p^3 \ . \quad (2.11)$$

Using instead of this expression the standard formula gives a wrong estimate r_p from the condition

$$a - b\hat{r}_p^2 + c\hat{r}_p^3 \rightarrow (1+|C|^2)a - br_p^2 + cr_p^3 \ . \quad (2.12)$$

This gives the equivalent conditions

$$\begin{aligned} \hat{r}_p^2 &= r_p^2 - \frac{|C|^2 a}{b} \ , \\ P_{tube} &\equiv |C|^2 \simeq 2\frac{b}{a} \times r_p^2 \times \frac{(r_p - \hat{r}_p}{r_p}) \ . \end{aligned} \quad (2.13)$$

The resulting estimate for the leakage probability is $P_{tube} \simeq .0015$. The model should be able to reproduce this probability.

2.3.2 A model for the coupling between standard states and flux tube states

Just for fun one can look whether the idea about confinement of muon to quark flux tube carrying electric flux could make sense.

1. Assume that the quark is accompanied by a flux tube carrying electric flux $\int EdS = -\int \nabla\Phi \cdot dS = q$, where $q = 2e/3 = ke$ is the u quark charge. The potential created by the u quark at the proton end of the flux tube with transversal area $S = \pi R^2$ idealized as effectively 1-D structure is

$$\Phi = -\frac{ke}{\pi R^2}|x| + \Phi_0 . \tag{2.14}$$

The normalization factor comes from the condition that the total electric flux is q. The value of the additive constant V_0 is fixed by the condition that the potential coincides with Coulomb potential at $r = r_u$, where r_u is u quark Compton length. This gives

$$e\Phi_0 = \frac{e^2}{r_u} + Kr_u , \quad K = \frac{ke^2}{\pi R^2} . \tag{2.15}$$

2. Parameter R should be of order of magnitude of charge radius $\alpha_K r_u$ of u quark is free parameter in some limits. $\alpha_K = \alpha$ is expected to hold true in excellent approximation. Therefore a convenient parametrization is

$$R = z\alpha r_u . \tag{2.16}$$

This gives

$$K = \frac{4k}{\alpha r_u^2} , \quad e\Phi_0 = 4(\pi\alpha + \frac{k}{\alpha})\frac{1}{r_u} . \tag{2.17}$$

3. The requirement that electron with four times larger charge radius that u quark can topologically condensed inside the flux tube without a change in the average radius of the flux tube (and thus in a reduction in p-adic length scale increasing its mass by a factor 4!) suggests that $z \geq 4$ holds true at least far away from proton. Near proton the condition that the radius of the flux tube is smaller than electron's charge radius is satisfied for $z = 1$.

1. Reduction of Schrödinger equation at flux tube to Airy equation

The 1-D Schrödinger equation at flux tube has as its solutions Airy functions and the related functions known as "Bairy" functions.

1. What one has is a one-dimensional Schrödinger equation of general form

$$-\frac{\hbar^2}{2m_\mu}\frac{d^2\Psi}{dx^2} + (Kx - e\Phi_0)\Psi = E\Psi , \quad K = \frac{ke^2}{\pi R^2} . \tag{2.18}$$

By performing a linear coordinate change

$$u = (\frac{2m_\mu K}{\hbar^2})^{1/3}(x - x_E) , \quad x_E = \frac{-|E| + e\Phi_0}{K} , \tag{2.19}$$

one obtains

$$\frac{d^2\Psi}{du^2} - u\Psi = 0 . \quad (2.20)$$

This differential equation is known as Airy equation (or Stokes equation) and defines special functions $Ai(x)$ known as Airy functions and related functions $Bi(x)$ referred to as "Bairy" functions [15] . Airy functions characterize the intensity near an optical directional caustic such as that of rainbow.

2. The explicit expressions for $A_i(u)$ and $Bi(u)$ are is given by

$$\begin{aligned} Ai(u) &= \frac{1}{\pi}\int_0^\infty cos(\frac{1}{3}t^3 + ut)dt , \\ Bi(u) &= \frac{1}{\pi}\int_0^\infty \left[exp(-\frac{1}{3}t^3) + sin(\frac{1}{3}t^3 + ut)dt\right] . \end{aligned} \quad (2.21)$$

$Ai(u)$ oscillates rapidly for negative values of u having interpretation in terms of real wave vector and goes exponentially to zero for $u > 0$. $Bi(u)$ oscillates also for negative values of x but increases exponentially for positive values of u. The oscillatory behavior and its character become obvious by noticing that stationary phase approximation is possible for $x < 0$.

The approximate expressions of $Ai(u)$ and $Bi(u)$ for $u > 0$ are given by

$$\begin{aligned} Ai(u) &\sim \frac{1}{2\pi^{1/2}}exp(-\frac{2}{3}u^{3/2})u^{-1/4} , \\ Bi(u) &\sim \frac{1}{\pi^{1/2}}exp(\frac{2}{3}u^{3/2})u^{-1/4} . \end{aligned} \quad (2.22)$$

For $u < 0$ one has

$$\begin{aligned} Ai(u) &\sim \frac{1}{\pi^{1/2}}sin(\frac{2}{3}(-u)^{3/2})(-u)^{-1/4} , \\ Bi(u) &\sim \frac{1}{\pi^{1/2}}cos(\frac{2}{3}(-u)^{3/2})(-u)^{-1/4} . \end{aligned} \quad (2.23)$$

3. $u = 0$ corresponds to the turning point of the classical motion where the kinetic energy changes sign. $x = 0$ and $x = r_u$ correspond to the points

$$\begin{aligned} u_{min} \equiv u(0) &= -(\frac{2m_\mu K}{\hbar^2})^{1/3}x_E , \\ u_{max} \equiv u(r_u) &= (\frac{2m_\mu K}{\hbar^2})^{1/3}(r_u - x_E) , \\ x_E &= \frac{-|E| + e\Phi_0}{K} . \end{aligned} \quad (2.24)$$

4. The general solution is

$$\Psi = aAi(u) + bBi(u) . \quad (2.25)$$

The natural boundary condition is the vanishing of Ψ at the lower end of the flux tube giving

$$\frac{b}{a} = -\frac{Ai(u(0)}{Bi(u(0))} . \tag{2.26}$$

A non-vanishing value of b implies that the solution increases exponentially for positive values of the argument and the solution can be regarded as being concentrated in an excellent approximation near the upper end of the flux tube.

Second boundary condition is perhaps most naturally the condition that the energy is same for the flux tube amplitude as for the standard solution. Alternative boundary conditions would require the vanishing of the solution at both ends of the flux tube and in this case one obtains very large number of solutions as WKB approximation demonstrates. The normalization of the state so that it has a unit norm fixes the magnitude of the coefficients a and b since one can choose them to be real.

2. Estimate for the probability that muon is caught to the flux tube

The simplest estimate for the muon to be caught to the flux tube state characterized by the same energy as standard state is the overlap integral of the ordinary hydrogen wave function of muon and of the effectively one-dimensional flux tube. What one means with overlap integral is however not quite obvious.

1. The basic condition is that the modified "standard" state is orthogonal to the flux tube state. One can write the expression of a general state as

$$\begin{aligned} \Psi_{nlm} &\rightarrow N \times (\Psi_{nlm} - C(E, nlm)\Phi_{nlm}) , \\ \Phi_{nlm} &= Y_{lm}\Psi_E , \\ C(E, nlm) &= \langle \Psi_E | \Psi_{nlm} \rangle . \end{aligned} \tag{2.27}$$

   Here Φ_{nlm} depends a flux tube state in which spherical harmonics is wave function in the space of orientations of the flux tube and Ψ_E is flux tube state with same energy as standard state. Here an inner product between standard states and flux tube states is introduced.

2. Assuming same energy for flux tube state and standard state, the expression for the total total probability for ending up to single flux tube would be determined from the orthogonality condition as

$$P_{nlm} = \frac{|C(E, nlm)|^2}{1 - |C(E, lmn)|^2} . \tag{2.28}$$

   Here E refers to the common energy of flux tube state and standard state. The fact that flux tube states vanish at the lower end of the flux tube implies that they do not contribute to the expression for average charge density. The reduced contribution of the standard part implies that the attempt to interpret the experimental results in "standard model" gives a reduced value of the charge radius. The size of the contribution is given by P_{nlm} whose value should be about 4 per cent.

One can consider two alternative forms for the inner product between standard states and flux tube states. Intuitively it is clear that an overlap between the two wave functions must be in question.

1. The simplest possibility is that one takes only overlap at the upper end of the flux tube which defines 2-D surface. Second possibility is that that the overlap is over entire flux tube projection at the space-time sheet of atom.

$$\begin{aligned} \langle\Psi_E|\Psi_{nlm}\rangle &= \int_{end} \overline{\Psi}_r\Psi_{nlm}dS \text{ (Option I)} \ , \\ \langle\Psi_E|\Psi_{nlm}\rangle &= \int_{tube} \overline{\Psi}_r\Psi_{nlm}dV \text{ (Option II)} \ . \end{aligned} \tag{2.29}$$

2. For option I the inner product is non-vanishing only if Ψ_E is non-vanishing at the end of the flux tube. This would mean that electron ends up to the flux tube through its end. The inner product is dimensionless without introduction of a dimensional coupling parameter if the inner product for flux tube states is defined by 1-dimensional integral: one might criticize this assumption as illogical. Unitarity might be a problem since the local behaviour of the flux tube wave function at the end of the flux tube could imply that the contribution of the flux tube state in the quantum state dominates and this does not look plausible. One can of course consider the introduction to the inner product a coefficient representing coupling constant but this would mean loss of predictivity. Schrödinger equation at the end of the flux tubes guarantees the conservation of the probability current only if the energy of flux tube state is same as that of standard state or if the flux tube Schrödinger amplitude vanishes at the end of the flux tube.

3. For option II there are no problems with unitary since the overlap probability is always smaller than unity. Option II however involves overlap between standard states and flux tube states even when the wave function at the upper end of the flux tube vanishes. One can however consider the possibility that the possible flux tube states are orthogonalized with respect to standard states with leakage to flux tubes. The interpretation for the overlap integral would be that electron ends up to the flux tube via the formation of wormhole contact.

3. Option I fails

The considerations will be first restricted to the simpler option I. The generalization of the results of calculation to option II is rather straighforward. It turns out that option II gives correct order of magnitude for the reduction of charge radius for reasonable parameter values.

1. In a good approximation one can express the overlap integrals over the flux tube end (option I) as

$$\begin{aligned} C(E,nlm) &= \int_{tube} \overline{\Psi}_E\Psi_{nlm}dS \simeq \pi R^2 \times Y_{lm} \times C(E,nl) \ , \\ C(E,nl) &= \overline{\Psi}_E(r_u)R_{nl}(r_u) \ . \end{aligned} \tag{2.30}$$

An explicit expression for the coefficients can be deduced by using expression for Ψ_E as a superposition of Airy and Bairy functions. This gives

$$\begin{aligned} C(E,nl) &= \overline{\Psi}_E(r_u)R_{nl}(r_u) \ , \\ \Psi_E(x) &= a_E Ai(u_E) + bBi(u_E) \ , \ \frac{a_E}{b_E} = -\frac{Bi(u_E(0))}{Ai(u_E(0)} \ , \\ u_E(x) &= (\frac{2m_\mu K}{\hbar^2})^{1/3}(x-x_E) \ , \ x_E = \frac{|E|-e\Phi_0}{K} \ , \\ K &= \frac{ke^2}{\pi R^2} \ , \ R = z\alpha_K r_u \ , \ k = \frac{2}{3} \ . \end{aligned} \tag{2.31}$$

The normalization of the coefficients is fixed from the condition that a and b chosen in such a manner that Ψ has unit norm. For these boundary conditions Bi is expected to dominate completely in the sum and the solution can be regarded as exponentially decreasing function concentrated around the upper end of the flux tube.

In order to get a quantitative view about the situation one can express the parameters u_{min} and u_{max} in terms of the basic dimensionless parameters of the problem.

1. One obtains

$$\begin{aligned} u_{min} \equiv u(0) &= -2(\frac{k}{z\alpha})^{1/3}\left[1+\pi\frac{z}{k}\alpha^2(1-\frac{1}{2}\alpha r)\right] \times r^{1/3} \ , \\ u_{max} \equiv u(r_u) &= u(0)+2\frac{k}{z\alpha}\times r^{1/3} \ , \\ r &= \frac{m_\mu}{m_u} \ , \quad R = z\alpha r_u \ . \end{aligned} \tag{2.32}$$

Using the numerical values of the parameters one obtains for $z = 1$ and $\alpha = 1/137$ the values $u_{min} = -33.807$ and $u_{max} = 651.69$. The value of u_{max} is so large that the normalization is in practice fixed by the exponential behavior of Bi for the suggested boundary conditions.

2. The normalization constant is in good approximation defined by the integral of the approximate form of Bi^2 over positive values of u and one has

$$N^2 \simeq \frac{dx}{du}\times\int_{u_{min}}^{u_{max}} Bi(u)^2du \ , \quad \frac{dx}{du} = \frac{1}{2}(\frac{z^2\alpha}{k})^{1/3}\times r^{1/3}r_u \ , \tag{2.33}$$

By taking $t = exp(\frac{4}{3}u^{3/2})$ as integration variable one obtains

$$\begin{aligned} \int_{u_{min}}^{u_{max}} Bi(u)^2du &\simeq \pi^{-1}\int_{u_{min}}^{u_{max}} exp(\frac{4}{3}u^{3/2})u^{-1/2}du \\ &= (\frac{4}{3})^{2/3}\pi^{-1}\int_{t_{min}}^{t_{max}}\frac{dt}{log(t)^{2/3}} \simeq \frac{1}{\pi}\frac{exp(\frac{4}{3}u_{max}^{3/2})}{u_{max}} \ . \end{aligned} \tag{2.34}$$

This gives for the normalization factor the expression

$$N \simeq \frac{1}{2}(\frac{z^2\alpha}{k})^{2/3}r^{1/3}r_u^{1/2}exp(\frac{2}{3}u_{max}^{3/2}) \ . \tag{2.35}$$

3. One obtains for the value of Ψ_E at the end of the flux tube the estimate

$$\Psi_E(r_u) = \frac{Bi(u_{max})}{N} \simeq 2\pi^{-1/2}\times(\frac{k}{z^2\alpha})^{2/3}r^{1/3}r_u^{-1/2} \ , \quad r = \frac{r_u}{r_\mu} \ . \tag{2.36}$$

4. The inner product defined as overlap integral gives for the ground state

$$\begin{aligned} C_{E,00} &= \Psi_E(r_u)\times\Psi_{1,0,0}(r_u)\times\pi R^2 \\ &= 2\pi^{-1/2}(\frac{k}{z^2\alpha})^{2/3}r^{1/3}r_u^{-1/2}\times(\frac{1}{\pi a(\mu)^3})^{1/2}\times exp(-\alpha r)\times\pi z^2\alpha^2r_u^2 \\ &= 2\pi^{1/2}k^{2/3}z^{2/3}r^{11/6}\alpha^{17/6}exp(-\alpha r) \ . \end{aligned} \tag{2.37}$$

The relative reduction of charge radius equals to $P = C_{E,00}^2$. For $z = 1$ one obtains $P = C_{E,00}^2 = 5.5\times10^{-6}$, which is by three orders of magnitude smaller than the value needed for $P_{tube} = C_{E,20}^2 = .0015$. The obvious explanation for the smallness is the α^2 factor coming from the area of flux tube in the inner product.

4. Option II could work

The failure of the simplest model is essentially due to the inner product. For option II the inner product for the flux tube states involves the integral over the area of flux tube so that the normalization factor for the state is obtained from the previous one by the replacement $N \rightarrow N/\sqrt{\pi R^2}$. In the integral over the flux tube the exponent function is is in the first approximation equal to constant since the wave function for ground state is at the end of the flux tube only by a factor .678 smaller than at the origin and the wave function is strongly concentrated near the end of the flux tube. The inner product defined by the overlap integral over the flux tube implies $N \rightarrow NS^{1/2}$, $S = \pi R^2 = z^2\alpha^2 r_u^2$. In good approximation the inner product for option II means the replacement

$$\begin{aligned} C_{E,n0} &\rightarrow A \times B \times C_{E,n0} \ , \\ A &= \frac{\frac{dx}{du}}{\sqrt{\pi R^2}} = \frac{1}{2\sqrt{\pi}} z^{-1/3} k^{-1/3} \alpha^{-2/3} r^{1/3} \ , \\ B &= \frac{\int Bi(u)du}{\sqrt{Bi(u_{max})}} = u_{max}^{-1/4} = 2^{-1/4} z^{1/2} k^{-1/4} \alpha^{1/4} r^{-1/12} \ . \end{aligned} \tag{2.38}$$

Using the expression

$$R_{20}(r_u) = \frac{1}{2\sqrt{2}} \times (\frac{1}{a_\mu})^{3/2} \times (2 - r\alpha) \times exp(-r\alpha) \ , \ r = \frac{r_u}{r_\mu} \tag{2.39}$$

one obtains for $C_{E,20}$ the expression

$$C_{E,20} = 2^{-3/4} z^{5/6} k^{1/12} \alpha^{29/12} r^{25/12} \times (2 - r\alpha) \times exp(-r\alpha) \ . \tag{2.40}$$

By the earlier general argument one should have $P_{tube} = |C_{E,20}|^2 \simeq .0015$. $P_{tube} = .0015$ is obtained for $z = 1$ and $N = 2$ corresponding to single flux tube per u quark. If the flux tubes are in opposite directions, the leakage into 2P state vanishes. Note that this leakage does not affect the value of the coefficient a in the general formula for the Lamb shift. The radius of the flux tube is by a factor 1/4 smaller than the classical radius of electron and one could argue that this makes it impossible for electron to topologically condense at the flux tube. For $z = 4$ one would have $P_{tube} = .015$ which is 10 times too large a value. Note that the nucleus possess a wave function for the orientation of the flux tube. If this corresponds to S-wave state then only the leakage beween S-wave states and standard states is possible.

2.3.3 Are exotic flux tube bound states possible?

There seems to be no deep reason forbidding the possibility of genuine flux tube states decoupling from the standard states completely. To get some idea about the energy eigenvalues one can apply WKB approximation. This approach should work now: in fact, the study on WKB approximation near turning point by using linearization of the the potential leads always to Airy equation so that the linear potential represents an ideal situation for WKB approximation. As noticed these states do not seem to be directly relevant for the recent situation. The fact that these states have larger binding energies than the ordinary states of hydrogen atom might make possible to liberate energy by inducing transitions to these states.

1. Assume that a bound state with a negative energy E is formed inside the flux tube. This means that the condition $p^2 = 2m(E - V) \geq 0$, $V = -e\Phi$, holds true in the region $x \leq x_{max} < r_u$ and $p^2 = 2m(E - V) < 0$ in the region $r_u > x \geq x_{max}$. The expression for x_{max} is

$$x_{max} = \frac{\pi R^2}{k}(-\frac{|E|}{e^2} + \frac{1}{r_u} + \frac{kr_u}{\pi R^2})\hbar \ . \tag{2.41}$$

   $x_{max} < r_u$ holds true if one has

$$|E| \ < \ \frac{e^2}{r_u} = E_{max} \ . \tag{2.42}$$

The ratio of this energy to the ground state energy of muonic hydrogen is from $E(1) = e^2/2a(\mu)$ and $a = \hbar/\alpha m$ given by

$$\frac{E_{max}}{E(n=1)} \ = \ \frac{2m_u}{\alpha m_\mu} \simeq 5.185 \ . \tag{2.43}$$

This encourages to think that the ground state energy could be reduced by the formation of this kind of bound state if it is possible to find a value of n in the allowed range. The physical state would of course contain only a small fraction of this state. In the case of electron the increase of the binding energy is even more dramatic since one has

$$\frac{E_{max}}{E(n=1)} \ = \ \frac{2m_u}{\alpha m_e} = \frac{8}{\alpha} \simeq 1096 \ . \tag{2.44}$$

Obviously the formation of this kind of states could provide a new source of energy. There have been claims about anomalous energy production in hydrogen [50] . I have discussed these claims from TGD viewpoint in [11]

2. One can apply WKB quantization in the region where the momentum is real to get the condition

$$I \ = \ \int_0^{x_{max}} \sqrt{2m(E+e\Phi)}\frac{dx}{\hbar} = n + \frac{1}{2} \ . \tag{2.45}$$

By performing the integral one obtains the quantization condition

$$\begin{aligned} I &= k^{-1}(8\pi\alpha)^{1/2} \times \frac{R^2}{r_u^{3/2} r_\mu} \times A^{3/2} = n + \frac{1}{2} \ , \\ A &= 1 + kx^2 - \frac{|E|r_u}{e^2} \ , \\ x &= \frac{r_u}{R} \ , \ k = \frac{2}{3\pi} \ , \ r_i = \frac{\hbar}{m_i} \ . \end{aligned} \tag{2.46}$$

3. Parameter R should be of order of magnitude of charge radius $\alpha_K r_u$ of u quark is free parameter in some limits. $\alpha_K = \alpha$ is expected to hold true in excellent approximation. Therefore a convenient parametrization is

$$R \ = \ z\alpha r_u \ . \tag{2.47}$$

This gives for the binding energy the general expression in terms of the ground state binging energy $E(1,\mu)$ of muonic hydrogen as

$$\begin{aligned} |E| &= C \times E(1,\mu) \ , \\ C &= D \times (1 + Kz^{-2}\alpha^{-2} - (\frac{y}{z^2})^{2/3} \times (n+1/2)^{2/3}) \ , \\ D &= 2y \times (\frac{K^2}{8\pi\alpha})^{1/3} \ , \\ y &= \frac{m_u}{m_\mu} \ , \ K = \frac{2}{3\pi} \ . \end{aligned} \tag{2.48}$$

4. There is a finite number of bound states. The above mentioned consistency conditions coming from $0 < x_{max} < r_\mu$ give $0 < C < C_{max} = 5.185$ restricting the allowed value of n to some interval. One obtains the estimates

$$n_{min} \simeq \frac{z^2}{y}(1+Kz^{-2}\alpha^{-2} - \frac{C_{max}}{D})^{3/2} - \frac{1}{2} ,$$
$$n_{max} = \frac{z^2}{y}(1+Kz^{-2}\alpha^{-2})^{3/2} - \frac{1}{2} . \quad (2.49)$$

Very large value of n is required by the consistency condition. The calculation gives $n_{min} \in \{1.22 \times 10^7, 4.59 \times 10^6, 1.48 \times 10^5\}$ and $n_{max} \in \{1.33 \times 10^7, 6.66 \times 10^6, 3.34 \times 10^6\}$ for $z \in \{1,2,4\}$. This would be a very large number of allowed bound states -about 3.2×10^6 for $z = 1$.

The WKB state behaves as a plane wave below x_{max} and sum of exponentially decaying and increasing amplitudes above x_{max}:

$$\frac{1}{\sqrt{k(x)}}\left[Aexp(i\int_0^x k(y)dy) + Bexp(-i\int_0^x k(y)dy)\right] ,$$
$$\frac{1}{\sqrt{\kappa(x)}}\left[Cexp(-\int_{x_{max}}^x \kappa(y)dy + Dexp(\int_{x_{max}}^x \kappa(y)dy\right] ,$$
$$k(x) = \sqrt{2m(-|E| + e\Phi} \ , \ \kappa(x)\sqrt{2m(|E| - e\Phi} . \quad (2.50)$$

At the classical turning point these two amplitudes must be identical.

The next task is to decide about natural boundary conditions. Two types of boundary conditions must be considered. The basic condition is that genuine flux tube states are in question. This requires that the inner product between flux tube states and standard states defined by the integral over flux tube ends vanishes. This is guaranteed if the Schrödinger amplitude for the flux tube state vanishes at the ends of the flux tube so that flux tube behaves like an infinite potential well. The condition $\Psi(0) = 0$ at the lower end of the flux tube would give $A = -B$. Combined with the continuity condition at the turning point these conditions imply that Ψ can be assumed to be real. The $\Psi(r_u) = 0$ gives a condition leading to the quantization of energy.

The wave function over the directions of flux tube with a given value of n is given by the spherical harmonics assigned to the state (n,l,m).

2.4 Explanation for the soft photon excess in hadron production

There is quite a recent article entitled Study of the Dependence of Direct Soft Photon Production on the Jet Characteristics in Hadronic Z^0 Decays discussing one particular manifestation of an anomaly of hadron physics known for two decades: the soft photon production rate in hadronic reactions is by an averge factor of about four higher than expected. In the article soft photons assignable to the decays of Z^0 to quark-antiquark pairs. This anomaly has not reached the attention of particle physics which seems to be the fate of anomalies quite generally nowadays: large extra dimensions and blackholes at LHC are much more sexy topics of study than the anomalies about which both existing and speculative theories must remain silent.

2.4.1 Soft photon anomaly

The general observations are summarized by the abstract of the paper.

An analysis of the direct soft photon production rate as a function of the par ent jet characteristics is presented, based on hadronic events collected by the DELPHI experiment at LEP1. The dependences of the photon rates on the jet kinematic characteristics (momentum, mass, etc.) and on the jet charged, neutral and total hadron multiplicities are reported. Up to a scale factor of about four, which characterizes the overall value of the soft photon excess, a similarity of the observed soft photon behaviour to that of the inner hadronic bremsstrahlung predictions is found for the momentum, mass, and jet charged multiplicity dependences. However for the dependence of the soft photon rate on the jet neutral and total hadron multiplicities a prominent difference is found for the observed soft photon signal as compared to the expected bremsstrahlung from final state hadrons. The observed linear increase of the soft photon production rate with the jet total hadron multiplicity and its strong dependence on the jet neutral multiplicity suggest that the rate is proportional to the number of quark pairs produced in the fragmentation process, with the neutral pairs being more effectively radiating than the charged ones.

I try to abstract the essentials of the article.

1. One considers soft photon production in kinematic range .2 GeV< E < 1 GeV, p_T < .08 GeV, where p_T is photon transverse momentum with respect to the parent jet direction. The soft photon excess is associated with hadron production only and does not appear in leptonic sector. As one subtracts the photon yield due to the decays of hadrons (mainly neutral pions), one finds that what remains is on the average 4 times larger than the photon yield by inner hadronic brehmstrahlung, which means bremsstrahlung by charged final state hadrons. This suggests that the description in terms of charged hadron bremstrahlung is not correct and one must go to quark level.

2. Up to the scale factor with average value four, the dependence of soft photon production on jet momentum, mass, and jet charged multiplicity is consistent with the inner hadronic bremsstrahlung predictions.

3. The dependence of the soft photon rate on jet neural and total hadron multiplicities differs from the expected bremsstrahlung from final state hadons. The linear increase of the rate with the jet total hadron multiplicity and strong dependence on the jet neutral multiplicity does not conform with internal hadron bremsstrahlung prediction which suggests that the anomalous soft photon production is proportional to the number of neutral quark pairs giving rise to neutral mesons. For some reason neutral pairs would thus radiate more effectively than the charged ones. Therefore the hypothesis that sea quarks alone are responsible for anomalous brehmstrahlung cannot hold true as such.

The article discusses the data and also the models that has been proposed. Incoherent production of photons by quarks predict satisfactorily the linear dependence of total intensity of bremmstrahling on total number of jet particle if the number of quarks in jet is assumed to be proportional to the number jet particles (see Fig. 7 of [26]). The model cannot however explain the deviations from the model based on charged hadron inner bremmstrahling: the problems are produced by the sensitive dependence on the number of neutral hadrons (see Fig. 6 of [26]).

The models assuming that jet acts as a coherent structure fail also and it is proposed that somehow neutral quark pairs must act as electric dipoles generating dipole radiation at low energies. The dipole moments assignable to neutral quark pairs $U\overline{U}$ and $D\overline{D}$. $U\overline{D}$, $D\overline{U}$ with given respect to center of mass are proportional to the difference of the quark charges $4/3, 2/3, 1/3, -1/3$ so that one might argue that the dipole radiation from neutral pairs is by a factor 16 *resp.* 4 stronger than from charged pair and authors argue that this might be part of the explanation. This would suggest that the excess radiation comes from dipole radiation from quarks inside neutral hadrons. The dipole radiation intensity is expected to be weaker than monopole radiation by a factor $1/\lambda^2$ roughly so that this line of thought does not look promising.

2.4.2 TGD based explanation of the anomaly

Could one find an explanation for the anomaly in TGD framework? The following model finds its inspiration from TGD inspired models for two other anomalies.

1. The first model explains the reported deviation of the charge radius of muonic hydrogen from the predicted radius. Key role is played by the electric flux tubes associated with quarks and having

size scale of order quark Compton radius and therefore extending up to the Bohr radius of muonic hydrogen in the case of u quark.

2. Second model explains the observed anomalous behavior of the quark-gluon plasma. What is observed is almost perfect fluid behavior instead of gas like behavior reflecting itself as small viscosity to entropy ratio. The findings suggest coherence in rather long length scales and also existence of string like objects. Color magnetic (or color electric or both) flux tubes containing quarks and antiquarks are proposed as a space-time correlate for the quark gluon plasma.

Electric flux tubes as basic objects provide a promising candidate for the counterparts of dipoles now. In the case of neutral hadrons color flux tubes and em flux tubes can be one and the same thing. In the case of charged hadrons this cannot be the case and em flux tubes connect oppositely charged hadrons. This could explain the difference between neutral and charged hadrons. If the production amplitude is coherent sum over amplitudes for quarks and antiquarks inside hadron and if also sea quarks contribute, only neutral hadrons would contribute to the brehmstrahlung at long wave length limit and the excess would correspond to the contribution of sea quarks insided neutral hadrons.

A more precise argument goes as follows.

1. The first guess would be that the production amplitude of photons is sum over incoherent contributions of valence and sea quarks. This cannot be the case since both charged and neutral hadrons would contribute equally.

2. Quantum classical correspondence requires some space-time correlate for the classical electric fields. In TGD electric flux is carred by flux tubes and this suggests that flux tubes serve as this correlate. These flux tubes must begin from quark and end to an anti-quark of opposite charge. One must distinguish between the flux tubes assignable to electric field and gluon field. The flux tubes connecting charged hadrons cannot correspond to color flux tubes. For electromagnetically neutral hadrons color flux tubes and em flux tubes can be one and the same thing: this conforms with the fact that classical color fields are proportional to the induced Kähler form as is also the U(1) part of the classical em field. This will be assumed so that only the flux tubes associated with neutral quark pairs (hadrons) can contribute to the coherent dipole radiation. In particular, the sea quarks at these flux tubes can contribute. The flux tubes connecting different hadrons of the final state would not carry color gauge flux making possible materialization of sea quarks from vacuum. If the sea quarks at flux electric flux tubes are responsible for the anomaly, the excess is present only for the neutral hadrons.

3. Low energy phenomenon is in question. This means that the description of quark pairs as coherently scattering pairs of charges (dipole approximation is not necessary) should make sense only when the photon wavelength is longer than the size scale of the dipole: the relevant length scale could be expressed in terms of the distance d between the quark and antiquark of the pair. The criterion can be written as $\lambda \geq xd/2$, where x is a numerical constant of order unity whose value, which should be fixed by the precise criterion of coherence length which should be few wave lengths. For higher energies description as incoherently radiating quarks should be a good approximation. The quark and antiquark with opposite charges can belong to the same to-be-hadron or different charged to-be-hadrons. In the first case there distance remainsmore or leass the same during fragmentation process. In the latter case it increases. In the first case the treatment of the flux tube as a coherently radiating unit makes sense for wavelengths $\lambda \geq xd/2$.

4. The assumption that the bremsstrahling amplitude is a coherent sum over the amplitudes for the quarks and antiquarks inside to-be-hadron gives a heuristic estimate for the radiation power. Consider first the situation in which the ends of the flux tube contain quark and antiquark. Denoting by A value of the photon emission amplitude for free quark, this would give amplitude squared $|A|^2|(1 - exp(exp(ik \cdot d))|^2$, whose maximum value is by a factor 4 larger than that for a single particle. The maxima would correspond to $\lambda = 2dcos(\theta)/(2n+1)$, where θ is the angle between the wave vector of photon and d. $n = 0$ would correspond to $\lambda = 2dcos(\theta)$. For given value of λ one would obtain a diffraction pattern with maxima at $cos(\theta) = (n + 1/2)\lambda/d$. This cannot however

give large enough radiation power: the angle average of the factor $|1-exp(i\phi)|^2$ is 2 instead of 4 and corresponds to the incoherent sum of production rates.

5. More complex model would assume that the flux tubes contain quarks and antiquarks also in their interior so that one would have coherent sum of a larger number of amplitudes which would give diffraction conditions for λ analogous to those above. In this case the maximum of the diffractive factor would be N^2, where $N=2n$ is total number of quarks and antiquarks for mesons. For neutral baryons flux tube would contain odd number of quarks. The angle average would be in this case be equal to N. If all quarks and antiquarks inside the flux tube appear as valence quarks of the final state hadron, one obtains just the result predicted by the independent quark model. Therefore the only possible interpretation for additional contribution is in terms of sea quarks.

Consider now a more detailed quantitative estimate. Assume that the emission inside flux tubes is incoherent. Assume that the sea quarks with charges $\pm 2/3$ and $\pm 1/3$ appear with same probablities and this is true also for valence quarks for energetic enough jets. Therefore the average quark charge squared is $\langle Q_q^2\rangle = 5/18$.

1. The model based on incoherent bremmstrahlung on quarks mentioned in [26] assumes that the number of partons in jet is proportional to the hadrons in the jet:

$$R \propto (N_{sea,neu} + N_{val,neu} + N_{sea,ch} + N_{val,ch}) \propto N_{tot} \ . \tag{2.51}$$

According to [26] the model explains the excess as a a linear function of jet total hadron multiplicity N_{tot} (see Fig. 7 of [26]). This behavior is obtained if the production rate satisifies

$$R \propto (N_{sea,neu} + N_{val,neu} + N_{sea,ch} + N_{val,ch})\langle Q_q^2\rangle \ .$$

One however considers inclusive distribution meaning integration over the various combinations (N_{neu}, N_{ch}) and also other jet variables weighted by differential cross section so that similar result is obtained under much weaker conditions.

2. Indeed, if sea quarks and valence quarks have same p-adic mass scale, one has

$$R \propto (N_{sea,neu} + N_{val,neu} + N_{val,ch})\langle Q_q^2\rangle \tag{2.52}$$

p-Adic length cale hypothesis however allows the sea quarks to be considerably lighter than valence quarks so that their contribution to the brehmstrahlung can be larger. This would mean the proportionality

$$\begin{aligned} R &\propto (xN_{sea,neu} + N_{val,neu} + N_{val,ch})\langle Q_q^2\rangle \ , \\ x &= (\frac{m_{val}}{m_{sea}})^2 \ . \end{aligned} \tag{2.53}$$

p-Adic length scale hypothesis predicts that x is power of two: $x = 2^k$, $k \in \{0,1,2,..\}$.

The above constraint gives rise to the consistency condition

$$\langle R\rangle \propto \langle xN_{sea,neu} + N_{val,neu} + N_{val,ch}\rangle \propto N_{tot} \ . \tag{2.54}$$

3. The data [26] support the the appearence of $N_{sea,neu}$ in the rate.

(a) The dependence on xN_{sea} could explain the exceptionally large deviation (by factor of 8, see Fig. 5 of [26]) from hadronic inner bremsstrahlung for smallest charged multiplicity meaning large number sea quarks assignable to neutral hadrons. For large values of charged multiplicity the contribution of $xN_{sea,neu} + N_{val,neu}$ becomes small and the one should obtain approximate factor 4.

(b) The linear fit of the distribution in the form $R = a_1N_{ch} + a_2N_{neu}$ gives $a_2/a_1 \simeq 6$ so that the dependence on neutral multiplicity is six time stronger than on charged multiplicity (see table 6 of [26]). This suggests that $xN_{sea,neu}$ dominates in the formula. The first possibility is that the parameter $r = N_{sea,neu}/N_{val,neu}$ is considerably larger than unity. Second possibility is that one has $x > 1$.

(c) The ratio of signal to bremsstrahlung prediction increases rapidly as a function of neutral jet multiplicity n_{neu} and increases from 2.5 to about 16 in the range $[0, 6]$ for the neutral multiplicity (see Fig. 6 of [26]). This conforms with the dependence on $N_{sea,neu}$. Also the dependence of the signal to bremsstralung ratio on the core charged multiplicity is non-trivial being largest for vanishing core charge and decreasing with core n_{ch}. Also this confirms with the proposal.

To sum up, the model depends crucially on the notion of induced gauge field and proportionality of the classical color fields and U(1) part of em field to the induced Kähler form and therefore the anomaly gives support for the basic prediction of TGD distinguishing it from QCD. It is possible that two times lighter p-adic mass scale for sea quarks than for valence quarks is needed in order to explain the findings.

3 Simulating Big Bang in laboratory

Ultra-high energy collisions of heavy nuclei at Relativistic Heavy Ion Collider (RHIC) can create so high temperatures that there are hopes of simulating Big Bang in laboratory. The experiment with PHOBOS detector [30] probed the nature of the strong nuclear force by smashing two Gold atoms together at ultrahigh energies. The analysis of the experimental data has been carried out by Prof. Manly and his collaborators at RHIC in Brookhaven, NY [29]. The surprise was that the hydrodynamical flow for non-head-on collisions did not possess the expected longitudinal boost invariance.

This finding stimulates in TGD framework the idea that something much deeper might be involved.

1. The quantum criticality of the TGD inspired very early cosmology predicts the flatness of 3-space as do also inflationary cosmologies. The TGD inspired cosmology is 'silent whisper amplified to big bang' since the matter gradually topologically condenses from decaying cosmic string to the space-time sheet representing the cosmology. This suggests that one could model also the evolution of the quark-gluon plasma in an analogous manner. Now the matter condensing to the quark-gluon plasma space-time sheet would flow from other space-time sheets. The evolution of the quark-gluon plasma would very literally look like the very early critical cosmology.

2. What is so remarkable is that critical cosmology is not a small perturbation of the empty cosmology represented by the future light cone. By perturbing this cosmology so that the spherical symmetry is broken, it might possible to understand qualitatively the findings of [29]. Maybe even the breaking of the spherical symmetry in the collision might be understood as a strong gravitational effect on distances transforming the spherical shape of the plasma ball to a non–spherical shape without affecting the spherical shape of its M^4_+ projection.

3. The model seems to work at qualitative level and predicts strong gravitational effects in elementary particle length scales so that TGD based gravitational physics would differ dramatically from that predicted by the competing theories. Standard cosmology cannot produce these effects without a large breaking of the cherished Lorentz and rotational symmetries forming the basis of elementary particle physics. Thus the the PHOBOS experiment gives direct support for the view that Poincare symmetry is symmetry of the imbedding space rather than that of the space-time.

4. This picture was completed a couple of years later by the progress made in hadronic mass calculations [7] . It has already earlier been clear that quarks are responsible only for a small part of the mass of baryons (170 GeV in case of nucleons). The assumption that hadronic $k = 107$ space-time sheet carries a many-particle state of super-symplectic particles with vanishing electro-weak quantum numbers (meaning darkness in the strongest sense of the word.)

5. TGD allows a model of hadrons predicting their masses with accuracy better than one per cent. In this framework color glass condensate can be identified as a state formed when the hadronic space-time sheets of colliding hadrons fuse to single long stringy object and collision energy is transformed to super-symplectic hadrons.

What I have written above reflects the situation around 2005 when RHIC was in blogs. After 5 years later (2010) LHC gave its first results suggesting similar phenomena in proton-proton collisions. These results provide support for the idea that the formation of long entangled hadronic strings by a fusion of hadronic strings forming a structure analogous to black hole or initial string dominated phase of the cosmology are responsible for the RHIC findings. In the LHC case the mechanism leading to this kind of strings must be different since initial state contains only two protons. I would not anymore distinguish between hadrons and super-symplectic hadrons since in the recent picture super-symplectic excitations are responsible for most of the mass of the hadron. The view about dark matter as macroscopic quantum phase with large Planck constant has also evolved a lot from what it was at that time and I have polished reference to some short lived ideas for the benefit of the reader and me. I did not speak about zero energy ontology at that time and the understanding of the general mathematical structure of TGD has improved dramatically during these years.

3.1 Experimental arrangement and findings

3.1.1 Heuristic description of the findings

In the experiments using PHOBOS detector ultrahigh energy Au+Au collisions at center of mass energy for which nucleon-nucleon center of mass energy is $\sqrt{s_{NN}} = 130$ GeV, were studied [30].

1. In the analyzed collisions the Au nuclei did not collide quite head-on. In classical picture the collision region, where quark gluon plasma is created, can be modelled as the intersection of two colliding balls, and its intersection with plane orthogonal to the colliding beams going through the center of mass of the system is defined by two pieces of circles, whose intersection points are sharp tips. Thus rotational symmetry is broken for the initial state in this picture.

2. The particles in quark-gluon plasma can be compared to a persons in a crowded room trying to get out. The particles collide many times with the particles of the quark gluon plasma before reaching the surface of the plasma. The distance $d(z,\phi)$ from the point $(z,0)$ at the beam axis to the point $(0,\phi)$ at the plasma surface depends on ϕ. Obviously, the distance is longest to the tips $\phi = \pm\pi/2$ and shortest to the points $\phi = 0, \phi = \phi$ of the surface at the sides of the collision region. The time $\tau(z,\phi)$ spent by a particle to the travel to the plasma surface should be a monotonically increasing function $f(d)$ of d:

$$\tau(z,\phi) = f(d(z,\phi)) \ .$$

For instance, for diffusion one would have $\tau \propto d^2$ and $\tau \propto d$ for a pure drift.

3. What was observed that for $z = 0$ the difference

$$\Delta\tau = \tau(z = 0, \pi/2) - \tau(z = 0, 0)$$

was indeed non-vanishing but that for larger values of z the difference tended to zero. Since the variation of z correspond that for the rapidity variable y for a given particle energy, this means that particle distributions depend on rapidity which means a breaking of the longitudinal boost invariance assumed in hydrodynamical models of the plasma. It was also found that the difference vanishes for large values of y: this finding is also important for what follows.

3.1.2 A more detailed description

Consider now the situation in a more quantitative manner.

1. Let z-axis be in the direction of the beam and ϕ the angle coordinate in the plane E^2 orthogonal to the beam. The kinematical variables are the rapidity of the detected particle defined as $y = log[E+p_z)/(E-p_z)]/2$ (E and p_z denote energy and longitudinal momentum), Feynman scaling variable $x_F \simeq 2E/\sqrt{s}$, and transversal momentum p_T.

2. By quantum-classical correspondence, one can translate the components of momentum to space-time coordinates since classically one has $x^\mu = p^\mu a/m$. Here a is proper time for a future light cone, whose tip defines the point where the quark gluon plasma begins to be generated, and $v^\mu = p^\mu/m$ is the four-velocity of the particle. Momentum space is thus mapped to an $a = constant$ hyperboloid of the future light cone for each value of a.

   In this correspondence the rapidity variable y is mapped to $y = log[(t+z)/(t-z)]$, $|z| \leq t$ and non-vanishing values for y correspond to particles which emerge, not from the collision point defining the origin of the plane E^2, but from a point above or below E^2. $|z| \leq t$ tells the coordinate along the beam direction for the vertex, where the particle was created. The limit $y \rightarrow 0$ corresponds to the limit $a \rightarrow \infty$ and the limit $y \rightarrow \pm\infty$ to $a \rightarrow 0$ (light cone boundary).

3. Quark-parton models predict at low energies an exponential cutoff in transverse momentum p_T; Feynman scaling $dN/dx_F = f(x_F)$ independent of s; and longitudinal boost invariance, that is rapidity plateau meaning that the distributions of particles do not depend on y. In the space-time picture this means that the space-time is effectively two-dimensional and that particle distributions are Lorentz invariant: string like space-time sheets provide a possible geometric description of this situation.

4. In the case of an ideal quark-gluon plasma, the system completely forgets that it was created in a collision and particle distributions do not contain any information about the beam direction. In a head-on collision there is a full rotational symmetry and even Lorentz invariance so that transverse momentum cutoff disappears. Rapidity plateau is predicted in all directions.

5. The collisions studied were not quite head-on collisions and were characterized by an impact parameter vector with length b and direction angle ψ_2 in the plane E^2. The particle distribution at the boundary of the plane E^2 was studied as a function of the angle coordinate $\phi - \psi_2$ and rapidity y which corresponds for given energy distance to a definite point of beam axis.

The hydrodynamical view about the situation looks like follows.

1. The particle distributions $N(p^\mu)$ as function of momentum components are mapped to space-time distributions $N(x^\mu, a)$ of particles. This leads to the idea that one could model the situation using Robertson-Walker type cosmology. Co-moving Lorentz invariant particle currents depending on the cosmic time only would correspond in this picture to Lorentz invariant momentum distributions.

2. Hydrodynamical models assign to the particle distribution $d^2N/dyd\phi$ a hydrodynamical flow characterized by four-velocity $v^\mu(y, \phi)$ for each value of the rapidity variable y. Longitudinal boost invariance predicting rapidity plateau states that the hydrodynamical flow does not depend on y at all. Because of the breaking of the rotational symmetry in the plane orthogonal to the beam, the hydrodynamical flow v depends on the angle coordinate $\phi - \psi_2$. It is possible to Fourier analyze this dependence and the second Fourier coefficient v_2 of $cos(2(\phi - \psi_2)$ in the expansion

$$\frac{dN}{d\phi} \simeq 1 + \sum_n v_n cos(n(\phi - \psi_2)) \tag{3.1}$$

   was analyzed in [29].

3. It was found that the Fourier component v_2 depends on rapidity y, which means a breaking of the longitudinal boost invariance. v_2 also vanishes for large values of y. If this is true for all Fourier coefficients v_n, the situation becomes effectively Lorentz invariant for large values of y since one has $v(y,\phi) \to 1$.

   Large values of y correspond to small values of a and to the initial moment of big bang in cosmological analogy. Hence the finding could be interpreted as a cosmological Lorentz invariance inside the light cone cosmology emerging from the collision point. Small values of y in turn correspond to large values of a so that the breaking of the spherical symmetry of the cosmology should be manifest only at $a \to \infty$ limit. These observations suggest a radical re-consideration of what happens in the collision: the breaking of the spherical symmetry would not be a property of the initial state but of the final state.

3.2 TGD based model for the quark-gluon plasma

Consider now the general assumptions the TGD based model for the quark gluon plasma region in the approximation that spherical symmetry is not broken.

1. Quantum-classical correspondence supports the mapping of the momentum space of a particle to a hyperboloid of future light cone. Thus the symmetries of the particle distributions with respect to momentum variables correspond directly to space-time symmetries.

2. The M^4_+ projection of a Robertson-Walker cosmology imbedded to $H = M^4_+ \times CP_2$ is future light cone. Hence it is natural to model the hydrodynamical flow as a mini-cosmology. Even more, one can assume that the collision quite literally creates a space-time sheet which locally obeys Robertson-Walker type cosmology. This assumption is sensible in many-sheeted space-time and conforms with the fractality of TGD inspired cosmology (cosmologies inside cosmologies).

3. If the space-time sheet containing the quark-gluon plasma is gradually filled with matter, one can quite well consider the possibility that the breaking of the spherical symmetry develops gradually, as suggested by the finding $v_2 \to 1$ for large values of $|y|$ (small values of a). To achieve Lorentz invariance at the limit $a \to 0$, one must assume that the expanding region corresponds to $r = constant$ "coordinate ball" in Robertson-Walker cosmology, and that the breaking of the spherical symmetry for the induced metric leads for large values of a to a situation described as a "not head-on collision".

4. Critical cosmology is by definition unstable, and one can model the Au+Au collision as a perturbation of the critical cosmology breaking the spherical symmetry. The shape of $r = constant$ sphere defined by the induced metric is changed by strong gravitational interactions such that it corresponds to the shape for the intersection of the colliding nuclei. One can view the collision as a spontaneous symmetry breaking process in which a critical quark-gluon plasma cosmology develops a quantum fluctuation leading to a situation described in terms of impact parameter. This kind of modelling is not natural for a hyperbolic cosmology, which is a small perturbation of the empty M^4_+ cosmology.

3.2.1 The imbedding of the critical cosmology

Any Robertson-Walker cosmology can be imbedded as a space-time sheet, whose M^4_+ projection is future light cone. The line element is

$$ds^2 = f(a)da^2 - a^2(K(r)dr^2 + r^2 d\Omega^2) . \tag{3.2}$$

Here a is the scaling factor of the cosmology and for the imbedding as surface corresponds to the future light cone proper time.

This light cone has its tip at the point, where the formation of quark gluon plasma starts. (θ, ϕ) are the spherical coordinates and appear in $d\Omega^2$ defining the line element of the unit sphere. a and r

are related to the spherical Minkowski coordinates (m^0, r_M, θ, ϕ) by $(a = \sqrt{(m^0)^2 - r_M^2}, r = r_M/a)$. If hyperbolic cosmology is in question, the function $K(r)$ is given by $K(r) = 1/(1+r^2)$. For the critical cosmology 3-space is flat and one has $K(r) = 1$.

1. The critical cosmologies imbeddable to $H = M_+^4 \times CP_2$ are unique apart from a single parameter defining the duration of this cosmology. Eventually the critical cosmology must transform to a hyperbolic cosmology. Critical cosmology breaks Lorentz symmetry at space-time level since Lorentz group is replaced by the group of rotations and translations acting as symmetries of the flat Euclidian space.

2. Critical cosmology replaces Big Bang with a silent whisper amplified to a big but not infinitely big bang. The silent whisper aspect makes the cosmology ideal for the space-time sheet associated with the quark gluon plasma: the interpretation is that the quark gluon plasma is gradually transferred to the plasma space-time sheet from the other space-time sheets. In the real cosmology the condensing matter corresponds to the decay products of cosmic string in 'vapor phase'. The density of the quark gluon plasma cannot increase without limit and after some critical period the transition to a hyperbolic cosmology occurs. This transition could, but need not, correspond to the hadronization.

3. The imbedding of the critical cosmology to $M_+^4 \times S^2$ is given by

$$\begin{aligned} sin(\Theta) &= \frac{a}{a_m} , \\ \Phi &= g(r) . \end{aligned} \quad (3.3)$$

Here Θ and Φ denote the spherical coordinates of the geodesic sphere S^2 of CP_2. One has

$$\begin{aligned} f(a) &= 1 - \frac{R^2k^2}{(1-(a/a_m)^2)} , \\ (\partial_r \Phi)^2 &= \frac{a_m^2}{R^2} \times \frac{r^2}{1+r^2} . \end{aligned} \quad (3.4)$$

Here R denotes the radius of S^2. From the expression for the gradient of Φ it is clear that gravitational effects are very strong. The imbedding becomes singular for $a = a_m$. The transition to a hyperbolic cosmology must occur before this.

This model for the quark-gluon plasma would predict Lorentz symmetry and $v = 1$ (and $v_n = 0$) corresponding to head-on collision so that it is not yet a realistic model.

3.2.2 TGD based model for the quark-gluon plasma without breaking of spherical symmetry

There is a highly unique deformation of the critical cosmology transforming metric spheres to highly non–spherical structures purely gravitationally. The deformation can be characterized by the following formula

$$sin^2(\Theta) = (\frac{a}{a_m})^2 \times (1 + \Delta(a,\theta,\phi)^2) . \quad (3.5)$$

1. This induces deformation of the g_{rr} component of the induced metric given by

$$g_{rr} = -a^2 \left[1 + \Delta^2(a,\theta,\phi)\frac{r^2}{1+r^2}\right] . \quad (3.6)$$

Remarkably, g_{rr} does not depend at all on CP_2 size and the parameter a_m determining the duration of the critical cosmology. The disappearance of the dimensional parameters can be understood to reflect the criticality. Thus a strong gravitational effect independent of the gravitational constant (proportional to R^2) results. This implies that the expanding plasma space-time sheet having sphere as M^4_+ projection differs radically from sphere in the induced metric for large values of a. Thus one can understand why the parameter v_2 is non-vanishing for small values of the rapidity y.

2. The line element contains also the components g_{ij}, $i, j \in \{a, \theta, \phi\}$. These components are proportional to the factor

$$\frac{1}{1-(a/a_m)^2(1+\Delta^2)} , \tag{3.7}$$

which diverges for

$$a_m(\theta, \phi) = \frac{a_m}{\sqrt{1+\Delta^2}} . \tag{3.8}$$

Presumably quark-gluon plasma phase begins to hadronize first at the points of the plasma surface for which $\Delta(\theta, \phi)$ is maximum, that is at the tips of the intersection region of the colliding nuclei. A phase transition producing string like objects is one possible space-time description of the process.

3.3 Further experimental findings and theoretical ideas

The interaction between experiment and theory is pure magic. Although experimenter and theorist are often working without any direct interaction (as in case of TGD), I have the strong feeling that this disjointness is only apparent and there is higher organizing intellect behind this coherence. Again and again it has turned out that just few experimental findings allow to organize separate and loosely related physical ideas to a consistent scheme. The physics done in RHIC has played completely unique role in this respect.

3.3.1 Super-symplectic matter as the TGD counterpart of CGC?

The model discussed above explained the strange breaking of longitudinal Lorentz invariance in terms of a hadronic mini bang cosmology. The next twist in the story was the shocking finding, compared to Columbus's discovery of America, was that, rather than behaving as a dilute gas, the plasma behaved like a liquid with strong correlations between partons, and having density 30-50 times higher than predicted by QCD calculations [47] . When I learned about these findings towards the end of 2004, I proposed how TGD might explain them in terms of what I called conformal confinement [4] . This idea - although not wrong for any obvious reason - did not however have any obvious implications. After the progress made in p-adic mass calculations of hadrons leading to highly successful model for both hadron and meson masses [7] , the idea was replaced with the hypothesis that the condensate in question is Bose-Einstein condensate like state of super-symplectic particles formed when the hadronic space-time sheets of colliding nucleons fuse together to form a long string like object.

A further refinement of the idea comes from the hypothesis that quark gluon plasma is formed by the topological condensation of quarks to hadronic strings identified as color flux tubes. This would explain the high density of the plasma. The highly entangled hadronic string would be analogous to the initial state of TGD inspired cosmology with the only difference that string tension is extremely small in the hadronic context. This structure would possess also characteristics of blackhole.

3.3.2 Fireballs behaving like black hole like objects

The latest discovery in RHIC is that fireball, which lasts a mere 10^{-23} seconds, can be detected because it absorbs jets of particles produced by the collision [46]. The association with the notion black hole is unavoidable and there indeed exists a rather esoteric M-theory inspired model "The RHIC fireball as a dual black hole" by Hortiu Nastase [45] for the strange findings.

The Physics Today article [39] "What Have We Learned From the Relativistic Heavy Ion Collider?" gives a nice account about experimental findings. Extremely high collision energies are in question: Gold nuclei contain energy of about 100 GeV per nucleon: 100 times proton mass. The expectation was that a large volume of thermalized Quark-Gluon Plasma (QCP) is formed in which partons lose rapidly their transverse momentum. The great surprise was the suppression of high transverse momentum collisions suggesting that in this phase strong collective interactions are present. This has inspired the proposal that quark gluon plasma is preceded by liquid like phase which has been christened as Color Glass Condensate (CGC) thought to contain Bose-Einstein condensate of gluons.

3.3.3 The theoretical ideas relating CGC to gravitational interactions

Color glass condensate relates naturally to several gravitation related theoretical ideas discovered during the last year.

1. Classical gravitation and color confinement

Just some time ago it became clear that strong classical gravitation might play a key role in the understanding of color confinement [10] . Whether the situation looks confinement or asymptotic freedom would be in the eyes of beholder: this is one example of dualities filling TGD Universe. If one looks the situation at the hadronic space-time sheet or one has asymptotic freedom, particles move essentially like free massless particles. But - and this is absolutely essential- in the induced metric of hadronic space-time sheet. This metric represents classical gravitational field becoming extremely strong near hadronic boundary. From the point of view of outsider, the motion of quarks slows down to rest when they approach hadronic boundary: confinement. The distance to hadron surface is infinite or at least very large since the induced metric becomes singular at the light-like boundary! Also hadronic time ceases to run near the boundary and finite hadronic time corresponds to infinite time of observer. When you look from outside you find that this light-like 3-surface is just static surface like a black hole horizon which is also a light-like 3-surface. This gives confinement.

2. Dark matter in TGD

The evidence for hadronic black hole like structures is especially fascinating. In TGD Universe dark matter can be (not always) ordinary matter at larger space-time sheets in particular magnetic flux tubes. The mere fact that the particles are at larger space-time sheets might make them more or less invisible.

Matter can be however dark in much stronger sense, should I use the word "black"! The findings suggesting that planetary orbits obey Bohr rules with a gigantic Planck constant [9] ,[51] would suggest quantum coherence of dark matter even in astrophysical length scales and this raises the fascinating possibility that Planck constant is dynamical so that fine structure constant. Dark matter would correspond to phases with non-standard value of Planck constant. This quantization saves from black hole collapse just as the quantization of hydrogen atom saves from the infrared catastrophe.

The basic criterion for the transition to this phase would be that it occurs when some coupling strength - say fine structure constant multiplied by appropriate charges or gravitational constant multiplied by masses- becomes so large that the perturbation series for scattering amplitudes fails to converge. The phase transition increases Planck constant so that convergence is achieved. The attempts to build a detailed view about what might happen led to a generalization of the imbedding space concept by replacing M^4 (or rather the causal diamond) and CP_2 with their singular coverings. During 2010 it turned out that this generalization could be regarded as a conventional manner to describe a situation in which space-time surface becomes analogous to a multi-sheeted Riemann surface. If so, then Planck constant would be replaced by its integer multiple only in effective sense.

The obvious questions are following. Could black hole like objects/magnetic flux tubes/cosmic strings consist of quantum coherent dark matter? Does this dark matter consist dominantly from hadronic space-time sheets which have fused together and contain super-symplectic bosons and their super-partners (with quantum numbers of right handed neutrino) having therefore no electro-weak interactions. Electro-weak charges would be at different space-time sheets.

1. Gravitational interaction cannot force the transition to dark phase in a purely hadronic system at RHIC energies since the product GM_1M_2 characterizing the interaction strength of two masses must be larger than unity ($\hbar = c = 1$) for the phase transition increasing Planck constant to occur. Hence the collision energy should be above Planck mass for the phase transition to occur if gravitational interactions are responsible for the transition.

2. The criterion for the transition to dark phase is however much more general and states that the system does its best to stay perturbative by increasing its Planck constant in discrete steps and applies thus also in the case of color interactions and governs the phase transition to the TGD counterpart of non-perturbative QCD. Criterion would be roughly $\alpha_s Q_s^2 > 1$ for two color charges of opposite sign. Hadronic string picture would suggests that the criterion is equivalent to the generalization of the gravitational criterion to its strong gravity analog $nL_p^2M^2 > 1$, where L_p is the p-adic length scale characterizing color magnetic energy density (hadronic string tension) and M is the mass of the color magnetic flux tube and n is a numerical constant. Presumably L_p, $p = M_{107} = 2^{107} - 1$, is the p-adic length scale since Mersenne prime M_{107} labels the space-time sheet at which partons feed their color gauge fluxes. The temperature during this phase could correspond to Hagedorn temperature (for the history and various interpretations of Hagedorn temperature see the CERN Courier article [17]) for strings and is determined by string tension and would naturally correspond also to the temperature during the critical phase determined by its duration as well as corresponding black-hole temperature. This temperature is expected to be somewhat higher than hadronization temperature found to be about $\simeq$ 176 MeV. The density of inertial mass would be maximal during this phase as also the density of gravitational mass during the critical phase.

Lepto-hadron physics [12] , one of the predictions of TGD, is one instance of a similar situation. In this case electromagnetic interaction strength defined in an analogous manner becomes larger than unity in heavy ion collisions just above the Coulomb wall and leads to the appearance of mysterious states having a natural interpretation in terms of lepto-pion condensate. Lepto-pions are pairs of color octet excitations of electron and positron.

3. Description of collisions using analogy with black holes

The following view about RHIC events represents my immediate reaction to the latest RHIC news in terms of black-hole physics instead of notions related to big bang. Since black hole collapse is roughly the time reversal of big bang, the description is complementary to the earliest one.

In TGD context one can ask whether the fireballs possibly detected at RHIC are produced when a portion of quark-gluon plasma in the collision region formed by to Gold nuclei separates from hadronic space-time sheets which in turn fuse to form a larger space-time sheet separated from the remaining collision region by a light-like 3-D surface (I have used to speak about light-like causal determinants) mathematically completely analogous to a black hole horizon. This larger space-time sheet would contain color glass condensate of super-symplectic gluons formed from the collision energy. A formation of an analog of black hole would indeed be in question.

The valence quarks forming structures connected by color bonds would in the first step of the collision separate from their hadronic space-time sheets which fuse together to form color glass condensate. Similar process has been observed experimentally in the collisions demonstrating the experimental reality of Pomeron, a color singlet state having no Regge trajectory [33] and identifiable as a structure formed by valence quarks connected by color bonds. In the collision it temporarily separates from the hadronic space-time sheet. Later the Pomeron and the new mesonic and baryonic Pomerons created in the collision suffer a topological condensation to the color glass condensate: this process would be analogous to a process in which black hole sucks matter from environment.

Of course, the relationship between mass and radius would be completely different with gravitational constant presumably replacement by the the square of appropriate p-adic length scale presumably of order pion Compton length: this is very natural if TGD counterparts of black-holes are formed by color magnetic flux tubes. This gravitational constant expressible in terms of hadronic string tension of .9 GeV^2 predicted correctly by super-symplectic picture would characterize the strong gravitational interaction assignable to super-symplectic $J = 2$ gravitons. I have long time ago in the context of p-adic mass calculations formulated quantitatively the notion of elementary particle black hole analogy making the notion of elementary particle horizon and generalization of Hawking-Bekenstein law [8] .

The size L of the "hadronic black hole" would be relatively large using protonic Compton radius as a unit of length. For instnce, for $\hbar = 26\hbar_0$ the size would be $26 \times L(107) = 46$ fm and correspond to a size of a heavy nucleus. This large size would fit nicely with the idea about nuclear sized color glass condensate. The density of partons (possibly gluons) would be very high and large fraction of them would have been materialized from the brehmstrahlung produced by the de-accelerating nuclei. Partons would be gravitationally confined inside this region. The interactions of partons would lead to a generation of a liquid like dense phase and a rapid thermalization would occur. The collisions of partons producing high transverse momentum partons occurring inside this region would yield no detectable high p_T jets since the matter coming out from this region would be somewhat like a thermal radiation from an evaporating black hole identified as a highly entangled hadronic string in Hagedorn temperature. This space-time sheet would expand and cool down to QQP and crystallize into hadrons.

4. Quantitative comparison with experimental data

Consider now a quantitative comparison of the model with experimental data. The estimated freeze-out temperature of quark gluon plasma is $T_f \simeq 175.76$ MeV [39, 45], not far from the total contribution of quarks to the mass of nucleon, which is 170 MeV [7] . Hagedorn temperature identified as black-hole temperature should be higher than this temperature. The experimental estimate for the hadronic Hagedorn temperature from the transversal momentum distribution of baryons is $\simeq$ 160 MeV. On the other hand, according to the estimates of hep-ph/0006020 the values of Hagedorn temperatures for mesons and baryons are $T_H(M) = 195$ MeV and $T_H(B) = 141$ MeV respectively.

D-dimensional bosonic string model for hadrons gives for the mesonic Hagedorn temperature the expression [17]

$$T_H = \frac{\sqrt{6}}{2\pi(D-2)\alpha'} , \tag{3.9}$$

For a string in $D = 4$-dimensional space-time and for the value $\alpha' \sim 1$ GeV^{-2} of Regge slope, this would give $T_H = 195$ MeV, which is slightly larger than the freezing out temperature as it indeed should be, and in an excellent agreement with the experimental value of [16] . It deserves to be noticed that in the model for fireball as a dual 10-D black-hole the rough estimate for the temperature of color glass condensate becomes too low by a factor 1/8 [45]. In light of this I would not yet rush to conclude that the fireball is actually a 10-dimensional black hole.

Note that the baryonic Hagedorn temperature is smaller than mesonic one by a factor of about $\sqrt{2}$. According to [16] this could be qualitatively understood from the fact that the number of degrees of freedom is larger so that the effective value of D in the mesonic formula is larger. $D_{eff} = 6$ would give $T_H = 138$ MeV to be compared with $T_H(B) = 141$ MeV. On the other hand, TGD based model for hadronic masses [7] assumes that quarks feed their color fluxes to $k = 107$ space-time sheets. For mesons there are two color flux tubes and for baryons three. Using the same logic as in [16] , one would have $D_{eff}(B)/D_{eff}(M) = 3/2$. This predicts $T_H(B) = 159$ MeV to be compared with 160 MeV deduced from the distribution of transversal momenta in p-p collisions.

3.4 Are ordinary black-holes replaced with super-symplectic black-holes in TGD Universe?

Some variants of super string model predict the production of small black-holes at LHC. I have never taken this idea seriously but in a well-defined sense TGD predicts black-holes associated with super-symplectic gravitons with strong gravitational constant defined by the hadronic string tension. The proposal is that super-symplectic black-holes have been already seen in Hera, RHIC, and the strange cosmic ray events.

Baryonic super-symplectic black-holes of the ordinary M_{107} hadron physics would have mass 934.2 MeV, very near to proton mass. The mass of their M_{89} counterparts would be 512 times higher, about 478 GeV if quark massses scale also by this factor. This need not be the case: if one has $k = 113 \rightarrow 103$ instead of 105 one has 434 GeV mass. "Ionization energy" for Pomeron, the structure formed by valence quarks connected by color bonds separating from the space-time sheet of super-symplectic black-hole in the production process, corresponds to the total quark mass and is about 170 MeV for ordinary proton and 87 GeV for M_{89} proton. This kind of picture about black-hole formation expected to occur in LHC differs from the stringy picture since a fusion of the hadronic mini black-holes to a larger black-hole is in question.

An interesting question is whether the ultrahigh energy cosmic rays having energies larger than the GZK cut-off of 5×10^{10} GeV are baryons, which have lost their valence quarks in a collision with hadron and therefore have no interactions with the microwave background so that they are able to propagate through long distances.

In neutron stars the hadronic space-time sheets could form a gigantic super-symplectic black-hole and ordinary black-holes would be naturally replaced with super-symplectic black-holes in TGD framework (only a small part of black-hole interior metric is representable as an induced metric). This obviously means a profound difference between TGD and string models.

1. Hawking-Bekenstein black-hole entropy would be replaced with its p-adic counterpart given by

$$S_p = (\frac{M}{m(CP_2)})^2 \times log(p) , \tag{3.10}$$

   where $m(CP_2)$ is CP_2 mass, which is roughly 10^{-4} times Planck mass. M is the contribution of p-adic thermodynamics to the mass. This contribution is extremely small for gauge bosons but for fermions and super-symplectic particles it gives the entire mass.

2. If p-adic length scale hypothesis $p \simeq 2^k$ holds true, one obtains

$$S_p = klog(2) \times (\frac{M}{m(CP_2)})^2, \tag{3.11}$$

   $m(CP_2) = \hbar/R$, R the "radius" of CP_2, corresponds to the standard value of $\hbar_0$ for all values of $\hbar$.

3. Hawking-Bekenstein area law gives in the case of Schwartschild black-hole

$$S = \frac{A}{4G} \times \hbar = \pi G M^2 \times \hbar . \tag{3.12}$$

   For the p-adic variant of the law Planck mass is replaced with CP_2 mass and $klog(2) \simeq log(p)$ appears as an additional factor. Area law is obtained in the case of elementary particles if k is prime and wormhole throats have M^4 radius given by p-adic length scale $L_k = \sqrt{k}R$ which is exponentially smaller than L_p. For macroscopic super-symplectic black-holes modified area law results if the radius of the large wormhole throat equals to Schwartschild radius. Schwartschild radius is indeed natural: in [13] I have shown that a simple deformation of the Schwartschild exterior metric to a metric representing rotating star transforms Schwartschild horizon to a light-like 3-surface at which the signature of the induced metric is transformed from Minkowskian to Euclidian.

4. The formula for the gravitational Planck constant appearing in the Bohr quantization of planetary orbits and characterizing the gravitational field body mediating gravitational interaction between masses M and m [9] reads as

$$\hbar_{gr} = \frac{GMm}{v_0}\hbar_0 .$$

   $v_0 = 2^{-11}$ is the preferred value of v_0. One could argue that the value of gravitational Planck constant is such that the Compton length $\hbar_{gr}/M$ of the black-hole equals to its Schwartshild radius. This would give

$$\hbar_{gr} = \frac{GM^2}{v_0}\hbar_0 \ , \ v_0 = 1/2 \ . \quad (3.13)$$

The requirement that $\hbar_{gr}$ is a ratio of ruler-and-compass integers expressible as a product of distinct Fermat primes (only four of them are known) and power of 2 would quantize the mass spectrum of black hole [9] . Even without this constraint M^2 is integer valued using p-adic mass squared unit and if p-adic length scale hypothesis holds true this unit is in an excellent approximation power of two.

5. The gravitational collapse of a star would correspond to a process in which the initial value of v_0 , say $v_0 = 2^{-11}$, increases in a stepwise manner to some value $v_0 \leq 1/2$. For a supernova with solar mass with radius of 9 km the final value of v_0 would be $v_0 = 1/6$. The star could have an onion like structure with largest values of v_0 at the core as suggested by the model of planetary system. Powers of two would be favored values of v_0. If the formula holds true also for Sun one obtains $1/v_0 = 3 \times 17 \times 2^{13}$ with 10 per cent error.

6. Black-hole evaporation could be seen as means for the super-symplectic black-hole to get rid of its electro-weak charges and fermion numbers (except right handed neutrino number) as the antiparticles of the emitted particles annihilate with the particles inside super-symplectic black-hole. This kind of minimally interacting state is a natural final state of star. Ideal super-symplectic black-hole would have only angular momentum and right handed neutrino number.

7. In TGD light-like partonic 3-surfaces are the fundamental objects and space-time interior defines only the classical correlates of quantum physics. The space-time sheet containing the highly entangled cosmic string might be separated from environment by a wormhole contact with size of black-hole horizon.

This looks the most plausible option but one can of course ask whether the large partonic 3-surface defining the horizon of the black-hole actually contains all super-symplectic particles so that super-symplectic black-hole would be single gigantic super-symplectic parton. The interior of super-symplectic black-hole would be a space-like region of space-time, perhaps resulting as a large deformation of CP_2 type vacuum extremal. Black-hole sized wormhole contact would define a gauge boson like variant of the black-hole connecting two space-time sheets and getting its mass through Higgs mechanism. A good guess is that these states are extremely light.

3.5 Very cautious conclusions

The model for quark-gluon plasma in terms of valence quark space-time sheets separated from hadronic space-time sheets forming a color glass condensate relies on quantum criticality and implies gravitation like effects due to the presence of super-symplectic strong gravitons. At space-time level the change of the distances due to strong gravitation affects the metric so that the breaking of spherical symmetry is caused by gravitational interaction. TGD encourages to think that this mechanism is quite generally at work in the collisions of nuclei. One must take seriously the possibility that strong gravitation is present also in longer length scales (say biological), in particular in processes in which new space-time sheets are generated. Critical cosmology might provide a universal model for the emergence of a new space-time sheet.

The model supports TGD based early cosmology and quantum criticality. In standard physics framework the cosmology in question is not sensible since it would predict a large breaking of the Lorentz invariance, and would mean the breakdown of the entire conceptual framework underlying elementary particle physics. In TGD framework Lorentz invariance is not lost at the level of imbedding space, and the experiments provide support for the view about space-time as a surface and for the notion of many-sheeted space-time.

The attempts to understand later strange events reported by RHIC have led to a dramatic increase of understanding of TGD and allow to fuse together separate threads of TGD.

1. The description of RHIC events in terms of the formation of hadronic black hole and its evaporation seems to be also possible and essentially identical with description as a mini bang.

2. It took some time to realize that scaled down TGD inspired cosmology as a model for quark gluon plasma predicts a new phase identifiable as color glass condensate and still a couple of years to realize the proper interpretation of it in terms of super-symplectic bosons having no counterpart in QCD framework.

3. There is also a connection with the dramatic findings suggesting that Planck constant for dark matter has a gigantic value.

4. Black holes and their scaled counterparts would not be merciless information destroyers in TGD Universe. The entanglement of particles having particle like integrity would make black hole like states ideal candidates for quantum computer like systems. One could even imagine that the galactic black hole is a highly tangled cosmic string in Hagedorn temperature performing quantum computations the complexity of which is totally out of reach of human intellect! Indeed, TGD inspired consciousness predicts that evolution leads to the increase of information and intelligence, and the evolution of stars should not form exception to this. Also the interpretation of black hole as consisting of dark matter follows from this picture.

Summarizing, it seems that thanks to some crucial experimental inputs the new physics predicted by TGD is becoming testable in laboratory.

3.6 Five years later

The emergence of the first interesting findings from LHC by CMS collaboration [24, 18] provide new insights to the TGD picture about the phase transition from QCD plasma to hadronic phase and inspired also the updating of the model of RHIC events (mainly elimination of some remnants from the time when the ideas about hierarchy of Planck constants had just born).

3.6.1 Anomalous behavior of quark gluon plasma is observed also in proton proton collisions

In some proton-proton collisions more than hundred particles are produced suggesting a single object from which they are produced. Since the density of matter approaches to that observed in heavy ion collisions for five years ago at RHIC, a formation of quark gluon plasma and its subsequent decay is what one would expect. The observations are not however quite what QCD plasma picture would allow to expect. Of course, already the RHIC results disagreed with what QCD expectations. What is so striking is the evolution of long range correlations between particles in events containing more than 90 particles as the transverse momentum of the particles increases in the range 1-3 GeV (see the excellent description of the correlations by Lubos Motl in his blog [42]).

One studies correlation function for two particles as a function of two variables. The first variable is the difference $\Delta\phi$ for the emission angles and second is essentially the difference for the velocities described relativistically by the difference $\Delta\eta$ for hyperbolic angles. As the transverse momentum p_T increases the correlation function develops structure. Around origin of $\Delta\eta$ axis a widening plateau develops near $\Delta\phi = 0$. Also a wide ridge with almost constant value as function of $\Delta\eta$ develops near $\Delta\phi = \pi$. The interpretation is that particles tend to move collinearly and or in opposite directions. In the latter case their velocity differences are large since they move in opposite directions so that a long ridge develops in $\Delta\eta$ direction in the graph.

Ideal QCD plasma would predict no correlations between particles and therefore no structures like this. The radiation of particles would be like blackbody radiation with no correlations between photons. The description in terms of string like object proposed also by Lubos on basis of analysis of the graph showing the distributions as an explanation of correlations looks attractive. The decay of a string like structure producing particles at its both ends moving nearly parallel to the string to opposite directions could be in question.

Since the densities of particles approach those at RHIC, I would bet that the explanation (whatever it is!) of the hydrodynamical behavior observed at RHIC for some years ago should apply also now. The introduction of string like objects in this model was natural since in TGD framework even ordinary nuclei are string like objects with nucleons connected by color flux tubes [6] ,[6] : this predicts a lot of new nuclear physics for which there is evidence. The basic idea was that in the high density hadronic color flux tubes associated with the colliding nucleon connect to form long highly entangled hadronic strings containing quark gluon plasma. The decay of these structures would explain the strange correlations. It must be however emphasized that in the recent case the initial state consists of two protons rather than heavy nuclei so that the long hadronic string could form from the QCD like quark gluon plasma at criticality when long range fluctuations emerge.

The main assumptions of the model for the RHIC events and those observed now deserve to be summarized. Consider first the "macroscopic description".

1. A critical system associated with confinement-deconfinement transition of the quark-gluon plasma formed in the collision and inhibiting long range correlations would be in question.

2. The proposed hydrodynamic space-time description was in terms of a scaled variant of what I call critical cosmology defining a universal space-time correlate for criticality: the specific property of this cosmology is that the mass contained by comoving volume approaches to zero at the the initial moment so that Big Bang begins as a silent whisper and is not so scaring;-). Criticality means flat 3-space instead of Lobatchevski space and means breaking of Lorentz invariance to SO(4). Breaking of Lorentz invariance was indeed observed for particle distributions but now I am not so sure whether it has much to do with this.

3. The system behaves like almost perfect fluid in the sense that the viscosity entropy ratio is near to its lower bound whose values is predicted by string theory considerations to be $\eta/s = \hbar/4\pi$.

The microscopic level the description would be like follows.

1. A highly entangled long hadronic string like object (color-magnetic flux tube) would be formed at high density of nucleons via the fusion of ordinary hadronic color-magnetic flux tubes to much longer one and containing quark gluon plasma. In QCD world plasma would not be at flux tube.

2. This geometrically (and perhaps also quantally!) entangled string like object would straighten and split to hadrons in the subsequent "cosmological evolution" and yield large numbers of almost collinear particles. The initial situation should be apart from scaling similar as in cosmology where a highly entangled soup of cosmic strings (magnetic flux tubes) precedes the space-time as we understand it. Maybe ordinary cosmology could provide analogy as galaxies arranged to form linear structures?

3. This structure would have also black hole like aspects but in totally different sense as the 10-D hadronic black-hole proposed by Nastase to describe the findings. Note that M-theorists identify black holes as highly entangled strings: in TGD 1-D strings are replaced by 3-D string like objects.

This picture leaves does not yet make the perfect fluid behavior obvious. The following argument relates it to the properties of the preferred extremals of Kähler action.

3.6.2 Preferred extremals as perfect fluids

Almost perfect fluids seems to be abundant in Nature. For instance, QCD plasma was originally thought to behave like gas and therefore have a rather high viscosity to entropy density ratio $x = \eta/s$. Already RHIC found that it however behaves like almost perfect fluid with x near to the minimum predicted by AdS/CFT. The findings from LHC gave additional conform the discovery [22]. Also Fermi gas is predicted on basis of experimental observations to have at low temperatures a low viscosity roughly 5-6 times the minimal value [49] . In the following the argument that the preferred extremals of Kähler action are perfect fluids apart from the symmetry breaking to space-time sheets is developed. The argument requires some basic formulas summarized first.

The detailed definition of the viscous part of the stress energy tensor linear in velocity (oddness in velocity relates directly to second law) can be found in [48] .

1. The symmetric part of the gradient of velocity gives the viscous part of the stress-energy tensor as a tensor linear in velocity. Velocity gardient decomposes to a term traceless tensor term and a term reducing to scalar.

$$\partial_i v_j + \partial_j v_i = \frac{2}{3}\partial_k v^k g_{ij} + (\partial_i v_j + \partial_j v_i - \frac{2}{3}\partial_k v^k g_{ij}) \ . \tag{3.14}$$

   The viscous contribution to stress tensor is given in terms of this decomposition as

$$\sigma_{visc;ij} = \zeta \partial_k v^k g_{ij} + \eta(\partial_i v_j + \partial_j v_i - \frac{2}{3}\partial_k v^k g_{ij}) \ . \tag{3.15}$$

   From $dF^i = T^{ij}S_j$ it is clear that bulk viscosity ζ gives to energy momentum tensor a pressure like contribution having interpretation in terms of friction opposing. Shear viscosity η corresponds to the traceless part of the velocity gradient often called just viscosity. This contribution to the stress tensor is non-diagonal and corresponds to momentum transfer in directions not parallel to momentum and makes the flow rotational. This termm is essential for the thermal conduction and thermal conductivity vanishes for ideal fluids.

2. The 3-D total stress tensor can be written as

$$\sigma_{ij} = \rho v_i v_j - p g_{ij} + \sigma_{visc;ij} \ . \tag{3.16}$$

   The generalization to a 4-D relativistic situation is simple. One just adds terms corresponding to energy density and energy flow to obtain

$$T^{\alpha\beta} = (\rho - p)u^\alpha u^\beta + p g^{\alpha\beta} - \sigma^{\alpha\beta}_{visc} \ . \tag{3.17}$$

   Here u^α denotes the local four-velocity satisfying $u^\alpha u_\alpha = 1$. The sign factors relate to the concentions in the definition of Minkowski metric ($(1,-1,-1,-1)$).

3. If the flow is such that the flow parameters associated with the flow lines integrate to a global flow parameter one can identify new time coordinate t as this flow parametger. This means a transition to a coordinate system in which fluid is at rest everywhere (comoving coordinates in cosmology) so that energy momentum tensor reduces to a diagonal term plus viscous term.

$$T^{\alpha\beta} = (\rho - p)g^{tt}\delta^\alpha_t\delta^\beta_t + p g^{\alpha\beta} - \sigma^{\alpha\beta}_{visc} \ . \tag{3.18}$$

   In this case the vanishing of the viscous term means that one has perfect fluid in strong sense.

   The existence of a global flow parameter means that one has

$$v_i = \Psi \partial_i \Phi \ . \tag{3.19}$$

   Ψ and Φ depend on space-time point. The proportionality to a gradient of scalar Φ implies that Φ can be taken as a global time coordinate. If this condition is not satisfied, the perfect fluid property makes sense only locally.

AdS/CFT correspondence allows to deduce a lower limit for the coefficient of shear viscosity as

$$x = \frac{\eta}{s} \geq \frac{\hbar}{4\pi} . \tag{3.20}$$

This formula holds true in units in which one has $k_B = 1$ so that temperature has unit of energy.

What makes this interesting from TGD view is that in TGD framework perfect fluid property in approriately generalized sense indeed characterizes locally the preferred extremals of Kähler action defining space-time surface.

1. Kähler action is Maxwell action with U(1) gauge field replaced with the projection of CP_2 Kähler form so that the four CP_2 coordinates become the dynamical variables at QFT limit. This means enormous reduction in the number of degrees of freedom as compared to the ordinary unifications. The field equations for Kähler action define the dynamics of space-time surfaces and this dynamics reduces to conservation laws for the currents assignable to isometries. This means that the system has a hydrodynamic interpretation. This is a considerable difference to ordinary Maxwell equations. Notice however that the "topological" half of Maxwell's equations (Faraday's induction law and the statement that no non-topological magnetic are possible) is satisfied.

2. Even more, the resulting hydrodynamical system allows an interpretation in terms of a perfect fluid. The general ansatz for the preferred extremals of field equations assumes that various conserved currents are proportional to a vector field characterized by so called Beltrami property. The coefficient of proportionality depends on space-time point and the conserved current in question. Beltrami fields by definition is a vector field such that the time parameters assignable to its flow lines integrate to single global coordinate. This is highly non-trivial and one of the implications is almost topological QFT property due to the fact that Kähler action reduces to a boundary term assignable to wormhole throats which are light-like 3-surfaces at the boundaries of regions of space-time with Euclidian and Minkowskian signatures. The Euclidian regions (or wormhole throats, depends on one's tastes) define what I identify as generalized Feynman diagrams.

   Beltrami property means that if the time coordinate for a space-time sheet is chosen to be this global flow parameter, all conserved currents have only time component. In TGD framework energy momentum tensor is replaced with a collection of conserved currents assignable to various isometries and the analog of energy momentum tensor complex constructed in this manner has no counterparts of non-diagonal components. Hence the preferred extremals allow an interpretation in terms of perfect fluid without any viscosity.

This argument justifies the expectation that TGD Universe is characterized by the presence of low-viscosity fluids. Real fluids of course have a non-vanishing albeit small value of x. What causes the failure of the exact perfect fluid property?

1. Many-sheetedness of the space-time is the underlying reason. Space-time surface decomposes into finite-sized space-time sheets containing topologically condensed smaller space-time sheets containing.... Only within given sheet perfect fluid property holds true and fails at wormhole contacts and because the sheet has a finite size. As a consequence, the global flow parameter exists only in given length and time scale. At imbedding space level and in zero energy ontology the phrasing of the same would be in terms of hierarchy of causal diamonds (CDs).

2. The so called eddy viscosity is caused by eddies (vortices) of the flow. The space-time sheets glued to a larger one are indeed analogous to eddies so that the reduction of viscosity to eddy viscosity could make sense quite generally. Also the phase slippage phenomenon of super-conductivity meaning that the total phase increment of the super-conducting order parameter is reduced by a multiple of 2π in phase slippage so that the average velocity proportional to the increment of the phase along the channel divided by the length of the channel is reduced by a quantized amount.

   The standard arrangement for measuring viscosity involves a lipid layer flowing along plane. The velocity of flow with respect to the surface increases from $v = 0$ at the lower boundary to v_{upper}

at the upper boundary of the layer: this situation can be regarded as outcome of the dissipation process and prevails as long as energy is feeded into the system. The reduction of the velocity in direction orthogonal to the layer means that the flow becomes rotational during dissipation leading to this stationary situation.

This suggests that the elementary building block of dissipation process corresponds to a generation of vortex identifiable as cylindrical space-time sheets parallel to the plane of the flow and orthogonal to the velocity of flow and carrying quantized angular momentum. One expects that vortices have a spectrum labelled by quantum numbers like energy and angular momentum so that dissipation takes in discrete steps by the generation of vortices which transfer the energy and angular momentum to environment and in this manner generate the velocity gradient.

3. The quantization of the parameter x is suggestive in this framework. If entropy density and viscosity are both proportional to the density n of the eddies, the value of x would equal to the ratio of the quanta of entropy and kinematic viscosity η/n for single eddy if all eddies are identical. The quantum would be $\hbar/4\pi$ in the units used and the suggestive interpretation is in terms of the quantization of angular momentum. One of course expects a spectrum of eddies so that this simple prediction should hold true only at temperatures for which the excitation energies of vortices are above the thermal energy. The increase of the temperature would suggest that gradually more and more vortices come into play and that the ratio increases in a stepwise manner bringing in mind quantum Hall effect. In TGD Universe the value of $\hbar$ can be large in some situations so that the quantal character of dissipation could become visible even macroscopically. Whether this a situation with large $\hbar$ is encountered even in the case of QCD plasma is an interesting question.

The following poor man's argument tries to make the idea about quantization a little bit more concrete.

1. The vortices transfer momentum parallel to the plane from the flow. Therefore they must have momentum parallel to the flow given by the total cm momentum of the vortex. Before continuing some notations are needed. Let the densities of vortices and absorbed vortices be n and n_{abs} respectively. Denote by $v_{||}$ *resp.* $v_{\perp}$ the components of cm momenta parallel to the main flow *resp.* perpendicular to the plane boundary plane. Let m be the mass of the vortex. Denote by S are parallel to the boundary plane.

2. The flow of momentum component parallel to the main flow due to the absorbed at S is

$$n_{abs} m v_{||} v_{\perp} S \ .$$

This momentum flow must be equal to the viscous force

$$F_{visc} = \eta \frac{v_{||}}{d} \times S \ .$$

From this one obtains

$$\eta = n_{abs} m v_{\perp} d \ .$$

If the entropy density is due to the vortices, it equals apart from possible numerical factors to

$$s = n$$

so that one has

$$\frac{\eta}{s} = m v_{\perp} d \ .$$

This quantity should have lower bound $x = \hbar/4\pi$ and perhaps even quantized in multiples of x, Angular momentum quantization suggests strongly itself as origin of the quantization.

3. Local momentum conservation requires that the comoving vortices are created in pairs with opposite momenta and thus propagating with opposite velocities $v_{\perp}$. Only one half of vortices is absorbed so that one has $n_{abs} = n/2$. Vortex has quantized angular momentum associated with its internal rotation. Angular momentum is generated to the flow since the vortices flowing downwards are absorbed at the boundary surface.

   Suppose that the distance of their center of mass lines parallel to plane is $D = \epsilon d$, ϵ a numerical constant not too far from unity. The vortices of the pair moving in opposite direction have same angular momentum $mv \quad D/2$ relative to their center of mass line between them. Angular momentum conservation requires that the sum these relative angular momenta cancels the sum of the angular momenta associated with the vortices themselves. Quantization for the total angular momentum for the pair of vortices gives

$$\frac{\eta}{s} = \frac{n\hbar}{\epsilon}$$

   Quantization condition would give

$$\epsilon = 4\pi \ .$$

   One should understand why $D = 4\pi d$ - four times the circumference for the largest circle contained by the boundary layer- should define the minimal distance between the vortices of the pair. This distance is larger than the distance d for maximally sized vortices of radius $d/2$ just touching. This distance obviously increases as the thickness of the boundary layer increasess suggesting that also the radius of the vortices scales like d.

4. One cannot of course take this detailed model too literally. What is however remarkable that quantization of angular momentum and dissipation mechanism based on vortices identified as space-time sheets indeed could explain why the lower bound for the ratio η/s is so small.

3.7 Evidence for TGD view about QCD plasma

The emergence of the first interesting findings from LHC by CMS collaboration [24, 18] provide new insights to the TGD picture about the phase transition from QCD plasma to hadronic phase and inspired also the updating of the model of RHIC events (mainly elimination of some remnants from the time when the ideas about hierarchy of Planck constants had just born).

In some proton-proton collisions more than hundred particles are produced suggesting a single object from which they are produced. Since the density of matter approaches to that observed in heavy ion collisions for five years ago at RHIC, a formation of quark gluon plasma and its subsequent decay is what one would expect. The observations are not however quite what QCD plasma picture would allow to expect. Of course, already the RHIC results disagreed with what QCD expectations. What is so striking is the evolution of long range correlations between particles in events containing more than 90 particles as the transverse momentum of the particles increases in the range 1-3 GeV (see the excellent description of the correlations by Lubos Motl in his blog [42]).

One studies correlation function for two particles as a function of two variables. The first variable is the difference $\Delta\phi$ for the emission angles and second is essentially the difference for the velocities described relativistically by the difference $\Delta\eta$ for hyperbolic angles. As the transverse momentum p_T increases the correlation function develops structure. Around origin of $\Delta\eta$ axis a widening plateau develops near $\Delta\phi = 0$. Also a wide ridge with almost constant value as function of $\Delta\eta$ develops near $\Delta\phi = \pi$. The interpretation is that particles tend to move collinearly and or in opposite directions. In the latter case their velocity differences are large since they move in opposite directions so that a long ridge develops in $\Delta\eta$ direction in the graph.

Ideal QCD plasma would predict no correlations between particles and therefore no structures like this. The radiation of particles would be like blackbody radiation with no correlations between photons. The description in terms of string like object proposed also by Lubos on basis of analysis of the graph showing the distributions as an explanation of correlations looks attractive. The decay of a string like structure producing particles at its both ends moving nearly parallel to the string to opposite directions could be in question.

Since the densities of particles approach those at RHIC, I would bet that the explanation (whatever it is!) of the hydrodynamical behavior observed at RHIC for some years ago should apply also now. The introduction of string like objects in this model was natural since in TGD framework even ordinary nuclei are string like objects with nucleons connected by color flux tubes [6] ,[6] : this predicts a lot of new nuclear physics for which there is evidence. The basic idea was that in the high density hadronic color flux tubes associated with the colliding nucleon connect to form long highly entangled hadronic strings containing quark gluon plasma. The decay of these structures would explain the strange correlations. It must be however emphasized that in the recent case the initial state consists of two protons rather than heavy nuclei so that the long hadronic string could form from the QCD like quark gluon plasma at criticality when long range fluctuations emerge.

The main assumptions of the model for the RHIC events and those observed now deserve to be summarized. Consider first the "macroscopic description".

1. A critical system associated with confinement-deconfinement transition of the quark-gluon plasma formed in the collision and inhibiting long range correlations would be in question.

2. The proposed hydrodynamic space-time description was in terms of a scaled variant of what I call critical cosmology defining a universal space-time correlate for criticality: the specific property of this cosmology is that the mass contained by comoving volume approaches to zero at the the initial moment so that Big Bang begins as a silent whisper and is not so scaring;-). Criticality means flat 3-space instead of Lobatchevski space and means breaking of Lorentz invariance to SO(4). Breaking of Lorentz invariance was indeed observed for particle distributions but now I am not so sure whether it has much to do with this.

The microscopic level the description would be like follows.

1. A highly entangled long hadronic string like object (color-magnetic flux tube) would be formed at high density of nucleons via the fusion of ordinary hadronic color-magnetic flux tubes to much longer one and containing quark gluon plasma. In QCD world plasma would not be at flux tube.

2. This geometrically (and perhaps also quantally!) entangled string like object would straighten and split to hadrons in the subsequent "cosmological evolution" and yield large numbers of almost collinear particles. The initial situation should be apart from scaling similar as in cosmology where a highly entangled soup of cosmic strings (magnetic flux tubes) precedes the space-time as we understand it. Maybe ordinary cosmology could provide analogy as galaxies arranged to form linear structures?

3. This structure would have also black hole like aspects but in totally different sense as the 10-D hadronic black-hole proposed by Nastase to describe the findings. Note that M-theorists identify black holes as highly entangled strings: in TGD 1-D strings are replaced by 3-D string like objects.

References

Books related to TGD

[1] M. Pitkänen. Construction of Quantum Theory: M-matrix. In *Towards M-Matrix*. Onlinebook. `http://tgd.wippiespace.com/public_html/tgdquant/tgdquant.html#towards`, 2006.

[2] M. Pitkänen. Does TGD Predict the Spectrum of Planck Constants? In *Towards M-Matrix*. Onlinebook. `http://tgd.wippiespace.com/public_html/tgdquant/tgdquant.html#Planck`, 2006.

[3] M. Pitkänen. Knots and TGD. In *Quantum Physics as Infinite-Dimensional Geometry*. Onlinebook. `http://tgd.wippiespace.com/public_html/tgdgeom/tgdgeom.html#knotstgd`, 2006.

[4] M. Pitkänen. Massless states and particle massivation. In *p-Adic Length Scale Hypothesis and Dark Matter Hierarchy*. Onlinebook. http://tgd.wippiespace.com/public_html/paddark/paddark.html#mless, 2006.

[5] M. Pitkänen. New Particle Physics Predicted by TGD: Part I. In *p-Adic Length Scale Hypothesis and Dark Matter Hierarchy*. Onlinebook. http://tgd.wippiespace.com/public_html/paddark/paddark.html#mass4, 2006.

[6] M. Pitkänen. Nuclear String Hypothesis. In *p-Adic Length Scale Hypothesis and Dark Matter Hierarchy*. Onlinebook. http://tgd.wippiespace.com/public_html/paddark/paddark.html#nuclstring, 2006.

[7] M. Pitkänen. p-Adic Particle Massivation: Hadron Masses. In *p-Adic Length Scale Hypothesis and Dark Matter Hierarchy*. Onlinebook. http://tgd.wippiespace.com/public_html/paddark/paddark.html#mass3, 2006.

[8] M. Pitkänen. p-Adic Physics: Physical Ideas. In *TGD as a Generalized Number Theory*. Onlinebook. http://tgd.wippiespace.com/public_html/tgdnumber/tgdnumber.html#phblocks, 2006.

[9] M. Pitkänen. TGD and Astrophysics. In *Physics in Many-Sheeted Space-Time*. Onlinebook. http://tgd.wippiespace.com/public_html/tgdclass/tgdclass.html#astro, 2006.

[10] M. Pitkänen. TGD as a Generalized Number Theory: Quaternions, Octonions, and their Hyper Counterparts. In *TGD as a Generalized Number Theory*. Onlinebook. http://tgd.wippiespace.com/public_html/tgdnumber/tgdnumber.html#visionb, 2006.

[11] M. Pitkänen. The Notion of Free Energy and Many-Sheeted Space-Time Concept. In *TGD and Fringe Physics*. Onlinebook. http://tgd.wippiespace.com/public_html/freenergy/freenergy.html#freenergy, 2006.

[12] M. Pitkänen. The Recent Status of Lepto-hadron Hypothesis. In *p-Adic Length Scale Hypothesis and Dark Matter Hierarchy*. Onlinebook. http://tgd.wippiespace.com/public_html/paddark/paddark.html#leptc, 2006.

[13] M. Pitkänen. The Relationship Between TGD and GRT. In *Physics in Many-Sheeted Space-Time*. Onlinebook. http://tgd.wippiespace.com/public_html/tgdclass/tgdclass.html#tgdgrt, 2006.

Articles about TGD

[14] M. Pitkänen. TGD inspired vision about entropic gravitation. http://tgd.wippiespace.com/public_html/articles/egtgd.pdf, 2011.

Theoretical Physics

[15] Airy functions. http://en.wikipedia.org/wiki/Airy_function.

[16] W. Broniowski. Two Hagedorn temperatures. http://arxiv.org/abs/hep-ph/0006020, 2002.

[17] T. Ericson and J. Rafelski. The tale of the Hagedorn temperature. *Cern Courier*, 43(7), 2002.

Particle and Nuclear Physics

[18] CMS observes a potentially new and interesting effect. http://user.web.cern.ch/user/news/2010/100921.html.

[19] Lamb shift. http://en.wikipedia.org/wiki/Lamb_shift.

[20] Massive High p_T Jets: Updates from CDF. http://indico.cern.ch/getFile.py/access?contribId=16&resId=0&materialId=slides&confId=113980.

[21] A. S. Antognini. The Lamb shift Experiment in Muonic Hydrogen. http://edoc.ub.uni-muenchen.de/5044/1/Antognini_Aldo.pdf, 2005.

[22] Alice Collaboration. Charged-particle multiplicity density at mid-rapidity in central Pb-Pb collisions at $\sqrt{s_{NN}}$= 2.76 TeV. http://arxiv.org/abs/1011.3916, 2010.

[23] CDF collaboration. Evidence for a Mass Dependent Forward-Backward Asymmetry in Top Quark Pair Production. http://arxiv.org/abs/1101.0034, 2011.

[24] CMS Collaboration. Observation of Long-Range, Near-Side Angular Correlations in Proton-Proton Collisions at the LHC. http://cms.web.cern.ch/cms/News/2010/QCD-10-002/QCD-10-002.pdf, 2010.

[25] CMS collaboration. Measurement of the Charge Asymmetry in Top Quark Pair Production. http://cdsweb.cern.ch/record/1369205/files/TOP-11-014-pas.pdf, 2011.

[26] Delphi Collaboration. Study of the Dependence of Direct Soft Photon Production on the Jet Characteristics in Hadronic Z^0 Decays. http://arxiv.org/abs/hep-ex/1004.1587, 2010.

[27] A. Brandt et al. *Phys. Lett. B*, 297:417, 1992.

[28] A. M. Smith et al. *Phys. Lett. B*, 163:267, 1985.

[29] B. B. Back et al. *Phys. Rev. Lett.*, 89(22), November 2002.

[30] B. B. Back et al. *Nucl. Phys. A*, 698:416, 2002.

[31] D. T. H. Davies et al. Precise Charm to Strange Mass Ratio and Light Quark Masses from Full Lattice QCD. *Phys. Rev.*, 104, 2010.

[32] J. Ashman et al. *Phys. Lett. B*, 206, 1988.

[33] M. Derrick et al. *Phys. Lett B*, 315:481, 1993.

[34] M. J. Alguard et al. *Phys. Rev.*, 41, 1978.

[35] R. Pohl et al. The size of proton. *Nature*, 466, 2010.

[36] Y.Takeuchi et al. Measurement of the Forward Backward Asymmetry in Top Pair Production in the Dilepton Decay Channel using 5.1 fb^{-1}. http://www-cdf.fnal.gov/physics/new/top/2011/DilAfb/, 2011.

[37] X. Zheng et al (The Jefferson Lab Hall Collaboration). Precision Measurement of the Neutron Spin Asymmetries and Spin-Dependent Structure Functions in the Valence Quark Region. http://arxiv.org/abs/nucl-ex/0405006, 2004.

[38] J. Flowers. Quantum electrodynamics: A chink in the armour? *Nature*, 466, 2010.

[39] T. Ludham and L. McLerran. What Have We Learned From the Relativistic Heavy Ion Collider? *Physics Today*, October 2003.

[40] Bo-Ciang Ma. The spin structure of the proton, 2000.

[41] K. McAlpine. Incredible shrinking proton raises eyebrows. http://www.newscientist.com/article/dn19141-incredible-shrinking-proton-raises-eyebrows.html, 2010.

[42] L. Motl. LHC probably sees new shocking physics. http://motls.blogspot.com/2010/09/lhc-probably-sees-new-shocking-physics.html, 2010.

[43] G. Violini N. M. Queen. *Dispersion Theory in High Energy Physics.* The Macmillan Press Limited, 1974.

[44] NASA. NASA announces dark matter discovery. http://www.nasa.gov/home/hqnews/2006/aug/HQ_M06128_dark_matter.html, 2006.

[45] H. Nastase. The RHIC fireball as a dual black hole. http://arxiv.org/abs/hep-th/0501068, 2005.

[46] E. S. Reich. Black hole like phenomenon created by collider. *New Scientist*, 19(2491), 2005.

[47] E. Samuel. Ghost in the Atom. *New Scientist*, (2366):30, October 2002.

Condensed Matter Physics

[48] Viscosity. http://en.wikipedia.org/wiki/Viscosity.

[49] C. Cao et al. Universal Quantum Viscosity in a Unitary Fermi Gas. http://www.sciencemag.org/content/early/2010/12/08/science.1195219, 2010.

[50] R. Mills et al. Spectroscopic and NMR identification of novel hybrid ions in fractional quantum energy states formed by an exothermic reaction of atomic hydrogen with certain catalysts. http://www.blacklightpower.com/techpapers.html, 2003.

Cosmology and Astro-Physics

[51] D. Da Roacha and L. Nottale. Gravitational Structure Formation in Scale Relativity. http://arxiv.org/abs/astro-ph/0310036, 2003.

Article

Particle Massivation in TGD Universe

Matti Pitkänen [1]

Abstract

This article represents the most recent view about particle massivation in TGD framework. This topic is necessarily quite extended since many several notions and new mathematics is involved. Indeed, the calculation of particle masses involves five chapters of [11]. In the following my goal is to provide an up-to-date summary whereas the chapters are unavoidably a story about evolution of ideas.

The identification of the spectrum of light particles reduces to two tasks: the construction of massless states and the identification of the states which remain light in p-adic thermodynamics. The latter task is relatively straightforward. The thorough understanding of the massless spectrum requires however a real understanding of quantum TGD. It would be also highly desirable to understand why p-adic thermodynamics combined with p-adic length scale hypothesis works. A lot of progress has taken place in these respects during last years.

Zero energy ontology providing a detailed geometric view about bosons and fermions, the generalization of S-matrix to what I call M-matrix, the notion of finite measurement resolution characterized in terms of inclusions of von Neumann algebras, the derivation of p-adic coupling constant evolution and p-adic length scale hypothesis from the first principles, the realization that the counterpart of Higgs mechanism involves generalized eigenvalues of the modified Dirac operator: these are represent important steps of progress during last years with a direct relevance for the understanding of particle spectrum and massivation although the predictions of p-adic thermodynamics are not affected.

During 2010 a further progress took place. These steps of progress relate closely to zero energy ontology, bosonic emergence, the realization of the importance of twistors in TGD, and to the discovery of the weak form of electric-magnetic duality. Twistor approach and the understanding of the Chern-Simons Dirac operator served as a midwife in the process giving rise to the birth of the idea that all particles at fundamental level are massless and that both ordinary elementary particles and string like objects emerge from them. Even more, one can interpret virtual particles as being composed of these massless on mass shell particles assignable to wormhole throats so that four-momentum conservation poses extremely powerful constraints on loop integrals and makes them manifestly finite.

The weak form of electric-magnetic duality led to the realization that elementary particles correspond to bound states of two wormhole throats with opposite Kähler magnetic charges with second throat carrying weak isospin compensating that of the fermion state at second wormhole throat. Both fermions and bosons correspond to wormhole contacts: in the case of fermions topological condensation generates the second wormhole throat. This means that altogether four wormhole throats are involved with both fermions, gauge bosons, and gravitons (for gravitons this is unavoidable in any case). For p-adic thermodynamics the mathematical counterpart of string corresponds to a wormhole contact with size of order CP_2 size with the role of its ends played by wormhole throats at which the signature of the induced 4-metric changes. The key observation is that for massless states the throats of spin 1 particle must have opposite three-momenta so that gauge bosons are necessarily massive, even photon and other particles usually regarded as massless must have small mass which in turn cancels infrared divergences and give hopes about exact Yangian symmetry generalizing that of $\mathcal{N} = 4$ SYM. Besides this there is weak "stringy" contribution to the mass assignable to the magnetic flux tubes connecting the two wormhole throats at the two space-time sheets.

[1] Correspondence: E-mail:matpitka@luukku.com

Contents

1 Introduction

This article represenst the most recent view about particle massivation in TGD framework. This topic is necessarily quite extended since many several notions and new mathematics is involved. Therefore the calculation of particle masses involves five chapters ([2, 8, 12, 9] of [11]. In the following my goal is to provide an up-to-date summary whereas the chapters are unavoidably a story about the evolution of ideas.

The identification of the spectrum of light particles reduces to two tasks: the construction of massless states and the identification of the states which remain light in p-adic thermodynamics. The latter task is relatively straightforward. The thorough understanding of the massless spectrum requires however a real understanding of quantum TGD. It would be also highly desirable to understand why p-adic thermodynamics combined with p-adic length scale hypothesis works. A lot of progress has taken place in these respects during last years.

Zero energy ontology providing a detailed geometric view about bosons and fermions, the generalization of S-matrix to what I call M-matrix, the notion of finite measurement resolution characterized in terms of inclusions of von Neumann algebras, the derivation of p-adic coupling constant evolution and p-adic length scale hypothesis from the first principles, the realization that the counterpart of Higgs mechanism involves generalized eigenvalues of the modified Dirac operator: these are represent important steps of progress during last years with a direct relevance for the understanding of particle spectrum and massivation although the predictions of p-adic thermodynamics are not affected.

During 2010 a further progress took place as I wrote articles about TGD to Prespacetime journal [22, 23, 26, 27, 24, 21, 25, 28]. These steps of progress relate closely to zero energy ontology, bosonic emergence, the realization of the importance of twistors in TGD, and to the discovery of the weak form of electric-magnetic duality. Twistor approach and the understanding of the Chern-Simons Dirac operator served as a midwife in the process giving rise to the birth of the idea that all particles at fundamental level are massless and that both ordinary elementary particles and string like objects emerge from them. Even more, one can interpret virtual particles as being composed of these massless on mass shell particles assignable to wormhole throats so that four-momentum conservation poses extremely powerful constraints on loop integrals and makes them manifestly finite.

The weak form of electric-magnetic duality led to the realization that elementary particles correspond to bound states of two wormhole throats with opposite Kähler magnetic charges with second throat carrying weak isospin compensating that of the fermion state at second wormhole throat. Both fermions and bosons correspond to wormhole contacts: in the case of fermions topological condensation generates the second wormhole throat. This means that altogether four wormhole throats are involved with both fermions, gauge bosons, and gravitons (for gravitons this is unavoidable in any case). For p-adic thermodynamics the mathematical counterpart of string corresponds to a wormhole contact with size of order CP_2 size with the role of its ends played by wormhole throats at which the signature of the induced 4-metric changes. The key observation is that for massless states the throats of spin 1 particle must have opposite three-momenta so that gauge bosons are necessarily massive, even photon and other particles usually regarded as massless must have small mass which in turn cancels infrared divergences and give hopes about exact Yangian symmetry generalizing that of $\mathcal{N} = 4$ SYM. Besides this there is weak "stringy" contribution to the mass assignable to the magnetic flux tubes connecting the two wormhole throats at the two space-time sheets.

1.1 Physical states as representations of super-symplectic and Super Kac-Moody algebras

Physical states belong to the representation of super-symplectic algebra and Super Kac-Moody algebra assignable $SO(2) \times SU(3) \times SU(2)_{rot} \times U(2)_{ew}$ associated with the 2-D surfaces X^2 defined by the intersections of 3-D light like causal determinants with $\delta M^4_\pm \times CP_2$. These 2-surfaces have interpretation as partons.

It has taken considerable effort to understand the relationship between super-symplectic and super Kac-Moody algebras and there are still many uncertainties involved. What looks like the most plausible option relies on the generalization of a coset construction proposed already for years ago but given up because of the lacking understanding of how SKM and SC algebras could be lifted to the level of imbedding space. The progress in the *Physics as generalized number theory* program provided finally a justification for the coset construction.

1. Assume a generalization of the coset construction in the sense that the differences of super Kac-Moody Virasoro generators (SKMV) and super-symplectic Virasoro generators (SSV) annihilate the physical states. The interpretation is in terms of TGD counterpart for Einstein's equations realizing Equivalence Principle. Mass squared is identified as the p-adic thermal expectation value of either $SKMV$ or SSV conformal weight (gravitational or inertial mass) in a superposition of states with $SKMV$ (SSV) conformal weight $n \geq 0$ annihilated by $SKMV - SSV$.

2. Construct first ground states with negative conformal weight annihilated by $SKMV$ and SSV generators G_n, L_n, $n < 0$. Apply to these states generators of tensor factors of Super Viraroso algebras to obtain states with vanishing SSV and $SKMV$ conformal weights. After this construct thermal states as superpositions of states obtained by applying $SKMV$ generators and corresponding SSV generators G_n,L_n, $n > 0$. Assume that these states are annihilated by SSV and $SKMV$ generators G_n, L_n,$n > 0$ and by the differences of all SSV and $SKMV$ generators.

3. Super-symplectic algebra represents a completely new element and in the case of hadrons the non-perturbative contribution to the mass spectrum is easiest to understand in terms of super-symplectic thermal excitations contributing roughly 70 per cent to the p-adic thermal mass of the hadron. It must be however emphasized that by SKMV-SSV duality one can regard these contributions equivalently as SKM or SC contributions.

Yangian algebras associated with the super-conformal algebras and motivated by twistorial approach generalize the super-conformal symmetry and make it multi-local in the sense that generators can act on several partonic 2-surfaces simultaneously. These partonic 2-surfaces generalize the vertices for the external massless particles in twistor Grassmann diagrams [19]. The implications of this symmetry are yet to be deduced but one thing is clear: Yangians are tailor made for the description of massive bound states formed from several partons identified as partonic 2-surfaces. The preliminary discussion of what is involved can be found in [19].

1.2 Particle massivation

Particle massivation can be regarded as a generation of thermal conformal weight identified as mass squared and due to a thermal mixing of a state with vanishing conformal weight with those having higher conformal weights. The observed mass squared is not p-adic thermal expectation of mass squared but that of conformal weight so that there are no problems with Lorentz invariance.

One can imagine several microscopic mechanisms of massivation. The following proposal is the winner in the fight for survival between several competing scenarios.

1. The original observation was that the pieces of CP_2 type vacuum extremals representing elementary particles have random light-like curve as an M^4 projection so that the average motion correspond to that of massive particle. Light-like randomness gives rise to classical Virasoro conditions. This picture generalizes since the basic dynamical objects are light-like but otherwise random 3-surfaces. The identification of elementary particles developed in three steps.

(a) Fermions are identified as light-like 3-surfaces at which the signature of induced metric of deformed CP_2 type extremals changes from Euclidian to the Minkowskian signature of the background space-time sheet. Gauge bosons and Higgs correspond to wormhole contacts with light-like throats carrying fermion and antifermion quantum numbers. Gravitons correspond to pairs of wormhole contacts bound to string like object by the fluxes connecting the wormhole contacts. The randomness of the light-like 3-surfaces and associated super-conformal symmetries justify the use of thermodynamics and the question remains why this thermodynamics can be taken to be p-adic. The proposed identification of bosons means enormous simplification in thermodynamical description since all calculations reduced to the calculations to fermion level. This picture generalizes to include super-symmetry. The fermionic oscillator operators associated with the partonic 2-surfaces act as generators of badly broken SUSY and right-handed neutrino gives to the not so badly broken $\mathcal{N} = 1$ SUSY consistent with empirical facts.

(b) The next step was to realize that the topological condensation of fermion generates second wormhole throat which carries momentum but no fermionic quantum numbers. This is also needed to the massivation by p-adic thermodynamics applied to the analogs of string like objects defined by wormhole throats with throats taking the role of string ends. p-Adic thermodynamics did not however allow a satisfactory understanding of the gauge bosons masses and it was clear that Higgsy contribution should be present and dominate for gauge bosons. Gauge bosons should also somehow obtain their longitudinal polarizations and here Higgs like particles indeed predicted by the basic picture suggests itself strongly.

(c) A further step was the discovery of the weak form of electric-magnetic duality, which led to the realization that wormhole throats possess Kähler magnetic charge so that a wormole throat with opposite magnetic charge is needed to compensate this charge. This wormhole throat can also compensate the weak isospin of the second wormhole throat so that weak confinement and massivation results. In the case of quarks magnetic confinement might take place in hadronic rather than weak length scale. Second crucial observation was that gauge bosons are necessarily massive since the light-like momenta at two throats must correspond to opposite three-momenta so that no Higgs potential is needed. This leads to a picture in which gauge bosons eat the Higgs scalars and also photon, gluons, and gravitons develop small mass.

2. The fundamental parton level description of TGD is based on almost topological QFT for light-like 3-surfaces. Dynamics is constrained by the requirement that CP_2 projection is for extremals of Chern-Simons action 2-dimensional and for off-shell states light-likeness is the only constraint. As a matter fact, the basic theory relies on the modified Dirac action associated with Chern-Simons action and Kähler action in the sense that the generalizes eigenmodes of Chern-Simons Dirac operator correspond to the zero modes of Kähler action localized to the light-like 3-surfaces representing partons. In this manner the data about the dynamics of Kähler action is feeded to the eigenvalue spectrum. Eigenvalues are interpreted as square roots of ground state conformal weights.

3. The symmetries respecting light-likeness property give rise to Kac-Moody type algebra and super-symplectic symmetries emerge also naturally as well as $\mathcal{N} = 4$ character of super-conformal invariance. The coset construction for super-symplectic Virasoro algebra and Super Kac-Moody algebra identified in physical sense as sub-algebra of former implies that the four-momenta assignable to the two algebras are identical. The interpretation is in terms of the identity of gravitational inertial masses and generalization of Equivalence Principle.

4. Instead of energy, the Super Kac-Moody Virasoro (or equivalently super-symplectic) generator L_0 (essentially mass squared) is thermalized in p-adic thermodynamics (and also in its real version assuming it exists). The fact that mass squared is thermal expectation of conformal weight guarantees Lorentz invariance. That mass squared, rather than energy, is a fundamental quantity at CP_2 length scale is also suggested by a simple dimensional argument (Planck mass squared is proportional to $\hbar$ so that it should correspond to a generator of some Lie-algebra (Virasoro generator L_0!)).

5. By Equivalence Principle the thermal average of mass squared can be calculated either in terms of thermodynamics for either super-symplectic of Super Kac-Moody Virasoro algebra and p-adic thermodynamics is consistent with conformal invariance.

6. There is also a modular contribution to the mass squared, which can be estimated using elementary particle vacuum functionals in the conformal modular degrees of freedom of the partonic 2-surface. It dominates for higher genus partonic 2-surfaces. For bosons both Virasoro and modular contributions seem to be negligible and could be due to the smallness of the p-adic temperature.

7. A long standing problem has been whether coupling to Higgs boson is needed to explain gauge boson masses via a generation of Higgs vacuum expectation having possibly interpretation in terms of a coherent state. Before the detailed model for elementary particles in terms of pairs of wormhole contacts at the ends of flux tubes the picture about the sitution was as follows. From the beginning it was clear that is that ground state conformal weight must be negative. Then it became clear that the ground state conformal weight need not be a negative integer. The deviation Δh of the total ground state conformal weight from negative integer gives rise to Higgs type contribution to the thermal mass squared and dominates in case of gauge bosons for which p-adic temperature is small. In the case of fermions this contribution to the mass squared is small. The possible Higgs vacuum expectation makes sense only at QFT limit and would be naturally proportional to Δh so that the coupling to Higgs would only apparently cause gauge boson massivation. It is natural to relate Δh to the generalized eigenvalues of Chern-Simons Dirac operator.

8. A natural identification of the non-integer contribution to the conformal weight is as Higgsy and stringy contributions to the vacuum conformal weight. In twistor approach the generalized eigenvalues of Chern-Simons Dirac operator for external particles indeed correspond to light-like momenta and when the three-momenta are opposite this gives rise to non-vanishing mass. Higgs is necessary to give longitudinal polarizations for gauge bosons and also gauge bosons usually regarded as exactly massless particles would naturally receive small mass in this manner so that Higgs would disappear completely from the spectrum. The theoretetical motivation for small mass would be exact Yangian symmetry. Higgs vacuum expectation assignable to coherent state of Higgs bosons is not needed to explain the boson masses.

An important question concerns the justification of p-adic thermodynamics.

1. The underlying philosophy is that real number based TGD can be algebraically continued to various p-adic number fields. This gives justification for the use of p-adic thermodynamics although the mapping of p-adic thermal expectations to real counterparts is not completely unique. The physical justification for p-adic thermodynamics is effective p-adic topology characterizing the 3-surface: this is the case if real variant of light-like 3-surface has large number of common algebraic points with its p-adic counterpart obeying same algebraic equations but in different number field. In fact, there is a theorem stating that for rational surfaces the number of rational points is finite and rational (more generally algebraic points) would naturally define the notion of number theoretic braid essential for the realization of number theoretic universality.

2. The most natural option is that the descriptions in terms of both real and p-adic thermodynamics make sense and are consistent. This option indeed makes if the number of generalized eigen modes of modified Dirac operator is finite. This is indeed the case if one accepts periodic boundary conditions for the Chern-Simons Dirac operator. In fact, the solutions are localized at the strands of braids [5]. This makes sense because the theory has hydrodynamic interpretation [5]. This reduces $\mathcal{N} = \infty$ to finite SUSY and realizes finite measurement resolution as an inherent property of dynamics.

   The finite number of fermionic oscillator operators implies an effective cutoff in the number conformal weights so that conformal algebras reduce to finite-dimensional algebras. The first guess would be that integer label for oscillator operators becomes a number in finite field for some prime. This means that one can calculate mass squared also by using real thermodynamics but the consistency with p-adic thermodynamics gives extremely strong number theoretical constraints on mass scale. This consistency condition allows also to solve the problem how to map a negative ground state conformal weight to its p-adic counterpart. Negative conformal weight is divided into a negative half odd integer part plus positive part Δh, and negative part corresponds as such to p-adic integer whereas positive part is mapped to p-adic number by canonical identification.

p-Adic thermodynamics is what gives to this approach its predictive power.

1. p-Adic temperature is quantized by purely number theoretical constraints (Boltzmann weight $exp(-E/kT)$ is replaced with p^{L_0/T_p}, $1/T_p$ integer) and fermions correspond to $T_p = 1$ whereas $T_p = 1/n$, $n > 1$, seems to be the only reasonable choice for gauge bosons.

2. p-Adic thermodynamics forces to conclude that CP_2 radius is essentially the p-adic length scale $R \sim L$ and thus of order $R \simeq 10^{3.5}\sqrt{\hbar G}$ and therefore roughly $10^{3.5}$ times larger than the naive guess. Hence p-adic thermodynamics describes the mixing of states with vanishing conformal weights with their Super Kac-Moody Virasoro excitations having masses of order $10^{-3.5}$ Planck mass.

1.3 What next?

The successes of p-adic mass calculations are basically due to the power of super-conformal symmetries and of number theory. One cannot deny that the description of the Higgsy aspects of massivation and of hadrons involves phenomenological elements. There are however excellent hopes that it might be possible some day to calculate everything from first principles. The non-local Yangian symmetry generalizing the super-conformal algebras suggests itself strongly as a fundamental symmetry of quantum TGD. The generalized of the Yangian symmetry replaces points with partonic 2-surfaces being multi-local with respect to them, and leads to general formulas for multi-local operators representing four-momenta and other conserved charges of composite states. In TGD framework even elementary particles involve two wormhole contacts having each two wormhole throats identified as the fundamental partonic entities. Therefore Yangian approach would naturally define the first principle approach to the understanding of masses of elementary particles and their bound states (say hadrons). The power of this extended symmetry might be enough to deduce universal mass formulas. One of the future challenges would therefore be the mathematical and physical understanding of Yangian symmetry. This would however require the contributions of professional mathematicians.

2 Identification of elementary particles

2.1 Partons as wormhole throats and particles as bound states of wormhole contacts

The assumption that partonic 2-surfaces correspond to representations of Super Virasoro algebra has been an unchallenged assumption of the p-adic mass calculations for a long time although one might argue that these objects do not possess stringy characteristics, in particular they do not possess two ends. The progress in the understanding of the modified Dirac equation and the introduction of the weak form of electric magnetic duality [5] however forces to modify the picture about the origin of the string mass spectrum.

1. The weak form of electric-magnetic duality, the basic facts about modified Dirac equation and the proposed twistorialization of quantum TGD [19] force to conclude that both strings and bosons and their super-counterparts emerge from massless fermions moving collinearly at partonic two-surfaces. Stringy mass spectrum is consistent with this only if p-adic thermodynamics describes wormhole contacts as analogs of stringy objects having quantum numbers at the throats playing the role of string ends. For instance, the three-momenta of massless wormhole throats could be in opposite direction so that wormhole contact would become massive. The fundamental string like objects would therefore correspond to the wormhole contacts with size scale of order CP_2 length. Already these objects must have a correct correlation between color and electroweak quantum numbers. The colored super-generators taking care that anomalous color is compensated can be assigned with purely bosonic quanta associated with the wormhole throats which carry no fermion number.

2. Second modification comes from the necessity to assume weak confinement in the sense that each wormhole throat carrying fermionic numbers is accompanied by a second wormhole throat carrying neutrino pair cancelling the net weak isospin so that only electromagnetic charge remains

unscreened. This screening must take place in weak length scale so that ordinary elementar particles are predicted to be string like objects. This string tension has however nothing to do with the fundamental string tension responsible for the mass spectrum. This picture is forced also by the fact that fermionic wormhole throats necessarily carry Kähler magnetic charge [5] so that in the case of leptons the second wormhole throat must carry a compensating Kähler magnetic charge. In the case of quarks one can consider the possibility that magnetic charges are not neutralized completely in weak scale and that the compensation occurs in QCD length scale so that Kähler magnetic confinement would accompany color confinement. This means color magnetic confinement since classical color gauge fields are proportional to induced Kähler field.

These modifications do not seem to appreciably affect the results of calculations, which depend only on the number of tensor factors in super Virasoro representation, they are not taken explicitly into account in the calculations. The predictions of the general theory are consistent with the earliest mass calculations, and the earlier ad hoc parameters disappear. In particular, optimal lowest order predictions for the charged lepton masses are obtained and photon, gluon and graviton appear as essentially massless particles. What is new is the possibility to describe the massivation of gauge bosons by including the contribution from the string tension of weak string like objects: weak boson masses have indeed been the trouble makers and have forced to conclude that Higgs expectation might be needed unless some other mechanism contributes to the conformal vacuum weight of the ground state.

2.2 Family replication phenomenon topologically

One of the basic ideas of TGD approach has been genus-generation correspondence: boundary components of the 3-surface should be carriers of elementary particle numbers and the observed particle families should correspond to various boundary topologies.

With the advent of zero energy ontology this picture changed somewhat. It is the wormhole throats identified as light-like 3-surfaces at with the induced metric of the space-time surface changes its signature from Minkowskian to Euclidian, which correspond to the light-like orbits of partonic 2-surfaces. One cannot of course exclude the possibility that also boundary components could allow to satisfy boundary conditions without assuming vacuum extremal property of nearby space-time surface. The intersections of the wormhole throats with the light-like boundaries of causal diamonds (CDs) identified as intersections of future and past directed light cones ($CD \times CP_2$ is actually in question but I will speak about CDs) define special partonic 2-surfaces and it is the moduli of these partonic 2-surfaces which appear in the elementary particle vacuum functionals naturally.

The first modification of the original simple picture comes from the identification of physical particles as bound states of pairs of wormhole contacts and from the assumption that for generalized Feynman diagrams stringy trouser vertices are replaced with vertices at which the ends of light-like wormhole throats meet. In this picture the interpretation of the analog of trouser vertex is in terms of propagation of same particle along two different paths. This interpretation is mathematically natural since vertices correspond to 2-manifolds rather than singular 2-manifolds which are just splitting to two disjoint components. Second complication comes from the weak form of electric-magnetic duality forcing to identify physical particles as weak strings with magnetic monopoles at their ends and one should understand also the possible complications caused by this generalization.

These modifications force to consider several options concerning the identification of light fermions and bosons and one can end up with a unique identification only by making some assumptions. Masslessness of all wormhole throats- also those appearing in internal lines- and dynamical $SU(3)$ symmetry for particle generations are attractive general enough assumptions of this kind. This means that bosons and their super-partners correspond to wormhole contacts with fermion and antifermion at the throats of the contact. Free fermions and their superpartners could correspond to CP_2 type vacuum extremals with single wormhole throat. It turns however that dynamical $SU(3)$ symmetry forces to identify massive (and possibly topologically condensed) fermions as (g, g) type wormhole contacts.

2.2.1 Do free fermions correspond to single wormhole throat or (g, g) wormhole?

The original interpretation of genus-generation correspondence was that free fermions correspond to wormhole throats characterized by genus. The idea of $SU(3)$ as a dynamical symmetry suggested that gauge bosons correspond to octet and singlet representations of $SU(3)$. The further idea that all lines of generalized Feynman diagrams are massless poses a strong additional constraint and it is not clear whether this proposal as such survives.

1. Twistorial program assumes that fundamental objects are massless wormhole throats carrying collinearly moving many-fermion states and also bosonic excitations generated by super-symplectic algebra. In the following consideration only purely bosonic and single fermion throats are considered since they are the basic building blocks of physical particles. The reason is that propagators for high excitations behave like p^{-n}, n the number of fermions associated with the wormhole throat. Therefore single throat allows only spins 0,1/2,1 as elementary particles in the usual sense of the word.

2. The identification of massive fermions (as opposed to free massless fermions) as wormhole contacts follows if one requires that fundamental building blocks are massless since at least two massless throats are required to have a massive state. Therefore the conformal excitations with CP_2 mass scale should be assignable to wormhole contacts also in the case of fermions. As already noticed this is not the end of the story: weak strings are required by the weak form of electric-magnetic duality.

3. If free fermions corresponding to single wormhole throat, topological condensation is an essential element of the formation of stringy states. The topological condensation of fermions by topological sum (fermionic CP_2 type vacuum extremal touches another space-time sheet) suggest $(g, 0)$ wormhole contact. Note however that the identification of wormhole throat is as 3-surface at which the signature of the induced metric changes so that this conclusion might be wrong. One can indeed consider also the possibility of (g, g) pairs as an outcome of topological conensation. This is suggested also by the idea that wormhole throats are analogous to string like objects and only this option turns out to be consistent with the BFF vertex based on the requirement of dynamical $SU(3)$ symmetry to be discussed later. The structure of reaction vertices makes it possible to interpret (g, g) pairs as $SU(3)$ triplet. If bosons are obtained as fusion of fermionic and antifermionic throats (touching of corresponding CP_2 type vacuum extremals) they correspond naturally to (g_1, g_2) pairs.

4. p-Adic mass calculations distinguish between fermions and bosons and the identification of fermions and bosons should be consistent with this difference. The maximal p-adic temperature $T = 1$ for fermions could relate to the weakness of the interaction of the fermionic wormhole throat with the wormhole throat resulting in topological condensation. This wormhole throat would however carry momentum and 3-momentum would in general be non-parallel to that of the fermion, most naturally in the opposite direction.

   p-Adic mass calculations suggest strongly that for bosons p-adic temperature $T = 1/n$, $n > 1$, so that thermodynamical contribution to the mass squared is negligible. The low p-adic temperature could be due to the strong interaction between fermionic and antifermionic wormhole throat leading to the "freezing" of the conformal degrees of freedom related to the relative motion of wormhole throats.

5. The weak form of electric-magnetic duality forces second wormhole throat with opposite magnetic charge and the light-like momenta could sum up to massive momentum. In this case string tension corresponds to electroweak length scale. Therefore p-adic thermodynamics must be assigned to wormhole contacts and these appear as basic units connected by Kähler magnetic flux tube pairs at the two space-time sheets involved. Weak stringy degrees of freedom are however expected to give additional contribution to the mass, perhaps by modifying the ground state conformal weight.

2.2.2 Dynamical $SU(3)$ fixes the identification of fermions and bosons and fundamental interaction vertices

For 3 light fermion families $SU(3)$ suggests itself as a dynamical symmetry with fermions in fundamental $N = 3$-dimensional representation and $N \times N = 9$ bosons in the adjoint representation and singlet representation. The known gauge bosons have same couplings to fermionic families so that they must correspond to the singlet representation. The first challenge is to understand whether it is possible to have dynamical $SU(3)$ at the level of fundamental reaction vertices.

This is a highly non-trivial constraint. For instance, the vertices in which n wormhole throats with same (g_1, g_2) glued along the ends of lines are not consistent with this symmetry. The splitting of the fermionic worm-hole contacts before the proper vertices for throats might however allow the realization of dynamical $SU(3)$. The condition of $SU(3)$ symmetry combined with the requirement that virtual lines resulting also in the splitting of wormhole contacts are always massless, leads to the conclusion that massive fermions correspond to (g, g) type wormhole contacts transforming naturally like $SU(3)$ triplet. This picture conformsl with the identification of free fermions as throats but not with the naive expectation that their topological condensation gives rise to $(g, 0)$ wormhole contact.

The argument leading to these conclusions runs as follows.

1. The question is what basic reaction vertices are allowed by dynamical $SU(3)$ symmetry. FFB vertices are in principle all that is needed and they should obey the dynamical symmetry. The meeting of entire wormhole contacts along their ends is certainly not possible. The splitting of fermionic wormhole contacts before the vertices might be however consistent with $SU(3)$ symmetry. This would give two a pair of 3-vertices at which three wormhole lines meet along partonic 2-surfaces (rather than along 3-D wormhole contacts).

2. Note first that crossing gives all possible reaction vertices of this kind from $F(g_1)\overline{F}(g_2) \rightarrow B(g_1, g_2)$ annihilation vertex, which is relatively easy to visualize. In this reaction $F(g_1)$ and $\overline{F}(g_2)$ wormhole contacts split first. If one requires that all wormhole throats involved are massless, the two wormhole throats resulting in splitting and carrying no fermion number must carry light-like momentum so that they cannot just disappear. The ends of the wormhole throats of the boson must glued together with the end of the fermionic wormhole throat and its companion generated in the splitting of the wormhole. This means that fermionic wormhole first splits and the resulting throats meet at the partonic 2-surface.

   his requires that topologically condensed fermions correspond to (g, g) pairs rather than $(g, 0)$ pairs. The reaction mechanism allows the interpretation of (g, g) pairs as a triplet of dynamical $SU(3)$. The fundamental vertices would be just the splitting of wormhole contact and 3-vertices for throats since $SU(3)$ symmetry would exclude more complex reaction vertices such as n-boson vertices corresponding the gluing of n wormhole contact lines along their 3-dimensional ends. The couplings of singlet representation for bosons would have same coupling to all fermion families so that the basic experimental constraint would be satisfied.

3. Both fermions and bosons cannot correspond to octet and singlet of $SU(3)$. In this case reaction vertices should correspond algebraically to the multiplication of matrix elements e_{ij}: $e_{ij}e_{kl} = \delta_{jk}e_{il}$ allowing for instance $F(g_1, g_2) + \overline{F}(g_2, g_3) \rightarrow B(g_1, g_3)$. Neither the fusion of entire wormhole contacts along their ends nor the splitting of wormhole throats before the fusion of partonic 2-surfaces allows this kind of vertices so that BFF vertex is the only possible one. Also the construction of QFT limit starting from bosonic emergence led to the formulation of perturbation theory in terms of Dirac action allowing only BFF vertex as fundamental vertex [6].

4. Weak electric-magnetic duality brings in an additional complication. $SU(3)$ symmetry poses also now strong constraints and it would seem that the reactions must involve copies of basic BFF vertices for the pairs of ends of weak strings. The string ends with the same Kähler magnetic charge should meet at the vertex and give rise to BFF vertices. For instance, $F\overline{F}B$ annihilation vertex would in this manner give rise to the analog of stringy diagram in which strings join along ends since two string ends disappear in the process.

If one accepts this picture the remaining question is why the number of genera is just three. Could this relate to the fact that $g \leq 2$ Riemann surfaces are always hyper-elliptic (have global Z_2 conformal symmetry) unlike $g > 2$ surfaces? Why the complete bosonic de-localization of the light families should be restricted inside the hyper-elliptic sector? Does the Z_2 conformal symmetry make these states light and make possible delocalization and dynamical $SU(3)$ symmetry? Could it be that for $g > 2$ elementary particle vacuum functionals vanish for hyper-elliptic surfaces? If this the case and if the time evolution for partonic 2-surfaces changing g commutes with Z_2 symmetry then the vacuum functionals localized to $g \leq 2$ surfaces do not disperse to $g > 2$ sectors.

2.2.3 The notion of elementary particle vacuum functional

Obviously one must know something about the dependence of the elementary particle state functionals on the geometric properties of the boundary component and in the sequel an attempt to construct what might be called elementary particle vacuum functionals, is made.

The basic assumptions underlying the construction are the following ones:

1. Elementary particle vacuum functionals depend on the geometric properties of the two-surface X^2 representing elementary particle.

2. Vacuum functionals possess extended Diff invariance: all 2-surfaces on the orbit of the 2-surface X^2 correspond to the same value of the vacuum functional. This condition is satisfied if vacuum functionals have as their argument, not X^2 as such, but some 2- surface Y^2 belonging to the unique orbit of X^2 (determined by the principle selecting preferred extremal of the Kähler action as a generalized Bohr orbit [7]) and determined in $Diff^3$ invariant manner.

3. Zero energy ontology allows to select uniquely the partonic two surface as the intersection of the wormhole throat at which the signature of the induced 4-metric changes with either the upper or lower boundary of $CD \times CP_2$. This is essential since otherwise one one could not specify the vacuum functional uniquely.

4. Vacuum functionals possess conformal invariance and therefore for a given genus depend on a finite number of variables specifying the conformal equivalence class of Y^2.

5. Vacuum functionals satisfy the cluster decomposition property: when the surface Y^2 degenerates to a union of two disjoint surfaces (particle decay in string model inspired picture), vacuum functional decomposes into a product of the vacuum functionals associated with disjoint surfaces.

6. Elementary particle vacuum functionals are stable against the decay $g \to g_1 + g_2$ and one particle decay $g \to g - 1$. This process corresponds to genuine particle decay only for stringy diagrams. For generalized Feynman diagrams the interpretation is in terms of propagation along two different paths simultaneously.

In [2] the construction of elementary particle vacuum functionals is described in more detail. This requires some basic concepts related to the description of the space of the conformal equivalence classes of Riemann surfaces and the concept of hyper-ellipticity. Since theta functions will play a central role in the construction of the vacuum functionals, also their basic properties are needed. Also possible explanations for the experimental absence of the higher fermion families are considered.

2.3 Basic facts about Riemann surfaces

In the following some basic aspects about Riemann surfaces will be summarized. The basic topological concepts, in particular the concept of the mapping class group, are introduced, and the Teichmueller parameters are defined as conformal invariants of the Riemann surface, which in fact specify the conformal equivalence class of the Riemann surface completely.

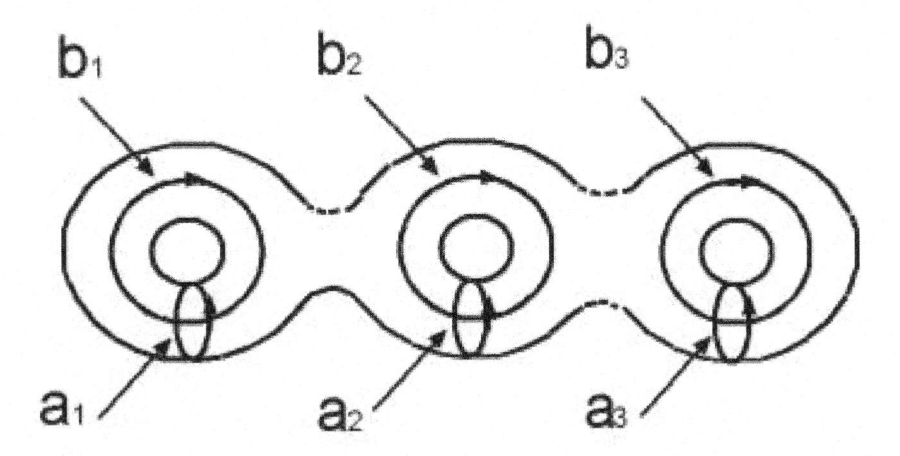

Figure 1: Definition of the canonical homology basis

2.3.1 Mapping class group

The first homology group $H_1(X^2)$ of a Riemann surface of genus g contains $2g$ generators [30, 31, 32] : this is easy to understand geometrically since each handle contributes two homology generators. The so called canonical homology basis can be identified as in Fig. 2.3.1.

One can define the so called intersection number $J(a,b)$ for two elements a and b of the homology group as the number of intersection points for the curves a and b counting the orientation. Since $J(a,b)$ depends on the homology classes of a and b only, it defines an antisymmetric quadratic form in $H_1(X^2)$. In the canonical homology basis the non-vanishing elements of the intersection matrix are:

$$J(a_i,b_j) \quad = \quad -J(b_j,a_i) = \delta_{i,j} \ . \tag{2.1}$$

J clearly defines symplectic structure in the homology group.

The dual to the canonical homology basis consists of the harmonic one-forms $\alpha_i, \beta_i, i = 1,..,g$ on X^2. These 1-forms satisfy the defining conditions

$$\begin{array}{ll} \int_{a_i} \alpha_j = \delta_{i,j} & \int_{b_i} \alpha_j = 0 \ , \\ \int_{a_i} \beta_j = 0 & \int_{b_i} \beta_j = \delta_{i,j} \ . \end{array} \tag{2.2}$$

The following identity helps to understand the basic properties of the Teichmueller parameters

$$\int_{X^2} \theta \wedge \eta \quad = \quad \sum_{i=1,..,g} [\int_{a_i} \theta \int_{b_i} \eta - \int_{b_i} \theta \int_{a_i} \eta] \ . \tag{2.3}$$

The existence of topologically nontrivial diffeomorphisms, when X^2 has genus $g > 0$, plays an important role in the sequel. Denoting by $Diff$ the group of the diffeomorphisms of X^2 and by $Diff_0$ the normal subgroup of the diffeomorphisms homotopic to identity, one can define the mapping class group M as the coset group

$$M = Diff/Diff_0 . \tag{2.4}$$

The generators of M are so called Dehn twists along closed curves a of X^2. Dehn twist is defined by excising a small tubular neighborhood of a, twisting one boundary of the resulting tube by 2π and gluing the tube back into the surface: see Fig. 2.3.1.

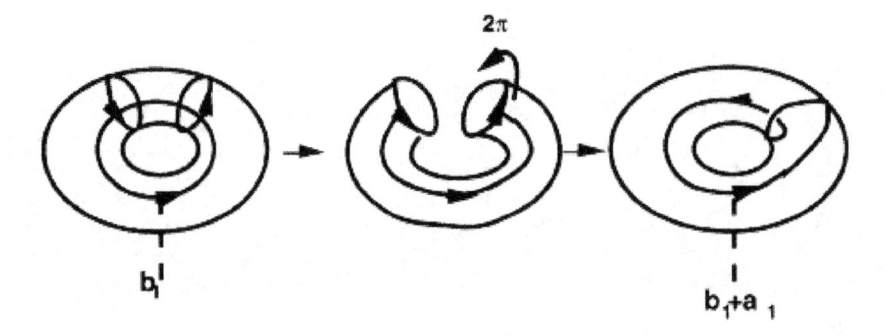

Figure 2: Definition of the Dehn twist

It can be shown that a minimal set of generators is defined by the following curves

$$a_1, b_1, a_1^{-1}a_2^{-1}, a_2, b_2, a_2^{-1}a_3^{-11}, ..., a_g, b_g . \tag{2.5}$$

The action of these transformations in the homology group can be regarded as a symplectic linear transformation preserving the symplectic form defined by the intersection matrix. Therefore the matrix representing the action of $Diff$ on $H_1(X^2)$ is $2g \times 2g$ matrix M with integer entries leaving J invariant: $MJM^T = J$. Mapping class group is often referred also as a symplectic modular group and denoted by $Sp(2g, Z)$. The matrix representing the action of M in the canonical homology basis decomposes into four $g \times g$ blocks A, B, C and D

$$M = \begin{pmatrix} A & B \\ C & D \end{pmatrix} , \tag{2.6}$$

where A and D operate in the subspaces spanned by the homology generators a_i and b_i respectively and C and D map these spaces to each other. The notation $D = [A, B; C, D]$ will be used in the sequel: in this notation the representation of the symplectic form J is $J = [0, 1; -1, 0]$.

2.3.2 Teichmueller parameters

The induced metric on the two-surface X^2 defines a unique complex structure. Locally the metric can always be written in the form

$$ds^2 = e^{2\phi} dz d\bar{z} . \tag{2.7}$$

where z is local complex coordinate. When one covers X^2 by coordinate patches, where the line element has the above described form, the transition functions between coordinate patches are holomorphic and therefore define a complex structure.

The conformal transformations ξ of X^2 are defined as the transformations leaving invariant the angles between the vectors of X^2 tangent space invariant: the angle between the vectors X and Y at point x is same as the angle between the images of the vectors under Jacobian map at the image point $\xi(x)$. These transformations need not be globally defined and in each coordinate patch they correspond to holomorphic (anti-holomorphic) mappings as is clear from the diagonal form of the metric in the local complex coordinates. A distinction should be made between local conformal transformations and globally defined conformal transformations, which will be referred to as conformal symmetries: for instance, for hyper-elliptic surfaces the group of the conformal symmetries contains two-element group Z_2.

Using the complex structure one can decompose one-forms to linear combinations of one-forms of type $(1,0)$ $(f(z,\bar{z})dz)$ and $(0,1)$ $(f(z,\bar{z})d\bar{z})$. $(1,0)$ form ω is holomorphic if the function f is holomorphic: $\omega = f(z)dz$ on each coordinate patch.

There are g independent holomorphic one forms ω_i known also as Abelian differentials of the first kind [30, 31, 32] and one can fix their normalization by the condition

$$\int_{a_i} \omega_j = \delta_{ij} \ . \tag{2.8}$$

This condition completely specifies ω_i.

Teichmueller parameters Ω_{ij} are defined as the values of the forms ω_i for the homology generators b_j

$$\Omega_{ij} = \int_{b_j} \omega_i \ . \tag{2.9}$$

The basic properties of Teichmueller parameters are the following:
i) The $g \times g$ matrix Ω is symmetric: this is seen by applying the formula (2.3) for $\theta = \omega_i$ and $\eta = \omega_j$.
ii) The imaginary part of Ω is positive: $Im(\Omega) > 0$. This is seen by the application of the same formula for $\theta = \eta$. The space of the matrices satisfying these conditions is known as Siegel upper half plane.
iii) The space of Teichmueller parameters can be regarded as a coset space $Sp(2g,R)/U(g)$ [32] : the action of $Sp(2g,R)$ is of the same form as the action of $Sp(2g,Z)$ and $U(g) \subset Sp(2g,R)$ is the isotropy group of a given point of Teichmueller space.
iv) Teichmueller parameters are conformal invariants as is clear from the holomorphy of the defining one-forms.
v) Teichmueller parameters specify completely the conformal structure of Riemann surface [31].

Although Teichmueller parameters fix the conformal structure of the 2-surface completely, they are not in one-to-one correspondence with the conformal equivalence classes of the two-surfaces:
i) The dimension for the space of the conformal equivalence classes is $D = 3g - 3$, when $g > 1$ and smaller than the dimension of Teichmueller space given by $d = (g \times g + g)/2$ for $g > 3$: all Teichmueller matrices do not correspond to a Riemann surface. In TGD approach this does not produce any problems as will be found later.
ii) The action of the topologically nontrivial diffeomorphisms on Teichmueller parameters is nontrivial and can be deduced from the action of the diffeomorphisms on the homology ($Sp(2g,Z)$ transformation) and from the defining condition $\int_{a_i} \omega_j = \delta_{i,j}$: diffeomorphisms correspond to elements $[A,B;C,D]$ of $Sp(2g,Z)$ and act as generalized Möbius transformations

$$\Omega \rightarrow (A\Omega + B)(C\Omega + D)^{-1} \ . \tag{2.10}$$

All Teichmueller parameters related by $Sp(2g,Z)$ transformations correspond to the same Riemann surface.
iii) The definition of the Teichmueller parameters is not unique since the definition of the canonical homology basis involves an arbitrary numbering of the homology basis. The permutation S of the handles is represented by same $g \times g$ orthogonal matrix both in the basis $\{a_i\}$ and $\{b_i\}$ and induces a similarity transformation in the space of the Teichmueller parameters

$$\Omega \rightarrow S\Omega S^{-1} . \quad (2.11)$$

Clearly, the Teichmueller matrices related by a similarity transformations correspond to the same conformal equivalence class. It is easy to show that handle permutations in fact correspond to $Sp(2g, Z)$ transformations.

2.3.3 Hyper-ellipticity

The motivation for considering hyper-elliptic surfaces comes from the fact, that $g > 2$ elementary particle vacuum functionals turn out to be vanishing for hyper-elliptic surfaces and this in turn will be later used to provide a possible explanation the non-observability of $g > 2$ particles.

Hyper-elliptic surface X can be defined abstractly as two-fold branched cover of the sphere having the group Z_2 as the group of conformal symmetries (see [29, 31, 32]. Thus there exists a map $\pi : X \rightarrow S^2$ so that the inverse image $\pi^{-1}(z)$ for a given point z of S^2 contains two points except at a finite number (say p) of points z_i (branch points) for which the inverse image contains only one point. Z_2 acts as conformal symmetries permuting the two points in $\pi^{-1}(z)$ and branch points are fixed points of the involution.

The concept can be generalized [29] : g-hyper-elliptic surface can be defined as a 2-fold covering of genus g surface with a finite number of branch points. One can consider also p-fold coverings instead of 2-fold coverings: a common feature of these Riemann surfaces is the existence of a discrete group of conformal symmetries.

A concrete representation for the hyper-elliptic surfaces [32] is obtained by studying the surface of C^2 determined by the algebraic equation

$$w^2 - P_n(z) = 0 , \quad (2.12)$$

where w and z are complex variables and $P_n(z)$ is a complex polynomial. One can solve w from the above equation

$$w_{\pm} = \pm\sqrt{P_n(z)} , \quad (2.13)$$

where the square root is determined so that it has a cut along the positive real axis. What happens that w has in general two roots (two-fold covering property), which coincide at the roots z_i of $P_n(z)$ and if n is odd, also at $z = \infty$: these points correspond to branch points of the hyper-elliptic surface and their number r is always even: $r = 2k$. w is discontinuous at the cuts associated with the square root in general joining two roots of $P_n(z)$ or if n is odd, also some root of P_n and the point $z = \infty$. The representation of the hyper-elliptic surface is obtained by identifying the two branches of w along the cuts. From the construction it is clear that the surface obtained in this manner has genus $k - 1$. Also it is clear that Z_2 permutes the different roots $w_{\pm}$ with each other and that $r = 2k$ branch points correspond to fixed points of the involution.

The following facts about the hyper-elliptic surfaces [31, 32] turn out to be important in the sequel:
i) All $g < 3$ surfaces are hyper-elliptic.
ii) $g \geq 3$ hyper-elliptic surfaces are not in general hyper-elliptic and form a set of codimension 2 in the space of the conformal equivalence classes [32].

2.3.4 Theta functions

An extensive and detailed account of the theta functions and their applications can be found in the book of Mumford [32]. Theta functions appear also in the loop calculations of string [57] [30]. In the following the so called Riemann theta function and theta functions with half integer characteristics will be defined as sections (not strictly speaking functions) of the so called Jacobian variety.

For a given Teichmueller matrix Ω, Jacobian variety is defined as the $2g$-dimensional torus obtained by identifying the points z of C^g (vectors with g complex components) under the equivalence

$$z \sim z + \Omega m + n \ , \tag{2.14}$$

where m and n are points of Z^g (vectors with g integer valued components) and Ω acts in Z^g by matrix multiplication.

The definition of Riemann theta function reads as

$$\Theta(z|\Omega) = \sum_n exp(i\pi n \cdot \Omega \cdot n + i2\pi n \cdot z) \ . \tag{2.15}$$

Here $\cdot$ denotes standard inner product in C^g. Theta functions with half integer characteristics are defined in the following manner. Let a and b denote vectors of C^g with half integer components (component either vanishes or equals to $1/2$). Theta function with characteristics $[a, b]$ is defined through the following formula

$$\Theta[a,b](z|\Omega) = \sum_n exp\left[i\pi(n+a)\cdot\Omega\cdot(n+a) + i2\pi(n+a)\cdot(z+b)\right] \ . \tag{2.16}$$

A brief calculation shows that the following identity is satisfied

$$\Theta[a,b](z|\Omega) = exp(i\pi a\cdot\Omega\cdot a + i2\pi a\cdot b)\times\Theta(z+\Omega a+b|\Omega) \tag{2.17}$$

Theta functions are not strictly speaking functions in the Jacobian variety but rather sections in an appropriate bundle as can be seen from the identities

$$\begin{aligned} \Theta[a,b](z+m|\Omega) &= exp(i2\pi a\cdot m)\Theta[a,b](z\Omega) \ , \\ \Theta[a,b](z+\Omega m|\Omega) &= exp(\alpha)\Theta[a,b](z|\Omega) \ , \\ exp(\alpha) &= exp(-i2\pi b\cdot m)exp(-i\pi m\cdot\Omega\cdot m - 2\pi m\cdot z) \ . \end{aligned} \tag{2.18}$$

The number of theta functions is 2^{2g} and same as the number of nonequivalent spinor structures defined on two-surfaces. This is not an accident [30] : theta functions with given characteristics turn out to be in a close relation to the functional determinants associated with the Dirac operators defined on the two-surface. It is useful to divide the theta functions to even and odd theta functions according to whether the inner product $4a \cdot b$ is even or odd integer. The numbers of even and odd theta functions are $2^{g-1}(2^g + 1)$ and $2^{g-1}(2^g - 1)$ respectively.

The values of the theta functions at the origin of the Jacobian variety understood as functions of Teichmueller parameters turn out to be of special interest in the following and the following notation will be used:

$$\Theta[a,b](\Omega) \equiv \Theta[a,b](0|\Omega) \ , \tag{2.19}$$

$\Theta[a,b](\Omega)$ will be referred to as theta functions in the sequel. From the defining properties of odd theta functions it can be found that they are odd functions of z and therefore vanish at the origin of the Jacobian variety so that only even theta functions will be of interest in the sequel.

An important result is that also some *even* theta functions vanish for $g > 2$ hyper-elliptic surfaces : in fact one can characterize $g > 2$ hyper-elliptic surfaces by the vanishing properties of the theta functions [31, 32]. The vanishing property derives from conformal symmetry (Z_2 in the case of hyper-elliptic surfaces) and the vanishing phenomenon is rather general [29] : theta functions tend to vanish for Riemann surfaces possessing discrete conformal symmetries. It is not clear (to the author) whether the presence of a conformal symmetry is in fact equivalent with the vanishing of some theta functions. As already noticed, spinor structures and the theta functions with half integer characteristics are in one-to-one correspondence and the vanishing of theta function with given half integer characteristics is equivalent with the vanishing of the Dirac determinant associated with the corresponding spinor structure or equivalently: with the existence of a zero mode for the Dirac operator [30]. For odd characteristics zero mode exists always: for even characteristics zero modes exist, when the surface is hyper-elliptic or possesses more general conformal symmetries.

2.4 Elementary particle vacuum functionals

The basic assumption is that elementary particle families correspond to various elementary particle vacuum functionals associated with the 2-dimensional boundary components of the 3-surface. These functionals need not be localized to a single boundary topology. Neither need their dependence on the boundary component be local. An important role in the following considerations is played by the fact that the minimization requirement of the Kähler action associates a unique 3-surface to each boundary component, the "Bohr orbit" of the boundary and this surface provides a considerable (and necessarily needed) flexibility in the definition of the elementary particle vacuum functionals. There are several natural constraints to be satisfied by elementary particle vacuum functionals.

2.4.1 Extended Diff invariance and Lorentz invariance

Extended Diff invariance is completely analogous to the extension of 3-dimensional Diff invariance to four-dimensional Diff invariance in the interior of the 3-surface. Vacuum functional must be invariant not only under diffeomorphisms of the boundary component but also under the diffeomorphisms of the 3-dimensional "orbit" Y^3 of the boundary component. In other words: the value of the vacuum functional must be same for any time slice on the orbit the boundary component. This is guaranteed if vacuum functional is functional of some two-surface Y^2 belonging to the orbit and defined in $Diff^3$ invariant manner.

An additional natural requirement is Poincare invariance. In the original formulation of the theory only Lorentz transformations of the light cone were exact symmetries of the theory. In this framework the definition of Y^2 as the intersection of the orbit with the hyperboloid $\sqrt{m_{kl}m^km^l} = a$ is $Diff^3$ and Lorentz invariant.

1. Interaction vertices as generalization of stringy vertices

For stringy diagrams Poincare invariance of conformal equivalence class and general coordinate invariance are far from being a trivial issues. Vertices are now not completely unique since there is an infinite number of singular 3-manifolds which can be identified as vertices even if one assumes space-likeness. One should be able to select a unique singular 3-manifold to fix the conformal equivalence class.

One might hope that Lorentz invariant invariant and general coordinate invariant definition of Y^2 results by introducing light cone proper time a as a height function specifying uniquely the point at which 3-surface is singular (stringy diagrams help to visualize what is involved), and by restricting the singular 3-surface to be the intersection of $a = constant$ hyperboloid of M^4 containing the singular point with the space-time surface. There would be non-uniqueness of the conformal equivalence class due to the choice of the origin of the light cone but the decomposition of the configuration space of 3-surfaces to a union of configuration spaces characterized by unions of future and past light cones could resolve this difficulty.

2. Interaction vertices as generalization of ordinary ones

If the interaction vertices are identified as intersections for the ends of space-time sheets representing particles, the conformal equivalence class is naturally identified as the one associated with the intersection of the boundary component or light like causal determinant with the vertex. Poincare invariance of the conformal equivalence class and generalized general coordinate invariance follow trivially in this case.

2.4.2 Conformal invariance

Conformal invariance implies that vacuum functionals depend on the conformal equivalence class of the surface Y^2 only. What makes this idea so attractive is that for a given genus g configuration space becomes effectively finite-dimensional. A second nice feature is that instead of trying to find coordinates for the space of the conformal equivalence classes one can construct vacuum functionals as functions of the Teichmueller parameters.

That one can construct this kind of functions as suitable functions of the Teichmueller parameters is not trivial. The essential point is that the boundary components can be regarded as submanifolds of $M_+^4 \times CP_2$: as a consequence vacuum functional can be regarded as a composite function:

2-surface $\rightarrow$ Teichmueller matrix Ω determined by the induced metric $\rightarrow \Omega_{vac}(\Omega)$

Therefore the fact that there are Teichmueller parameters which do not correspond to any Riemann surface, doesn't produce any trouble. It should be noticed that the situation differs from that in the Polyakov formulation of string models, where one doesn't assume that the metric of the two-surface is induced metric (although classical equations of motion imply this).

2.4.3 Diff invariance

Since several values of the Teichmueller parameters correspond to the same conformal equivalence class, one must pose additional conditions on the functions of the Teichmueller parameters in order to obtain single valued functions of the conformal equivalence class.

The first requirement of this kind is the invariance under topologically nontrivial Diff transformations inducing $Sp(2g, Z)$ transformation $(A, B; C, D)$ in the homology basis. The action of these transformations on Teichmueller parameters is deduced by requiring that holomorphic one-forms satisfy the defining conditions in the transformed homology basis. It turns out that the action of the topologically nontrivial diffeomorphism on Teichmueller parameters can be regarded as a generalized Möbius transformation:

$$\Omega \rightarrow (A\Omega + B)(C\Omega + D)^{-1} \ . \tag{2.20}$$

Vacuum functional must be invariant under these transformations. It should be noticed that the situation differs from that encountered in the string models. In TGD the integration measure over the configuration space is Diff invariant: in string models the integration measure is the integration measure of the Teichmueller space and this is not invariant under $Sp(2g, Z)$ but transforms like a density: as a consequence the counterpart of the vacuum functional must be also modular covariant since it is the product of vacuum functional and integration measure, which must be modular invariant.

It is possible to show that the quantities

$$(\Theta[a, b]/\Theta[c, d])^4 \ . \tag{2.21}$$

and their complex conjugates are $Sp(2g, Z)$ invariants [32] and therefore can be regarded as basic building blocks of the vacuum functionals.

Teichmueller parameters are not uniquely determined since one can always perform a permutation of the g handles of the Riemann surface inducing a redefinition of the canonical homology basis (permutation of g generators). These transformations act as similarities of the Teichmueller matrix:

$$\Omega \rightarrow S\Omega S^{-1} \ , \tag{2.22}$$

where S is the $g \times g$ matrix representing the permutation of the homology generators understood as orthonormal vectors in the g- dimensional vector space. Therefore the Teichmueller parameters related by these similarity transformations correspond to the same conformal equivalence class of the Riemann surfaces and vacuum functionals must be invariant under these similarities.

It is easy to find out that these similarities permute the components of the theta characteristics: $[a, b] \rightarrow [S(a), S(b)]$. Therefore the invariance requirement states that the handles of the Riemann surface behave like bosons: the vacuum functional constructed from the theta functions is invariant under the permutations of the theta characteristics. In fact, this requirement brings in nothing new. Handle permutations can be regarded as $Sp(2g, Z)$ transformations so that the modular invariance alone guarantees invariance under handle permutations.

2.4.4 Cluster decomposition property

Consider next the behavior of the vacuum functional in the limit, when boundary component with genus g splits to two separate boundary components of genera g_1 and g_2 respectively. The splitting into two separate boundary components corresponds to the reduction of the Teichmueller matrix Ω^g to a direct sum of $g_1 \times g_1$ and $g_2 \times g_2$ matrices ($g_1 + g_2 = g$):

$$\Omega^g = \Omega^{g_1} \oplus \Omega^{g_2} \ , \tag{2.23}$$

when a suitable definition of the Teichmueller parameters is adopted. The splitting can also take place without a reduction to a direct sum: the Teichmueller parameters obtained via $Sp(2g, Z)$ transformation from $\Omega^g = \Omega^{g_1} \oplus \Omega^{g_2}$ do not possess direct sum property in general.

The physical interpretation is obvious: the non-diagonal elements of the Teichmueller matrix describe the geometric interaction between handles and at this limit the interaction between the handles belonging to the separate surfaces vanishes. On the physical grounds it is natural to require that vacuum functionals satisfy cluster decomposition property at this limit: that is they reduce to the product of appropriate vacuum functionals associated with the composite surfaces.

Theta functions satisfy cluster decomposition property [30, 32]. Theta characteristics reduce to the direct sums of the theta characteristics associated with g_1 and g_2 ($a = a_1 \oplus a_2$, $b = b_1 \oplus b_2$) and the dependence on the Teichmueller parameters is essentially exponential so that the cluster decomposition property indeed results:

$$\Theta[a, b](\Omega^g) = \Theta[a_1, b_1](\Omega^{g_1})\Theta[a_2, b_2](\Omega^{g_2}) \ . \tag{2.24}$$

Cluster decomposition property holds also true for the products of theta functions. This property is also satisfied by suitable homogenous polynomials of thetas. In particular, the following quantity playing central role in the construction of the vacuum functional obeys this property

$$Q_0 = \sum_{[a,b]} \Theta[a, b]^4 \bar{\Theta}[a, b]^4 \ , \tag{2.25}$$

where the summation is over all even theta characteristics (recall that odd theta functions vanish at the origin of C^g).

Together with the $Sp(2g, Z)$ invariance the requirement of cluster decomposition property implies that the vacuum functional must be representable in the form

$$\Omega_{vac} = P_{M,N}(\Theta^4, \bar{\Theta}^4)/Q_{MN}(\Theta^4, \bar{\Theta}^4) \tag{2.26}$$

where the homogenous polynomials $P_{M,N}$ and $Q_{M,N}$ have same degrees (M and N as polynomials of $\Theta[a, b]^4$ and $\bar{\Theta}[a, b]^4$.

2.4.5 Finiteness requirement

Vacuum functional should be finite. Finiteness requirement is satisfied provided the numerator $Q_{M,N}$ of the vacuum functional is real and positive definite. The simplest quantity of this type is the quantity Q_0

defined previously and its various powers. $Sp(2g, Z)$ invariance and finiteness requirement are satisfied provided vacuum functionals are of the following general form

$$\Omega_{vac} = \frac{P_{N,N}(\Theta^4, \bar{\Theta}^4)}{Q_0^N} , \tag{2.27}$$

where $P_{N,N}$ is homogenous polynomial of degree N with respect to $\Theta[a,b]^4$ and $\bar{\Theta}[a,b]^4$. In addition $P_{N,N}$ is invariant under the permutations of the theta characteristics and satisfies cluster decomposition property.

2.4.6 Stability against the decay $g \rightarrow g_1 + g_2$

Elementary particle vacuum functionals must be stable against the genus conserving decays $g \rightarrow g_1 + g_2$. This decay corresponds to the limit at which Teichmueller matrix reduces to a direct sum of the matrices associated with g_1 and g_2 (note however the presence of $Sp(2g, Z)$ degeneracy). In accordance with the topological description of the particle reactions one expects that this decay doesn't occur if the vacuum functional in question vanishes at this limit.

In general the theta functions are non-vanishing at this limit and vanish provided the theta characteristics reduce to a direct sum of the odd theta characteristics. For $g < 2$ surfaces this condition is trivial and gives no constraints on the form of the vacuum functional. For $g = 2$ surfaces the theta function $\Theta(a, b)$, with $a = b = (1/2, 1/2)$ satisfies the stability criterion identically (odd theta functions vanish identically), when Teichmueller parameters separate into a direct sum. One can however perform $Sp(2g, Z)$ transformations giving new points of Teichmueller space describing the decay. Since these transformations transform theta characteristics in a nontrivial manner to each other and since all even theta characteristics belong to same $Sp(2g, Z)$ orbit [30, 32] , the conclusion is that stability condition is satisfied provided $g = 2$ vacuum functional is proportional to the product of fourth powers of all even theta functions multiplied by its complex conjugate.

If $g > 2$ there always exists some theta functions, which vanish at this limit and the minimal vacuum functional satisfying this stability condition is of the same form as in $g = 2$ case, that is proportional to the product of the fourth powers of all even Theta functions multiplied by its complex conjugate:

$$\Omega_{vac} = \prod_{[a,b]} \Theta[a,b]^4 \bar{\Theta}[a,b]^4 / Q_0^N , \tag{2.28}$$

where N is the number of even theta functions. The results obtained imply that genus-generation correspondence is one to one for $g > 1$ for the minimal vacuum functionals. Of course, the multiplication of the minimal vacuum functionals with functionals satisfying all criteria except stability criterion gives new elementary particle vacuum functionals: a possible physical identification of these vacuum functionals is most naturally as some kind of excited states.

One of the questions posed in the beginning was related to the experimental absence of $g > 0$, possibly massless, elementary bosons. The proposed stability criterion suggests a nice explanation. The point is that elementary particles are stable against decays $g \rightarrow g_1 + g_2$ but not with respect to the decay $g \rightarrow g + \ sphere$. As a consequence the direct emission of $g > 0$ gauge bosons is impossible unlike the emission of $g = 0$ bosons: for instance the decay muon $\rightarrow$ electron $+(g = 1)$ photon is forbidden.

2.4.7 Stability against the decay $g \rightarrow g - 1$

This stability criterion states that the vacuum functional is stable against single particle decay $g \rightarrow g - 1$ and, if satisfied, implies that vacuum functional vanishes, when the genus of the surface is smaller than g. In stringy framework this criterion is equivalent to a separate conservation of various lepton numbers: for instance, the spontaneous transformation of muon to electron is forbidden. Notice that this condition doesn't imply that that the vacuum functional is localized to a single genus: rather the vacuum functional of genus g vanishes for all surfaces with genus smaller than g. This hierarchical structure should have a close relationship to Cabibbo-Kobayashi-Maskawa mixing of the quarks.

The stability criterion implies that the vacuum functional must vanish at the limit, when one of the handles of the Riemann surface suffers a pinch. To deduce the behavior of the theta functions at this limit, one must find the behavior of Teichmueller parameters, when i:th handle suffers a pinch. Pinch implies that a suitable representative of the homology generator a_i or b_i contracts to a point.

Consider first the case, when a_i contracts to a point. The normalization of the holomorphic one-form ω_i must be preserved so that that ω_i must behaves as $1/z$, where z is the complex coordinate vanishing at pinch. Since the homology generator b_i goes through the pinch it seems obvious that the imaginary part of the Teichmueller parameter $\Omega_{ii} = \int_{b_i} \omega_i$ diverges at this limit (this conclusion is made also in [32]): $Im(\Omega_{ii}) \rightarrow \infty$.

Of course, this criterion doesn't cover all possible manners the pinch can occur: pinch might take place also, when the components of the Teichmueller matrix remain finite. In the case of torus topology one finds that $Sp(2g, Z)$ element $(A, B; C, D)$ takes $Im(\Omega) = \infty$ to the point C/D of real axis. This suggests that pinch occurs always at the boundary of the Teichmueller space: the imaginary part of Ω_{ij} either vanishes or some matrix element of $Im(\Omega)$ diverges.

Consider next the situation, when b_i contracts to a point. From the definition of the Teichmueller parameters it is clear that the matrix elements Ω_{kl}, with $k, l \neq i$ suffer no change. The matrix element Ω_{ki} obviously vanishes at this limit. The conclusion is that i:th row of Teichmueller matrix vanishes at this limit. This result is obtained also by deriving the $Sp(2g, Z)$ transformation permuting a_i and b_i with each other: in case of torus this transformation reads $\Omega \rightarrow -1/\Omega$.

Consider now the behavior of the theta functions, when pinch occurs. Consider first the limit, when $Im(\Omega_{ii})$ diverges. Using the general definition of $\Theta[a, b]$ it is easy to find out that all theta functions for which the i:th component a_i of the theta characteristic is non-vanishing (that is $a_i = 1/2$) are proportional to the exponent $exp(-\pi\Omega_{ii}/4)$ and therefore vanish at the limit. The theta functions with $a_i = 0$ reduce to $g - 1$ dimensional theta functions with theta characteristic obtained by dropping i:th components of a_i and b_i and replacing Teichmueller matrix with Teichmueller matrix obtained by dropping i:th row and column. The conclusion is that all theta functions of type $\Theta(a, b)$ with $a = (1/2, 1/2, ..., 1/2)$ satisfy the stability criterion in this case.

What happens for the $Sp(2g, Z)$ transformed points on the real axis? The transformation formula for theta function is given by [30, 32]

$$\Theta[a, b]((A\Omega + B)(C\Omega + D)^{-1}) = exp(i\phi)det(C\Omega + D)^{1/2}\Theta[c, d](\Omega) , \tag{2.29}$$

where

$$\begin{pmatrix} c \\ d \end{pmatrix} = \begin{pmatrix} A & B \\ C & D \end{pmatrix} \left(\begin{pmatrix} a \\ b \end{pmatrix} - \begin{pmatrix} (CD^T)_d/2 \\ (AB^T)_d/2 \end{pmatrix} \right) . \tag{2.30}$$

Here ϕ is a phase factor irrelevant for the recent purposes and the index d refers to the diagonal part of the matrix in question.

The first thing to notice is the appearance of the diverging square root factor, which however disappears from the vacuum functionals (P and Q have same degree with respect to thetas). The essential point is that theta characteristics transform to each other: as already noticed all even theta characteristics belong to the same $Sp(2g, Z)$ orbit. Therefore the theta functions vanishing at $Im(\Omega_{ii}) = \infty$ do not vanish at the transformed points. It is however clear that for a given Teichmueller parametrization of pinch some theta functions vanish always.

Similar considerations in the case $\Omega_{ik} = 0$, i fixed, show that all theta functions with $b = (1/2, ..., 1/2)$ vanish identically at the pinch. Also it is clear that for $Sp(2g, Z)$ transformed points one can always find some vanishing theta functions. The overall conclusion is that the elementary particle vacuum functionals obtained by using $g \rightarrow g_1 + g_2$ stability criterion satisfy also $g \rightarrow g - 1$ stability criterion since they are proportional to the product of all even theta functions. Therefore the only nontrivial consequence of

$g \to g-1$ criterion is that also $g=1$ vacuum functionals are of the same general form as $g>1$ vacuum functionals.

A second manner to deduce the same result is by restricting the consideration to the hyper-elliptic surfaces and using the representation of the theta functions in terms of the roots of the polynomial appearing in the definition of the hyper-elliptic surface [32]. When the genus of the surface is smaller than three (the interesting case), this representation is all what is needed since all surfaces of genus $g<3$ are hyper-elliptic.

Since hyper-elliptic surfaces can be regarded as surfaces obtained by gluing two compactified complex planes along the cuts connecting various roots of the defining polynomial it is obvious that the process $g \to g-1$ corresponds to the limit, when two roots of the defining polynomial coincide. This limit corresponds either to disappearance of a cut or the fusion of two cuts to a single cut. Theta functions are expressible as the products of differences of various roots (Thomae's formula [32])

$$\Theta[a,b]^4 \propto \prod_{i<j\in T}(z_i-z_j)\prod_{k<l\in CT}(z_k-z_l) \ , \tag{2.31}$$

where T denotes some subset of $\{1,2,...,2g\}$ containing $g+1$ elements and CT its complement. Hence the product of all even theta functions vanishes, when two roots coincide. Furthermore, stability criterion is satisfied only by the product of the theta functions.

Lowest dimensional vacuum functionals are worth of more detailed consideration.

i) $g=0$ particle family corresponds to a constant vacuum functional: by continuity this vacuum functional is constant for all topologies.

ii) For $g=1$ the degree of P and Q as polynomials of the theta functions is 24: the critical number of transversal degrees of freedom in bosonic string model! Probably this result is not an accident.

ii) For $g=2$ the corresponding degree is 80 since there are 10 even genus 2 theta functions.

There are large numbers of vacuum functionals satisfying the relevant criteria, which do not satisfy the proposed stability criteria. These vacuum functionals correspond either to many particle states or to unstable single particle states.

2.4.8 Continuation of the vacuum functionals to higher genus topologies

From continuity it follows that vacuum functionals cannot be localized to single boundary topology. Besides continuity and the requirements listed above, a natural requirement is that the continuation of the vacuum functional from the sector g to the sector $g+k$ reduces to the product of the original vacuum functional associated with genus g and $g=0$ vacuum functional at the limit when the surface with genus $g+k$ decays to surfaces with genus g and k: this requirement should guarantee the conservation of separate lepton numbers although different boundary topologies suffer mixing in the vacuum functional. These requirements are satisfied provided the continuation is constructed using the following rule:

Perform the replacement

$$\Theta[a,b]^4 \quad \to \quad \sum_{c,d}\Theta[a\oplus c,b\oplus d]^4 \tag{2.32}$$

for each fourth power of the theta function. Here c and d are Theta characteristics associated with a surface with genus k. The same replacement is performed for the complex conjugates of the theta function. It is straightforward to check that the continuations of elementary particle vacuum functionals indeed satisfy the cluster decomposition property and are continuous.

To summarize, the construction has provided hoped for answers to some questions stated in the beginning: stability requirements explain the separate conservation of lepton numbers and the experimental absence of $g>0$ elementary bosons. What has not not been explained is the experimental absence of $g>2$ fermion families. The vanishing of the $g>2$ elementary particle vacuum functionals for the hyper-elliptic surfaces however suggest a possible explanation: under some conditions on the surface X^2 the surfaces Y^2 are hyper-elliptic or possess some conformal symmetry so that elementary particle vacuum functionals vanish for them. This conjecture indeed might make sense since the surfaces Y^2 are determined by the asymptotic dynamics and one might hope that the surfaces Y^2 are analogous to the final states of a dissipative system.

2.5 Explanations for the absence of the $g > 2$ elementary particles from spectrum

The decay properties of the intermediate gauge bosons [47] are consistent with the assumption that the number of the light neutrinos is $N = 3$. Also cosmological considerations pose upper bounds on the number of the light neutrino families and $N = 3$ seems to be favored [47]. It must be however emphasized that p-adic considerations [9] encourage the consideration the existence of higher genera with neutrino masses such that they are not produced in the laboratory at present energies. In any case, for TGD approach the finite number of light fermion families is a potential difficulty since genus-generation correspondence suggests that the number of the fermion (and possibly also boson) families is infinite. Therefore one had better to find a good argument showing that the number of the observed neutrino families, or more generally, of the observed elementary particle families, is small also in the world described by TGD.

It will be later found that also TGD inspired cosmology requires that the number of the effectively massless fermion families must be small after Planck time. This suggests that boundary topologies with handle number $g > 2$ are unstable and/or very massive so that they, if present in the spectrum, disappear from it after Planck time, which correspond to the value of the light cone proper time $a \simeq 10^{-11}$ seconds.

In accordance with the spirit of TGD approach it is natural to wonder whether some geometric property differentiating between $g > 2$ and $g < 3$ boundary topologies might explain why only $g < 3$ boundary components are observable. One can indeed find a good candidate for this kind of property: namely hyper-ellipticity, which states that Riemann surface is a two-fold branched covering of sphere possessing two-element group Z_2 as conformal automorphisms. All $g < 3$ Riemann surfaces are hyper-elliptic unlike $g > 2$ Riemann surfaces, which in general do not posses this property. Thus it is natural to consider the possibility that hyper-ellipticity or more general conformal symmetries might explain why only $g < 2$ topologies correspond to the observed elementary particles.

As regards to the present problem the crucial observation is that some even theta functions vanish for the hyper-elliptic surfaces with genus $g > 2$ [32]. What is essential is that these surfaces have the group Z_2 as conformal symmetries. Indeed, the vanishing phenomenon is more general. Theta functions tend to vanish for $g > 2$ two-surfaces possessing discrete group of conformal symmetries [29] : for instance, instead of sphere one can consider branched coverings of higher genus surfaces.

From the general expression of the elementary particle vacuum functional it is clear that elementary particle vacuum functionals vanish, when Y^2 is hyper-elliptic surface with genus $g > 2$ and one might hope that this is enough to explain why the number of elementary particle families is three.

2.5.1 Hyper-ellipticity implies the separation of $g \leq 2$ and $g > 2$ sectors to separate worlds

If the vertices are defined as intersections of space-time sheets of elementary particles and if elementary particle vacuum functionals are required to have Z_2 symmetry, the localization of elementary particle vacuum functionals to $g \leq 2$ topologies occurs automatically. Even if one allows as limiting case vertices for which 2-manifolds are pinched to topologies intermediate between $g > 2$ and $g \leq 2$ topologies, Z_2 symmetry present for both topological interpretations implies the vanishing of this kind of vertices. This applies also in the case of stringy vertices so that also particle propagation would respect the effective number of particle families. $g > 2$ and $g \leq 2$ topologies would behave much like their own worlds in this approach. This is enough to explain the experimental findings if one can understand why the $g > 2$ particle families are absent as incoming and outgoing states or are very heavy.

2.5.2 What about $g > 2$ vacuum functionals which do not vanish for hyper-elliptic surfaces?

The vanishing of all $g \geq 2$ vacuum functionals for hyper-elliptic surfaces cannot hold true generally. There must exist vacuum functionals which do satisfy this condition. This suggest that elementary particle vacuum functionals for $g > 2$ states have interpretation as bound states of g handles and that the more general states which do not vanish for hyper-elliptic surfaces correspond to many-particle states composed of bound states $g \leq 2$ handles and cannot thus appear as incoming and outgoing states. Thus $g > 2$ elementary particles would decouple from $g \leq 2$ states.

2.5.3 Should higher elementary particle families be heavy?

TGD predicts an entire hierarchy of scaled up variants of standard model physics for which particles do not appear in the vertices containing the known elementary particles and thus behave like dark matter [18]. Also $g > 2$ elementary particles would behave like dark matter and in principle there is no absolute need for them to be heavy.

The safest option would be that $g > 2$ elementary particles are heavy and the breaking of Z_2 symmetry for $g \geq 2$ states could guarantee this. p-Adic considerations lead to a general mass formula for elementary particles such that the mass of the particle is proportional to $\frac{1}{\sqrt{p}}$ [11]. Also the dependence of the mass on particle genus is completely fixed by this formula. What remains however open is what determines the p-adic prime associated with a particle with given quantum numbers. Of course, it could quite well occur that p is much smaller for $g > 2$ genera than for $g \leq 2$ genera.

3 Non-topological contributions to particle masses from p-adic thermodynamics

In TGD framework p-adic thermodynamics provides a microscopic theory of particle massivation in the case of fermions. The idea is very simple. The mass of the particle results from a thermal mixing of the massless states with CP_2 mass excitations of super-conformal algebra. In p-adic thermodynamics the Boltzmann weight $exp(-E/T)$ does not exist in general and must be replaced with p^{L_0/T_p} which exists for Virasoro generator L_0 if the inverse of the p-adic temperature is integer valued $T_p = 1/n$. The expansion in powers of p converges extremely rapidly for physical values of p, which are rather large. Therefore the three lowest terms in expansion give practically exact results. Thermal massivation does not not necessarily lead to light states and this drops a large number of exotic states from the spectrum of light particles. The partition functions of N-S and Ramond type representations are not changed in TGD framework despite the fact that fermionic super generators carry fermion numbers and are not Hermitian. Thus the practical calculations are relatively straightforward.

In free fermion picture the p-adic thermodynamics in the boson sector is for fermion-antifermion states associated with the two throats of the bosonic wormhole. The question is whether the thermodynamical mass squared is just the sum of the two independent fermionic contributions for Ramond representations or should one use N-S type representation resulting as a tensor product of Ramond representations.

The overall conclusion about p-adic mass calculations is that fermionic mass spectrum is predicted in an excellent accuracy but that the thermal masses of the intermediate gauge bosons come 20-30 per cent to large for $T_p = 1$ and are completely negligible for $T_p = 1/2$. This forces to consider very seriously the possibility that thermal contribution to the bosonic mass is negligible and that TGD can, contrary to the original expectations, provide dynamical Higgs field as a fundamental field. The identification of Higgs as wormhole contact would provide this field. The bound state character of the boson states could be responsible for $T_p < 1$. For this option the Higgs contribution to fermion masses would be negligible.

A more plausible option is based on the identification of the Higgs like contribution in terms of the deviation of the ground state conformal weight from negative half integer. The negative ground state conformal weights in turn correspond to the squares of the generalized eigenvalues of the modified Dirac operator determined by the dynamics of Kähler action for preferred extremals.

A microscopic theory explaining the non-half integer contribution to the conformal weight follows from the identification of the physical elementary particles in terms of pairs of wormhole contacts with upper and lower throat pairs connected by Kähler magnetic flux tubes. This requires zero energy ontology, weak form of electric-magnetic duality, and twistor approach as theoretical ingredients. This gives also a nice connection with Higgs mechanism. TGD predicts scalar and pseudo scalar Higgs which correspond to SU(2) triplet and singlet and therefore same representations of SU(2) as electroweak gauge bosons. Higgs vacuum expectation is not needed and there are strong reasons to believe that also gauge bosons regarded usually as massless receive a small mass and scalar Higgs boson disappears completely from the spectrum. This could happen also for Higgsinos if they combine with gauginos to form massive fermions. Only pseudo scalar Higgs and its super partner would remain in the spectrum for this option unless they combine with possibly existing massless axial gauge bosons and their super partners to form massive states.

3.1 Partition functions are not changed

One must write Super Virasoro conditions for L_n and *both* G_n and $G_n^\dagger$ rather than for L_n and G_n as in the case of the ordinary Super Virasoro algebra, and it is a priori not at all clear whether the partition functions for the Super Virasoro representations remain unchanged. This requirement is however crucial for the construction to work at all in the fermionic sector, since even the slightest changes for the degeneracies of the excited states can change light state to a state with mass of order m_0 in the p-adic thermodynamics.

3.1.1 Super conformal algebra

Super Virasoro algebra is generated by the bosonic the generators L_n (n is an integer valued index) and by the fermionic generators G_r, where r can be either integer (Ramond) or half odd integer (NS). G_r creates quark/lepton for $r > 0$ and antiquark/antilepton for $r < 0$. For $r = 0$, G_0 creates lepton and its Hermitian conjugate anti-lepton. The defining commutation and anti-commutation relations are the following:

$$\begin{aligned}
[L_m, L_n] &= (m-n)L_{m+n} + \frac{c}{2}m(m^2-1)\delta_{m,-n} \ , \\
[L_m, G_r] &= (\frac{m}{2} - r)G_{m+r} \ , \\
[L_m, G_r^\dagger] &= (\frac{m}{2} - r)G_{m+r}^\dagger \ , \\
\{G_r, G_s^\dagger\} &= 2L_{r+s} + \frac{c}{3}(r^2 - \frac{1}{4})\delta_{m,-n} \ , \\
\{G_r, G_s\} &= 0 \ , \\
\{G_r^\dagger, G_s^\dagger\} &= 0 \ .
\end{aligned} \tag{3.1}$$

By the inspection of these relations one finds some results of a great practical importance.

1. For the Ramond algebra G_0, G_1 and their Hermitian conjugates generate the $r \geq 0, n \geq 0$ part of the algebra via anti-commutations and commutations. Therefore all what is needed is to assume that Super Virasoro conditions are satisfied for these generators in case that G_0 and $G_0^\dagger$ annihilate the ground state. Situation changes if the states are *not* annihilated by G_0 and $G_0^\dagger$ since then one must assume the gauge conditions for both L_1, G_1 and $G_1^\dagger$ besides the mass shell conditions associated with G_0 and $G_0^\dagger$, which however do not affect the number of the Super Virasoro excitations but give mass shell condition and constraints on the state in the cm spin degrees of freedom. This will be assumed in the following. Note that for the ordinary Super Virasoro only the gauge conditions for L_1 and G_1 are needed.

2. NS algebra is generated by $G_{1/2}$ and $G_{3/2}$ and their Hermitian conjugates (note that $G_{3/2}$ cannot be expressed as the commutator of L_1 and $G_{1/2}$) so that only the gauge conditions associated with these generators are needed. For the ordinary Super Virasoro only the conditions for $G_{1/2}$ and $G_{3/2}$ are needed.

3.1.2 Conditions guaranteing that partition functions are not changed

The conditions guaranteing the invariance of the partition functions in the transition to the modified algebra must be such that they reduce the number of the excitations and gauge conditions for a given conformal weight to the same number as in the case of the ordinary Super Virasoro.

1. The requirement that physical states are invariant under $G \leftrightarrow G^\dagger$ corresponds to the charge conjugation symmetry and is very natural. As a consequence, the gauge conditions for G and $G^\dagger$ are not independent and their number reduces by a factor of one half and is the same as in the case of the ordinary Super Virasoro.

2. As far as the number of the thermal excitations for a given conformal weight is considered, the only remaining problem are the operators $G_nG_n^\dagger$, which for the ordinary Super Virasoro reduce to $G_nG_n = L_{2n}$ and do not therefore correspond to independent degrees of freedom. In present case this situation is achieved only if one requires

$$(G_nG_n^\dagger - G_n^\dagger G_n)|phys\rangle = 0 . \tag{3.2}$$

It is not clear whether this condition must be posed separately or whether it actually follows from the representation of the Super Virasoro algebra automatically.

3.1.3 Partition function for Ramond algebra

Under the assumptions just stated, the partition function for the Ramond states not satisfying any gauge conditions

$$Z(t) = 1 + 2t + 4t^2 + 8t^3 + 14t^4 + ... , \tag{3.3}$$

which is identical to that associated with the ordinary Ramond type Super Virasoro.

For a Super Virasoro representation with $N = 5$ sectors, of main interest in TGD, one has

$$\begin{aligned} Z_N(t) &= Z^{N=5}(t) = \sum D(n)t^n \\ &= 1 + 10t + 60t^2 + 280t^3 + ... \end{aligned} \tag{3.4}$$

The degeneracies for the states satisfying gauge conditions are given by

$$d(n) = D(n) - 2D(n-1) . \tag{3.5}$$

corresponding to the gauge conditions for L_1 and G_1. Applying this formula one obtains for $N = 5$ sectors

$$d(0) = 1 \ , \quad d(1) = 8 \ , \quad d(2) = 40 \ , \quad d(3) = 160 \ . \tag{3.6}$$

The lowest order contribution to the p-adic mass squared is determined by the ratio

$$r(n) = \frac{D(n+1)}{D(n)} \ ,$$

where the value of n depends on the effective vacuum weight of the ground state fermion. Light state is obtained only provided the ratio is integer. The remarkable result is that for lowest lying states the ratio is integer and given by

$$r(1) = 8 \ , \quad r(2) = 5 \ , \quad r(3) = 4 \ . \tag{3.7}$$

It turns out that $r(2) = 5$ gives the best possible lowest order prediction for the charged lepton masses and in this manner one ends up with the condition $h_{vac} = -3$ for the tachyonic vacuum weight of Super Virasoro.

3.1.4 Partition function for NS algebra

For NS representations the calculation of the degeneracies of the physical states reduces to the calculation of the partition function for a single particle Super Virasoro

$$Z_{NS}(t) = \sum_n z(n/2)t^{n/2} . \tag{3.8}$$

Here $z(n/2)$ gives the number of Super Virasoro generators having conformal weight $n/2$. For a state with N active sectors (the sectors with a non-vanishing weight for a given ground state) the degeneracies can be read from the N-particle partition function expressible as

$$Z_N(t) = Z^N(t) . \tag{3.9}$$

Single particle partition function is given by the expression

$$Z(t) = 1+t^{1/2}+t+2t^{3/2}+3t^2+4t^{5/2}+5t^3+... . \tag{3.10}$$

Using this representation it is an easy task to calculate the degeneracies for the operators of conformal weight Δ acting on a state having N active sectors.

One can also derive explicit formulas for the degeneracies and calculation gives

$$\begin{array}{ll} D(0,N)=1 \ , & D(1/2,N)=N \ , \\ D(1,N)=\frac{N(N+1)}{2} \ , & D(3/2,N)=\frac{N}{6}(N^2+3N+8) \ , \\ D(2,N)=\frac{N}{2}(N^2+2N+3) \ , & D(5/2,N)=9N(N-1) \ , \\ D(3,N)=12N(N-1)+2N(N-1) & . \end{array} \tag{3.11}$$

as a function of the conformal weight $\Delta = 0, 1/2, ..., 3$.

The number of states satisfying Super Virasoro gauge conditions created by the operators of a conformal weight Δ, when the number of the active sectors is N, is given by

$$d(\Delta,N) = D(\Delta,N)-D(\Delta-1/2,N)-D(\Delta-3/2,N) . \tag{3.12}$$

The expression derives from the observation that the physical states satisfying gauge conditions for $G^{1/2}$, $G^{3/2}$ satisfy the conditions for all Super Virasoro generators. For $T_p = 1$ light bosons correspond to the integer values of $d(\Delta+1,N)/d(\Delta,N)$ in case that massless states correspond to thermal excitations of conformal weight Δ: they are obtained for $\Delta = 0$ only (massless ground state). This is what is required since the thermal degeneracy of the light boson ground state would imply a corresponding factor in the energy density of the black body radiation at very high temperatures. For the physically most interesting nontrivial case with $N = 2$ two active sectors the degeneracies are

$$d(0,2)=1 \ , \quad d(1,2)=1 \ , \quad d(2,2)=3 \ , \quad d(3,2)=4 \ . \tag{3.13}$$

N,Δ	0	1/2	1	3/2	2	5/2	3
2	1	1	1	3	3	4	4
3	1	2	3	9	11		
4	1	3	5	19	26		
5	1	4	10	24	150		

Table 3. Degeneracies $d(\Delta,N)$ of the operators satisfying NS type gauge conditions as a function of the number N of the active sectors and of the conformal weight Δ of the operator. Only those degeneracies, which are needed in the mass calculation for bosons assuming that they correspond to N-S representations are listed.

3.2 Fundamental length and mass scales

The basic difference between quantum TGD and super-string models is that the size of CP_2 is not of order Planck length but much larger: of order $10^{3.5}$ Planck lengths. This conclusion is forced by several consistency arguments, the mass scale of electron, and by the cosmological data allowing to fix the string tension of the cosmic strings which are basic structures in TGD inspired cosmology.

3.2.1 The relationship between CP_2 radius and fundamental p-adic length scale

One can relate CP_2 'cosmological constant' to the p-adic mass scale: for $k_L = 1$ one has

$$m_0^2 = \frac{m_1^2}{k_L} = m_1^2 = 2\Lambda \ . \tag{3.14}$$

$k_L = 1$ results also by requiring that p-adic thermodynamics leaves charged leptons light and leads to optimal lowest order prediction for the charged lepton masses. Λ denotes the 'cosmological constant' of CP_2 (CP_2 satisfies Einstein equations $G^{\alpha\beta} = \Lambda g^{\alpha\beta}$ with cosmological term).

The real counterpart of the p-adic thermal expectation for the mass squared is sensitive to the choice of the unit of p-adic mass squared which is by definition mapped as such to the real unit in canonical identification. Thus an important factor in the p-adic mass calculations is the correct identification of the p-adic mass squared scale, which corresponds to the mass squared unit and hence to the unit of the p-adic numbers. This choice does not affect the spectrum of massless states but can affect the spectrum of light states in case of intermediate gauge bosons.

1. For the choice

$$M^2 = m_0^2 \leftrightarrow 1 \tag{3.15}$$

the spectrum of L_0 is integer valued.

2. The requirement that all sufficiently small mass squared values for the color partial waves are mapped to real integers, would fix the value of p-adic mass squared unit to

$$M^2 = \frac{m_0^2}{3} \leftrightarrow 1 \ . \tag{3.16}$$

For this choice the spectrum of L_0 comes in multiples of 3 and it is possible to have a first order contribution to the mass which cannot be of thermal origin (say $m^2 = p$). This indeed seems to happen for electro-weak gauge bosons.

p-Adic mass calculations allow to relate m_0 to electron mass and to Planck mass by the formula

$$\begin{aligned} \frac{m_0}{m_{Pl}} &= \frac{1}{\sqrt{5+Y_e}} \times 2^{127/2} \times \frac{m_e}{m_{Pl}} \ , \\ m_{Pl} &= \frac{1}{\sqrt{\hbar G}} \ . \end{aligned} \tag{3.17}$$

For $Y_e = 0$ this gives $m_0 = .2437 \times 10^{-3} m_{Pl}$.

This means that CP_2 radius R defined by the length $L = 2\pi R$ of CP_2 geodesic is roughly $10^{3.5}$ times the Planck length. More precisely, using the relationship

$$\Lambda = \frac{3}{2R^2} = M^2 = m_0^2 \ ,$$

one obtains for

$$L = 2\pi R = 2\pi\sqrt{\frac{3}{2}\frac{1}{m_0}} \simeq 3.1167 \times 10^4 \sqrt{\hbar G} \ \text{for} \ Y_e = 0 \ . \tag{3.18}$$

The result came as a surprise: the first belief was that CP_2 radius is of order Planck length. It has however turned out that the new identification solved elegantly some long standing problems of TGD.

Y_e	0	.5	.7798
$(m_0/m_{Pl})10^3$	.2437	.2323	.2266
$K_R \times 10^{-7}$	2.5262	2.7788	2.9202
$(L_R/\sqrt{\hbar G}) \times 10^{-4}$	3.1580	3.3122	3.3954
$K \times 10^{-7}$	2.4606	2.4606	2.4606
$(L/\sqrt{\hbar G}) \times 10^{-4}$	3.1167	3.1167	3.1167
K_R/K	1.0267	1.1293	1.1868

Table 1. Table gives the values of the ratio $K_R = R^2/G$ and CP_2 geodesic length $L = 2\pi R$ for $Y_e \in \{0, 0.5, 0.7798\}$. Also the ratio of K_R/K, where $K = 2\times3\times5\times7\times11\times13\times17\times19\times23\times2^{-3}*(15/17)$ is rational number producing R^2/G approximately is given.

The value of top quark mass favors $Y_e = 0$ and $Y_e = .5$ is largest value of Y_e marginally consistent with the limits on the value of top quark mass.

3.2.2 CP_2 radius as the fundamental p-adic length scale

The identification of CP_2 radius as the fundamental p-adic length scale is forced by the Super Virasoro invariance. The pleasant surprise was that the identification of the CP_2 size as the fundamental p-adic length scale rather than Planck length solved many long standing problems of older TGD.

1. The earliest formulation predicted cosmic strings with a string tension larger than the critical value giving the angle deficit 2π in Einstein's equations and thus excluded by General Relativity. The corrected value of CP_2 radius predicts the value k/G for the cosmic string tension with k in the range $10^{-7} - 10^{-6}$ as required by the TGD inspired model for the galaxy formation solving the galactic dark matter problem.

2. In the earlier formulation there was no idea as how to derive the p-adic length scale $L \sim 10^{3.5}\sqrt{\hbar G}$ from the basic theory. Now this problem becomes trivial and one has to predict gravitational constant in terms of the p-adic length scale. This follows in principle as a prediction of quantum TGD. In fact, one can deduce G in terms of the p-adic length scale and the action exponential associated with the CP_2 extremal and gets a correct value if α_K approaches fine structure constant at electron length scale (due to the fact that electromagnetic field equals to the Kähler field if Z^0 field vanishes).

   Besides this, one obtains a precise prediction for the dependence of the Kähler coupling strength on the p-adic length scale by requiring that the gravitational coupling does not depend on the p-adic length scale. p-Adic prime p in turn has a nice physical interpretation: the critical value of α_K is same for the zero modes with given p. As already found, the construction of graviton state allows to understand the small value of the gravitational constant in terms of a de-coherence caused by multi-p fractality reducing the value of the gravitational constant from L_p^2 to G.

3. p-Adic length scale is also the length scale at which super-symmetry should be restored in standard super-symmetric theories. In TGD this scale corresponds to the transition to Euclidian field theory for CP_2 type extremals. There are strong reasons to believe that sparticles are however absent and that super-symmetry is present only in the sense that super-generators have complex conformal weights with $Re(h) = \pm 1/2$ rather than $h = 0$. The action of this super-symmetry changes the mass of the state by an amount of order CP_2 mass.

3.3 Color degrees of freedom

The ground states for the Super Virasoro representations correspond to spinor harmonics in $M^4 \times CP_2$ characterized by momentum and color quantum numbers. The correlation between color and electro-weak quantum numbers is wrong for the spinor harmonics and these states would be also hyper-massive. The super-symplectic generators allow to build color triplet states having negative vacuum conformal weights, and their values are such that p-adic massivation is consistent with the predictions of the earlier model differing from the recent one in the quark sector. In the following the construction and the properties of the color partial waves for fermions and bosons are considered. The discussion follows closely to the discussion of [33].

3.3.1 SKM algebra and counterpart of Super Virasoro conditions

The geometric part of SKM algebra is defined as an algebra respecting the light-likeness of the partonic 3-surface. It consists of X^3-local conformal transformations of $M^4_\pm$ and SU(3)-local SU(3) rotations. The requirement that generators have well defined radial conformal weight with respect to the lightlike coordinate r of X^3 restricts M^4 conformal transformations to the group $SO(3) \times E^3$. This involves choice of preferred time coordinate. If the preferred M^4 coordinate is chosen to correspond to a preferred light-like direction in $\delta M^4_\pm$ characterizing the theory, a reduction to $SO(2) \times E^2$ more familiar from string models occurs. The algebra decomposes into a direct sum of sub-algebras mapped to themselves by the Kac-Moody algebra generated by functions depending on r only. SKM algebra contains also $U(2)_{ew}$ Kac-Moody algebra acting as holonomies of CP_2 and having no bosonic counterpart.

p-Adic mass calculations require $N = 5$ sectors of super-conformal algebra. These sectors correspond to the 5 tensor factors for the $SO(3) \times E^3 \times SU(3) \times U(2)_{ew}$ (or $SO(2) \times E^2 \times SU(3) \times U(2)_{ew}$) decomposition of the SKM algebra to gauge symmetries of gravitation, color and electro-weak interactions. These symmetries act on the intersections $X^2 = X^3_l \cap X^7$ of 3-D light like causal determinants (CDs) X^3_l and 7-D light like CDs $X^7 = \delta M^4_+ \times CP_2$. This constraint leaves only the 2 transversal M^4 degrees of freedom since the translations in light like directions associated with X^3_l and δM^4_+ are eliminated.

The algebra differs from the standard one in that super generators $G(z)$ carry lepton and quark numbers are not Hermitian as in super-string models (Majorana conditions are not satisfied). The counterparts of Ramond representations correspond to zero modes of a second quantized spinor field with vanishing radial conformal weight. Non-zero modes with generalized eigenvalues $\lambda = 1/2 + iy$, $y = \sum_k n_k y_k$, $n_k \geq 0$, of the modified Dirac operator with $s_k = 1/2 + iy_k$ a zero or Rieman Zeta, define ground states of N-S type super Virasoro representations.

What is new is the imaginary part of conformal weight which means that the arrow of geometric time manifests itself via the sign of the imaginary part y already at elementary particle level. More concretely, positive energy particle propagating to the geometric future is not equivalent with negative energy particle propagating to the geometric past. The strange properties of the phase conjugate provide concrete physical demonstration of this difference. p-Adic mass calculations suggest the interpretation of y in terms of a decay width of the particle.

The Ramond or N-S type Virasoro conditions satisfied by the physical states in string model approach are replaced by the formulas expressing mass squared as a conformal weight. The condition is not equivalent with super Virasoro conditions since four-momentum does not appear in super Virasoro generators. It seems possible to assume that the commutator algebra $[SKM, SC]$ and the commutator of $[SKMV, SSV]$ of corresponding Super Virasoro algebras annihilate physical states. This would give rise to the analog of Super Virasoro conditions which could be seen as a Dirac equation in the world of classical worlds.

1. CP_2 CM degrees of freedom

Important element in the discussion are center of mass degrees of freedom parameterized by imbedding space coordinates. By the effective 2-dimensionality it is indeed possible to assign to partons momenta and color partial waves and they behave effectively as free particles. In fact, the technical problem of the earlier scenario was that it was not possible to assign symmetry transformations acting only on on the boundary components of 3-surface.

One can assign to each eigen state of color quantum numbers a color partial wave in CP_2 degrees of freedom. Thus color quantum numbers are not spin like quantum numbers in TGD framework except effectively in the length scales much longer than CP_2 length scale. The correlation between color partial waves and electro-weak quantum numbers is not physical in general: only the covariantly constant right handed neutrino has vanishing color.

2. Mass formula, and condition determining the effective string tension

Mass squared eigenvalues are given by

$$M^2 = m^2_{CP_2} + kL_0 \ . \tag{3.19}$$

The contribution of CP_2 spinor Laplacian to the mass squared operator is in general not integer valued.

The requirement that mass squared spectrum is integer valued for color partial waves possibly representing light states fixes the possible values of k determining the effective string tension modulo integer. The value $k = 1$ is the only possible choice. The earlier choice $k_L = 1$ and $k_q = 2/3$, $k_B = 1$ gave integer conformal weights for the lowest possible color partial waves. The assumption that the total vacuum weight h_{vac} is conserved in particle vertices implied $k_B = 1$.

3.3.2 General construction of solutions of Dirac operator of H

The construction of the solutions of massless spinor and other d'Alembertians in $M^4_+ \times CP_2$ is based on the following observations.

1. d'Alembertian corresponds to a massless wave equation $M^4 \times CP_2$ and thus Kaluza-Klein picture applies, that is M^4_+ mass is generated from the momentum in CP_2 degrees of freedom. This implies mass quantization:

$$M^2 = M^2_n \ ,$$

   where M^2_n are eigenvalues of CP_2 Laplacian. Here of course, ordinary field theory is considered. In TGD the vacuum weight changes mass squared spectrum.

2. In order to get a respectable spinor structure in CP_2 one must couple CP_2 spinors to an odd integer multiple of the Kähler potential. Leptons and quarks correspond to $n = 3$ and $n = 1$ couplings respectively. The spectrum of the electromagnetic charge comes out correctly for leptons and quarks.

3. Right handed neutrino is covariantly constant solution of CP_2 Laplacian for $n = 3$ coupling to Kähler potential whereas right handed 'electron' corresponds to the covariantly constant solution for $n = -3$. From the covariant constancy it follows that all solutions of the spinor Laplacian are obtained from these two basic solutions by multiplying with an appropriate solution of the scalar Laplacian coupled to Kähler potential with such a coupling that a correct total Kähler charge results. Left handed solutions of spinor Laplacian are obtained simply by multiplying right handed solutions with CP_2 Dirac operator: in this operation the eigenvalues of the mass squared operator are obviously preserved.

4. The remaining task is to solve scalar Laplacian coupled to an arbitrary integer multiple of Kähler potential. This can be achieved by noticing that the solutions of the massive CP_2 Laplacian can be regarded as solutions of S^5 scalar Laplacian. S^5 can indeed be regarded as a circle bundle over CP_2 and massive solutions of CP_2 Laplacian correspond to the solutions of S^5 Laplacian with $exp(is\tau)$ dependence on S^1 coordinate such that s corresponds to the coupling to the Kähler potential:

$$s = n/2 \ .$$

   Thus one obtains

$$D^2_5 = (D_\mu - iA_\mu\partial_\tau)(D^\mu - iA^\mu\partial_\tau) + \partial^2_\tau \tag{3.20}$$

   so that the eigen values of CP_2 scalar Laplacian are

$$m^2(s) = m^2_5 + s^2 \tag{3.21}$$

   for the assumed dependence on τ.

5. What remains to do, is to find the spectrum of S^5 Laplacian and this is an easy task. All solutions of S^5 Laplacian can be written as homogenous polynomial functions of C^3 complex coordinates Z^k and their complex conjugates and have a decomposition into the representations of $SU(3)$ acting in natural manner in C^3.

6. The solutions of the scalar Laplacian belong to the representations $(p, p+s)$ for $s \geq 0$ and to the representations $(p+|s|, p)$ of $SU(3)$ for $s \leq 0$. The eigenvalues $m^2(s)$ and degeneracies d are

$$\begin{aligned} m^2(s) &= \frac{2\Lambda}{3}[p^2 + (|s|+2)p + |s|] \ , \ p > 0 \ , \\ d &= \frac{1}{2}(p+1)(p+|s|+1)(2p+|s|+2) \ . \end{aligned} \tag{3.22}$$

Λ denotes the 'cosmological constant' of CP_2 ($R_{ij} = \Lambda s_{ij}$).

3.3.3 Solutions of the leptonic spinor Laplacian

Right handed solutions of the leptonic spinor Laplacian are obtained from the asatz of form

$$\nu_R = \Phi_{s=0}\nu_R^0 \ ,$$

where u_R is covariantly constant right handed neutrino and Φ scalar with vanishing Kähler charge. Right handed 'electron' is obtained from the ansats

$$e_R = \Phi_{s=3}e_R^0 \ ,$$

where e_R^0 is covariantly constant for $n = -3$ coupling to Kähler potential so that scalar function must have Kähler coupling $s = n/2 = 3$ a in order to get a correct Kähler charge. The d'Alembert equation reduces to

$$\begin{aligned} (D_\mu D^\mu - (1-\epsilon)\Lambda)\Phi = -m^2\Phi \ , \\ \epsilon(\nu) = 1 \ , \quad \epsilon(e) = -1 \ . \end{aligned} \tag{3.23}$$

The two additional terms correspond to the curvature scalar term and $J_{kl}\Sigma^{kl}$ terms in spinor Laplacian. The latter term is proportional to Kähler coupling and of different sign for ν and e, which explains the presence of the sign factor ϵ in the formula.

Right handed neutrinos correspond to (p, p) states with $p \geq 0$ with mass spectrum

$$\begin{aligned} m^2(\nu) &= \frac{m_1^2}{3}\left[p^2 + 2p\right] \ , \ p \geq 0 \ , \\ m_1^2 &\equiv 2\Lambda \ . \end{aligned} \tag{3.24}$$

Right handed 'electrons' correspond to $(p, p+3)$ states with mass spectrum

$$m^2(e) = \frac{m_1^2}{3}\left[p^2 + 5p + 6\right] \ , \ p \geq 0 \ . \tag{3.25}$$

Left handed solutions are obtained by operating with CP_2 Dirac operator on right handed solutions and have the same mass spectrum and representational content as right handed leptons with one exception: the action of the Dirac operator on the covariantly constant right handed neutrino ($(p = 0, p = 0)$ state) annihilates it.

3.3.4 Quark spectrum

Quarks correspond to the second conserved H-chirality of H-spinors. The construction of the color partial waves for quarks proceeds along similar lines as for leptons. The Kähler coupling corresponds to $n = 1$ (and $s = 1/2$) and right handed U type quark corresponds to a right handed neutrino. U quark type solutions are constructed as solutions of form

$$U_R = u_R \Phi_{s==1} ,$$

where u_R possesses the quantum numbers of covariantly constant right handed neutrino with Kähler charge $n = 3$ ($s = 3/2$). Hence Φ_s has $s = -1$. For D_R one has

$$D_R = d_r \Phi_{s=2} .$$

d_R has $s = -3/2$ so that one must have $s = 2$. For U_R the representations $(p+1, p)$ with triality one are obtained and $p = 0$ corresponds to color triplet. For D_R the representations $(p, p+2)$ are obtained and color triplet is missing from the spectrum ($p = 0$ corresponds to $\bar{6}$).

The CP_2 contributions to masses are given by the formula

$$\begin{aligned} m^2(U,p) &= \frac{m_1^2}{3}\left[p^2+3p+2\right] \ , \ p \geq 0 \ , \\ m^2(D,p) &= \frac{m_1^2}{3}\left[p^2+4p+4\right] \ , \ p \geq 0 \ . \end{aligned} \tag{3.26}$$

Left handed quarks are obtained by applying Dirac operator to right handed quark states and mass formulas and color partial wave spectrum are the same as for right handed quarks.

The color contributions to p-adic mass squared are integer valued if $m_0^2/3$ is taken as a fundamental p-adic unit of mass squared. This choice has an obvious relevance for p-adic mass calculations since canonical identification does not commute with a division by integer. More precisely, the images of number xp in canonical identification has a value of order 1 when x is a non-trivial rational whereas for $x = np$ the value is n/p and extremely is small for physically interesting primes. This choice does not however affect the spectrum of massless states but can affect the spectrum of light states in case of electro-weak gauge bosons.

3.4 Spectrum of elementary particles

The assumption that $k = 1$ holds true for all particles forces to modify the earlier construction of quark states. This turns out to be possible without affecting the p-adic mass calculations whose outcome depend in an essential manner on the ground state conformal weights h_{gr} of the fermions (which can be negative).

3.4.1 Leptonic spectrum

For $k = 1$ the leptonic mass squared is integer valued in units of m_0^2 only for the states satisfying

$$p \ mod \ 3 \neq 2 \ .$$

Only these representations can give rise to massless states. Neutrinos correspond to (p, p) representations with $p \geq 1$ whereas charged leptons correspond to $(p, p+3)$ representations. The earlier mass calculations demonstrate that leptonic masses can be understood if the ground state conformal weight is $h_{gr} = -1$ for charged leptons and $h_{gr} = -2$ for neutrinos.

The contribution of color partial wave to conformal weight is $h_c = (p^2+2p)/3$, $p \geq 1$, for neutrinos and $p = 1$ gives $h_c = 1$ (octet). For charged leptons $h_c = (p^2+5p+6)/3$ gives $h_c = 2$ for $p = 0$ (decuplet). In both cases super-symplectic operator O must have a net conformal weight $h_{sc} = -3$ to produce a correct conformal weight for the ground state. p-adic considerations suggests the use of operators O with super-symplectic conformal weight $z = -1/2 - i\sum n_k y_k$, where $s_k = 1/2 + iy_k$ corresponds to zero of Riemann ζ. If the operators in question are color Hamiltonians in octet representation net super-symplectic conformal weight $h_{sc} = -3$ results. The tensor product of two octets with conjugate super-symplectic conformal weights contains both octet and decuplet so that singlets are obtained. What strengthens the hopes that the construction is not adhoc is that the same operator appears in the construction of quark states too.

Right handed neutrino remains essentially massless. $p = 0$ right handed neutrino does not however generate $N = 1$ space-time (or rather, imbedding space) super symmetry so that no sparticles are predicted. The breaking of the electro-weak symmetry at the level of the masses comes out basically from the anomalous color electro-weak correlation for the Kaluza-Klein partial waves implying that the weights for the ground states of the fermions depend on the electromagnetic charge of the fermion. Interestingly, TGD predicts leptohadron physics based on color excitations of leptons and color bound states of these excitations could correspond topologically condensed on string like objects but not fundamental string like objects.

3.4.2 Spectrum of quarks

Earlier arguments [12] related to a model of CKM matrix as a rational unitary matrix suggested that the string tension parameter k is different for quarks, leptons, and bosons. The basic mass formula read as

$$M^2 = m_{CP_2}^2 + kL_0 \ .$$

The values of k were $k_q = 2/3$ and $k_L = k_B = 1$. The general theory however predicts that $k = 1$ for all particles.

1. By earlier mass calculations and construction of CKM matrix the ground state conformal weights of U and D type quarks must be $h_{gr}(U) = -1$ and $h_{gr}(D) = 0$. The formulas for the eigenvalues of CP_2 spinor Laplacian imply that if m_0^2 is used as unit, color conformal weight $h_c \equiv m_{CP_2}^2$ is integer for $p \ mod \ = \pm 1$ for U type quark belonging to $(p+1, p)$ type representation and obeying $h_c(U) = (p^2+3p+2)/3$ and for $p \ mod \ 3 = 1$ for D type quark belonging $(p, p+2)$ type representation and obeying $h_c(D) = (p^2+4p+4)/3$. Only these states can be massless since color Hamiltonians have integer valued conformal weights.

2. In the recent case $p = 1$ states correspond to $h_c(U) = 2$ and $h_c(D) = 3$. $h_{gr}(U) = -1$ and $h_{gr}(D) = 0$ reproduce the previous results for quark masses required by the construction of CKM matrix. This forces the super-symplectic operator O to compensate the anomalous color to have a net conformal weight $h_{sc} = -3$ just as in the leptonic case. The facts that the values of p are minimal for spinor harmonics and the super-symplectic operator is same for both quarks and leptons suggest that the construction is not had hoc. The real justification would come from the demonstration that $h_{sc} = -3$ defines null state for SSV: this would also explain why h_{sc} would be same for all fermions.

3. It would seem that the tensor product of the spinor harmonic of quarks (as also leptons) with Hamiltonians gives rise to a large number of exotic colored states which have same thermodynamical mass as ordinary quarks (and leptons). Why these states have smaller values of p-adic prime that ordinary quarks and leptons, remains a challenge for the theory. Note that the decay widths of intermediate gauge bosons pose strong restrictions on the possible color excitations of quarks. On the other hand, the large number of fermionic color exotics can spoil the asymptotic freedom, and it is possible to have and entire p-adic length scale hierarchy of QCDs existing only in a finite length scale range without affecting the decay widths of gauge bosons.

The following table summarizes the color conformal weights and super-symplectic vacuum conformal weights for the elementary particles.

	L	ν_L	U	D	W	γ, G, g
h_{vac}	-3	-3	-3	-3	-2	0
h_c	2	1	2	3	2	0

Table 2. The values of the parameters h_{vac} and h_c assuming that $k = 1$. The value of $h_{vac} \leq -h_c$ is determined from the requirement that p-adic mass calculations give best possible fit to the mass spectrum.

3.4.3 Photon, graviton and gluon

For photon, gluon and graviton the conformal weight of the $p = 0$ ground state is $h_{gr} = h_{vac} = 0$. The crucial condition is that $h = 0$ ground state is non-degenerate: otherwise one would obtain several physically more or less identical photons and this would be seen in the spectrum of black-body radiation. This occurs if one can construct several ground states not expressible in terms of the action of the Super Virasoro generators.

Masslessness or approximate masslessness requires low enough temperature $T_p = 1/n$, $n > 1$ at least and small enough value of the possible contribution coming from the ground state conformal weight.

In NS thermodynamics the only possibility to get exactly massless states in thermal sense is to have $\Delta = 0$ state with one active sector so that NS thermodynamics becomes trivial due to the absence of the thermodynamical excitations satisfying the gauge conditions. For neutral gauge bosons this is indeed achieved. For $T_p = 1/2$, which is required by the mass spectrum of intermediate gauge bosons, the thermal contribution to the mass squared is however extremely small even for W boson.

4 Modular contribution to the mass squared

The success of the p-adic mass calculations gives convincing support for the generation-genus correspondence. The basic physical picture is following.

1. Fermionic mass squared is dominated by partonic contribution, which is sum of cm and modular contributions: $M^2 = M^2(cm) + M^2(mod)$. Here 'cm' refers to the thermal contribution. Modular contribution can be assumed to depend on the genus of the boundary component only.

2. If Higgs contribution for diagonal (g, g) bosons (singlets with respect to "topological" $SU(3)$) dominates, the genus dependent contribution can be assumed to be negligible. This should be due to the bound state character of the wormhole contacts reducing thermal motion and thus the p-adic temperature.

3. Modular contribution to the mass squared can be estimated apart from an overall proportionality constant. The mass scale of the contribution is fixed by the p-adic length scale hypothesis. Elementary particle vacuum functionals are proportional to a product of all even theta functions and their conjugates, the number of even theta functions and their conjugates being $2N(g) = 2^g(2^g + 1)$. Also the thermal partition function must also be proportional to $2N(g)$:th power of some elementary partition function. This implies that thermal/ quantum expectation $M^2(mod)$ must be proportional to $2N(g)$. Since single handle behaves effectively as particle, the contribution must be proportional to genus g also. The success of the resulting mass formula encourages the belief that the argument is essentially correct.

The challenge is to construct theoretical framework reproducing the modular contribution to mass squared. There are two alternative manners to understand the origin modular contribution.

1. The realization that super-symplectic algebra is relevant for elementary particle physics leads to the idea that two thermodynamics are involved with the calculation of the vacuum conformal weight as a thermal expectation. The first thermodynamics corresponds to Super Kac-Moody algebra and second thermodynamics to super-symplectic algebra. This approach allows a first principle understanding of the origin and general form of the modular contribution without any need to introduce additional structures in modular degrees of freedom. The very fact that super-symplectic algebra does not commute with the modular degrees of freedom explains the dependence of the super-symplectic contribution on moduli.

2. The earlier approach was based on the idea that he modular contribution could be regarded as a quantum mechanical expectation value of the Virasoro generator L_0 for the elementary particle vacuum functional. Quantum treatment would require generalization the concepts of the moduli space and theta function to the p-adic context and finding an acceptable definition of the Virasoro generator L_0 in modular degrees of freedom. The problem with this interpretation is that it forces

to introduce, not only Virasoro generator L_0, but the entire super Virasoro algebra in modular degrees of freedom. One could also consider of interpreting the contribution of modular degrees of freedom to vacuum conformal weight as being analogous to that of CP_2 Laplacian but also this would raise the challenge of constructing corresponding Dirac operator. Obviously this approach has become obsolete.

The thermodynamical treatment taking into account the constraints from that p-adicization is possible might go along following lines.

1. In the real case the basic quantity is the thermal expectation value $h(M)$ of the conformal weight as a function of moduli. The average value of the deviation $\Delta h(M) = h(M) - h(M_0)$ over moduli space $\mathcal{M}$ must be calculated using elementary particle vacuum functional as a modular invariant partition function. Modular invariance is achieved if this function is proportional to the logarithm of elementary particle vacuum functional: this reproduces the qualitative features basic formula for the modular contribution to the conformal weight. p-Adicization leads to a slight modification of this formula.

2. The challenge of algebraically continuing this calculation to the p-adic context involves several sub-tasks. The notions of moduli space $\mathcal{M}_p$ and theta function must be defined in the p-adic context. An appropriately defined logarithm of the p-adic elementary particle vacuum functional should determine $\Delta h(M)$. The average of $\Delta h(M)$ requires an integration over $\mathcal{M}_p$. The problems related to the definition of this integral could be circumvented if the integral in the real case could be reduced to an algebraic expression, or if the moduli space is discrete in which case integral could be replaced by a sum.

3. The number theoretic existence of the p-adic Θ function leads to the quantization of the moduli so that the p-adic moduli space is discretized. Accepting the sharpened form of Riemann hypothesis [14] , the quantization means that the imaginary *resp.* real parts of the moduli are proportional to integers *resp.* combinations of imaginary parts of zeros of Riemann Zeta. This quantization could occur also for the real moduli for the maxima of Kähler function. This reduces the problematic p-adic integration to a sum and the resulting sum defining $\langle \Delta h \rangle$ converges extremely rapidly for physically interesting primes so that only the few lowest terms are needed.

4.1 Conformal symmetries and modular invariance

The full SKM invariance means that the super-conformal fields depend only on the conformal moduli of 2-surface characterizing the conformal equivalence class of the 2-surface. This means that all induced metrics differing by a mere Weyl scaling have same moduli. This symmetry is extremely powerful since the space of moduli is finite-dimensional and means that the entire infinite-dimensional space of deformations of parton 2-surface X^2 degenerates to a finite-dimensional moduli spaces under conformal equivalence. Obviously, the configurations of given parton correspond to a fiber space having moduli space as a base space. Super-symplectic degrees of freedom could break conformal invariance in some appropriate sense.

4.1.1 Conformal and SKM symmetries leave moduli invariant

Conformal transformations and super Kac Moody symmetries must leave the moduli invariant. This means that they induce a mere Weyl scaling of the induced metric of X^2 and thus preserve its non-diagonal character $ds^2 = g_{z\overline{z}}dzd\overline{z}$. This is indeed true if

1. the Super Kac Moody symmetries are holomorphic isometries of $X^7 = \delta M^4_{\pm} \times CP_2$ made local with respect to the complex coordinate z of X^2, and

2. the complex coordinates of X^7 are holomorphic functions of z.

Using complex coordinates for X^7 the infinitesimal generators can be written in the form

$$J^{An} = z^n j^{Ak} D_k + \overline{z}^n j^{A\overline{k}} D_{\overline{k}} . \quad (4.1)$$

The intuitive picture is that it should be possible to choose X^2 freely. It is however not always possible to choose the coordinate z of X^2 in such a manner that X^7 coordinates are holomorphic functions of z since a consistency of inherent complex structure of X^2 with that induced from X^7 is required. Geometrically this is like meeting of two points in the space of moduli.

Lorentz boosts produce new inequivalent choices of S^2 with their own complex coordinate: this set of complex structures is parameterized by the hyperboloid of future light cone (Lobatchevski space or mass shell), but even this is not enough. The most plausible manner to circumvent the problem is that only the maxima of Kähler function correspond to the holomorphic situation so that super-symplectic algebra representing quantum fluctuations would induce conformal anomaly.

4.1.2 The isometries of δM^4_+ are in one-one correspondence with conformal transformations

For CP_2 factor the isometries reduce to $SU(3)$ group acting also as symplectic transformations. For $\delta M^4_+ = S^2 \times R_+$ one might expect that isometries reduce to Lorentz group containing rotation group of $SO(3)$ as conformal isometries. If r_M corresponds to a macroscopic length scale, then X^2 has a finite sized S^2 projection which spans a rather small solid angle so that group $SO(3)$ reduces in a good approximation to the group $E^2 \times SO(2)$ of translations and rotations of plane.

This expectation is however wrong! The light-likeness of δM^4_+ allows a dramatic generalization of the notion of isometry. The point is that the conformal transformations of S^2 induce a conformal factor $|df/dw|^2$ to the metric of δM^4_+ and the local radial scaling $r_M \rightarrow r_M/|df/dw|$ compensates it. Hence the group of conformal isometries consists of conformal transformations of S^2 with compensating radial scalings. This compensation of two kinds of conformal transformations is the deep geometric phenomenon which translates to the condition $L_{SC} - L_{SKM} = 0$ in the sub-space of physical states. Note that an analogous phenomenon occurs also for the light-like CDs X^3_l with respect to the metrically 2-dimensional induced metric.

The X^2-local radial scalings $r_M \rightarrow r_M(z, \overline{z})$ respect the conditions $g_{zz} = g_{\overline{z}\overline{z}} = 0$ so that a mere Weyl scaling leaving moduli invariant results. By multiplying the conformal isometries of δM^4_+ by z^n (z is used as a complex coordinate for X^2 and w as a complex coordinate for S^2) a conformal localization of conformal isometries would result. Kind of double conformal transformations would be in question. Note however that this requires that X^7 coordinates are holomorphic functions of X^2 coordinate. These transformations deform X^2 unlike the conformal transformations of X^2. For X^3_l similar local scalings of the light like coordinate leave the moduli invariant but lead out of X^7.

4.1.3 Symplectic transformations break the conformal invariance

In general, infinitesimal symplectic transformations induce non-vanishing components $g_{zz}, g_{\overline{z}\overline{z}}$ of the induced metric and can thus change the moduli of X^2. Thus the quantum fluctuations represented by super-symplectic algebra and contributing to the configuration space metric are in general moduli changing. It would be interesting to know explicitly the conditions (the number of which is the dimension of moduli space for a given genus), which guarantee that the infinitesimal symplectic transformation is moduli preserving.

4.2 The physical origin of the genus dependent contribution to the mass squared

Different p-adic length scales are not enough to explain the charged lepton mass ratios and an additional genus dependent contribution in the fermionic mass formula is required. The general form of this contribution can be guessed by regarding elementary particle vacuum functionals in the modular degrees of freedom as an analog of partition function and the modular contribution to the conformal weight as an analog of thermal energy obtained by averaging over moduli. p-Adic length scale hypothesis determines the overall scale of the contribution.

The exact physical origin of this contribution has remained mysterious but super-symplectic degrees of freedom represent a good candidate for the physical origin of this contribution. This would mean a sigh of relief since there would be no need to assign conformal weights, super-algebra, Dirac operators, Laplacians, etc.. with these degrees of freedom.

4.2.1 Thermodynamics in super-symplectic degrees of freedom as the origin of the modular contribution to the mass squared

The following general picture is the simplest found hitherto.

1. Elementary particle vacuum functionals are defined in the space of moduli of surfaces X^2 corresponding to the maxima of Kähler function. There some restrictions on X^2. In particular, p-adic length scale poses restrictions on the size of X^2. There is an infinite hierarchy of elementary particle vacuum functionals satisfying the general constraints but only the lowest elementary particle vacuum functionals are assumed to contribute significantly to the vacuum expectation value of conformal weight determining the mass squared value.

2. The contribution of Super-Kac Moody thermodynamics to the vacuum conformal weight h coming from Virasoro excitations of the $h = 0$ massless state is estimated in the previous calculations and does not depend on moduli. The new element is that for a partonic 2-surface X^2 with given moduli, Virasoro thermodynamics is present also in super-symplectic degrees of freedom.

   Super-symplectic thermodynamics means that, besides the ground state with $h_{gr} = -h_{SC}$ with minimal value of super-symplectic conformal weight h_{SC}, also thermal excitations of this state by super-symplectic Virasoro algebra having $h_{gr} = -h_{SC} - n$ are possible. For these ground states the SKM Virasoro generators creating states with net conformal weight $h = h_{SKM} - h_{SC} - n \geq 0$ have larger conformal weight so that the SKM thermal average h depends on n. It depends also on the moduli M of X^2 since the Beltrami differentials representing a tangent space basis for the moduli space $\mathcal{M}$ do not commute with the super-symplectic algebra. Hence the thermally averaged SKM conformal weight h_{SKM} for given values of moduli satisfies

$$h_{SKM} = h(n, M) . \quad (4.2)$$

3. The average conformal weight induced by this double thermodynamics can be expressed as a super-symplectic thermal average $\langle\cdot\rangle_{SC}$ of the SKM thermal average $h(n, M)$:

$$h(M) = \langle h(n, M)\rangle_{SC} = \sum p_n(M)h(n) , \quad (4.3)$$

   where the moduli dependent probability $p_n(M)$ of the super-symplectic Virasoro excitation with conformal weight n should be consistent with the p-adic thermodynamics. It is convenient to write $h(M)$ as

$$h(M) = h_0 + \Delta h(M) , \quad (4.4)$$

   where h_0 is the minimum value of $h(M)$ in the space of moduli. The form of the elementary particle vacuum functionals suggest that h_0 corresponds to moduli with $Im(\Omega_{ij}) = 0$ and thus to singular configurations for which handles degenerate to one-dimensional lines attached to a sphere.

4. There is a further averaging of $\Delta h(M)$ over the moduli space $\mathcal{M}$ by using the modulus squared of elementary particle vacuum functional so that one has

$$h = h_0 + \langle\Delta h(M)\rangle_{\mathcal{M}} . \quad (4.5)$$

   Modular invariance allows to pose very strong conditions on the functional form of $\Delta h(M)$. The simplest assumption guaranteing this and thermodynamical interpretation is that $\Delta h(M)$ is proportional to the logarithm of the vacuum functional Ω:

$$\Delta h(M) \propto -log(\frac{\Omega(M)}{\Omega_{max}}) . \quad (4.6)$$

Here Ω_{max} corresponds to the maximum of Ω for which $\Delta h(M)$ vanishes.

4.2.2 Justification for the general form of the mass formula

The proposed general ansatz for $\Delta h(M)$ provides a justification for the general form of the mass formula deduced by intuitive arguments.

1. The factorization of the elementary particle vacuum functional Ω into a product of $2N(g) = 2^g(2^g + 1)$ terms and the logarithmic expression for $\Delta h(M)$ imply that the thermal expectation values is a sum over thermal expectation values over $2N(g)$ terms associated with various even characteristics (a, b), where a and b are g-dimensional vectors with components equal to $1/2$ or 0 and the inner product $4a \cdot b$ is an even integer. If each term gives the same result in the averaging using Ω_{vac} as a partition function, the proportionality to $2N_g$ follows.

2. For genus $g \geq 2$ the partition function defines an average in $3g - 3$ complex-dimensional space of moduli. The analogy of $\langle \Delta h \rangle$ and thermal energy suggests that the contribution is proportional to the complex dimension $3g - 3$ of this space. For $g \leq 1$ the contribution the complex dimension of moduli space is g and the contribution would be proportional to g.

$$\begin{aligned} \langle \Delta h \rangle &\propto g \times X(g) \text{ for } g \leq 1 , \\ \langle \Delta h \rangle &\propto (3g-3) \times X(g) \text{ for } g \geq 2 , \\ X(g) &= 2^g(2^g + 1) . \end{aligned} \quad (4.7)$$

   If X^2 is hyper-elliptic for the maxima of Kähler function, this expression makes sense only for $g \leq 2$ since vacuum functionals vanish for hyper-elliptic surfaces.

3. The earlier argument, inspired by the interpretation of elementary particle vacuum functional as a partition function, was that each factor of the elementary particle vacuum functional gives the same contribution to $\langle \Delta h \rangle$, and that this contribution is proportional to g since each handle behaves like a particle:

$$\langle \Delta h \rangle \propto g \times X(g) . \quad (4.8)$$

   The prediction following from the previous differs by a factor $(3g - 3)/g$ for $g \geq 2$. This would scale up the dominant modular contribution to the masses of the third $g = 2$ fermionic generation by a factor $\sqrt{3/2} \simeq 1.22$. One must of course remember, that these rough arguments allow $g-$dependent numerical factors of order one so that it is not possible to exclude either argument.

4.3 Generalization of Θ functions and quantization of p-adic moduli

The task is to find p-adic counterparts for theta functions and elementary particle vacuum functionals. The constraints come from the p-adic existence of the exponentials appearing as the summands of the theta functions and from the convergence of the sum. The exponentials must be proportional to powers of p just as the Boltzmann weights defining the p-adic partition function. The outcome is a quantization of moduli so that integration can be replaced with a summation and the average of $\Delta h(M)$ over moduli is well defined.

It is instructive to study the problem for torus in parallel with the general case. The ordinary moduli space of torus is parameterized by single complex number τ. The points related by $SL(2, Z)$ are equivalent, which means that the transformation $\tau \rightarrow (A\tau+B)/(C\tau+D)$ produces a point equivalent with τ. These transformations are generated by the shift $\tau \rightarrow \tau+1$ and $\tau \rightarrow -1/\tau$. One can choose the fundamental domain of moduli space to be the intersection of the slice $Re(\tau) \in [-1/2, 1/2]$ with the exterior of unit circle $|\tau| = 1$. The idea is to start directly from physics and to look whether one might some define p-adic version of elementary particle vacuum functionals in the p-adic counter part of this set or in some modular invariant subset of this set.

Elementary particle vacuum functionals are expressible in terms of theta functions using the functions $\Theta^4[a,b]\overline{\Theta}^4[a,b]$ as a building block. The general expression for the theta function reads as

$$\Theta[a,b](\Omega) = \sum_n exp(i\pi(n+a)\cdot\Omega\cdot(n+a))exp(2i\pi(n+a)\cdot b) \ . \tag{4.9}$$

The latter exponential phase gives only a factor $\pm i$ or ± 1 since $4a \cdot b$ is integer. For $p \ mod \ 4 = 3$ imaginary unit exists in an algebraic extension of p-adic numbers. In the case of torus (a,b) has the values $(0,0)$, $(1/2,0)$ and $(0,1/2)$ for torus since only even characteristics are allowed.

Concerning the p-adicization of the first exponential appearing in the summands in Eq. 4.9, the obvious problem is that π does not exists p-adically unless one allows infinite-dimensional extension.

1. Consider first the real part of Ω. In this case the proper manner to treat the situation is to introduce and algebraic extension involving roots of unity so that $Re(\Omega)$ rational. This approach is proposed as a general approach to the p-adicization of quantum TGD in terms of harmonic analysis in symmetric spaces allowing to define integration also in p-adic context in a physically acceptable manner by reducing it to Fourier analysis. The simplest situation corresponds to integer values for $Re(\Omega)$ and in this case the phase are equal to $\pm i$ or ± 1 since a is half-integer valued. One can consider a hierarchy of variants of moduli space characterized by the allowed roots of unity. The physical interpretation for this hierarchy would be in terms of a hierarchy of measurement resolutions. Note that the real parts of Ω can be assumed to be rationals of form m/n where n is constructed as a product of finite number of primes and therefore the allowed rationals are linear combinations of inverses $1/p_i$ for a subset $\{p_i\}$ of primes.

2. For the imaginary part of Ω different approach is required. One wants a rapid convergence of the sum formula and this requires that the exponents reduces in this case to positive powers of p. This is achieved if one has

$$Im(\Omega) = -n\frac{log(p)}{\pi)} \ , \tag{4.10}$$

   Unfortunately this condition is not consistent with the condition $Im(\Omega) > 0$. A manner to circumvent the difficulty is to replace Ω with its complex conjugate. Second approach is to define the real discretized variant of theta function first and then map it by canonical identification to its p-adic counterpart: this would map phase to phases and powers of p to their inverses. Note that a similar change of sign must be performed in p-adic thermodynamics for powers of p to map p-adic probabilities to real ones. By rescaling $Im(\Omega) \rightarrow \frac{log(p)}{\pi)} Im(\Omega)$ one has non-negative integer valued spectrum for $Im(\Omega)$ making possible to reduce integration in moduli space to a summation over finite number of rationals associated with the real part of Ω and powers of p associated with the imaginary part of Ω.

3. Since the exponents appearing in

$$p^{(n+a)\cdot Im(\Omega_{ij,p})\cdot(n+a)} = p^{a\cdot Im(\Omega)\cdot a} \times p^{2a\cdot Im(\Omega\cdot n} \times p^{+n\cdot Im(\Omega_{ij,p})\cdot n}$$

   are positive integers valued, $\Theta_{[a,b]}$ exist in R_p and converges. The problematic factor is the first exponent since the components of the vector a can have values $1/2$ and 0 and its existence implies a quantization of $Im(\Omega_{ij})$ as

$$Im(\Omega) = -Kn\frac{log(p)}{p} \ , \ n \in Z \ , \ n \geq 1 \ , \tag{4.11}$$

In p-adic context this condition mustbe formulated for the exponent of Ω defining the natural coordinate. $K = 4$ guarantees the existence of Θ functions and $K = 1$ the existence of the building blocks $\Theta^4[a,b]\overline{\Theta}^4[a,b]$ of elementary particle vacuum functionals in R_p. The extension to higher genera means only replacement of Ω with the elements of a matrix.

4. One can criticize this approach for the loss of the full modular covariance in the definition of theta functions. The modular transformations $\Omega \rightarrow \Omega + n$ are consistent with the number theoretic constraints but the transformations $\Omega \rightarrow -1/\Omega$ do not respect them. It seem that one can circumvent the difficulty by restricting the consideration to a fundamental domain satisfying the number theoretic constraints.

This variant of moduli space is discrete and p-adicity is reflected only in the sense that the moduli space makes sense also p-adically. One can consider also a continuum variant of the p-adic moduli space using the same prescription as in the construction of p-adic symmetric spaces [17].

1. One can introduce $exp(i\pi Re(\Omega))$ as the counterpart of $Re(\Omega)$ as a coordinate of the Teichmueller space. This coordinate makes sense only as a local coordinate since it does not differentiate between $Re(\Omega)$ and $Re(\Omega+2n)$. On the other hand, modular invariance states that Ω abd $\Omega+n$ correspond to the same moduli so that nothing is lost. In the similar manner one can introduce $exp(\pi Im(\Omega)) \in \{p^n, n > 0\}$ as the counterpart of discretized version of $Im(\Omega)$.

2. The extension to continuum would mean in the case of $Re(\Omega)$ the extension of the phase $exp(i\pi Re(\Omega))$ to a product $exp(i\pi Re(\Omega))exp(ipx) = exp(i\pi Re(\Omega) + exp(ipx)$, where x is p-adic integer which can be also infinite as a real integer. This would mean that each root of unity representing allowed value $Re(\Omega)$ would have a p-adic neighborhood consisting of p-adic integers. This neighborhood would be the p-adic counterpart for the angular integral $\Delta\phi$ for a given root of unity and would not make itself visible in p-adic integration.

3. For the imaginary part one can also consider the extension of $exp(\pi Im(\Omega))$ to $p^n \times exp(npx)$ where x is a p-adic integer. This would assign to each point p^n a p-adic neighborhood defined by p-adic integers. This neighborhood is same all integers n with same p-adic norm. When n is proportional to p^k one has $exp(npx) - 1 \propto p^k$.

The quantization of moduli characterizes precisely the conformal properties of the partonic 2-surfaces corresponding to different p-adic primes. In the real context -that is in the intersection of real and p-adic worlds- the quantization of moduli of torus would correspond to

$$\tau = K\left[\sum q + i \times n\frac{log(p)}{\pi}\right] \ , \tag{4.12}$$

where q is a rational number expressible as linear combination of inverses of a finite fixed set of primes defining the allowed roots of unity. $K = 1$ guarantees the existence of elementary particle vacuum functionals and $K = 4$ the existence of Theta functions. The ratio for the complex vectors defining the sides of the plane parallelogram defining torus via the identification of the parallel sides is quantized. In other words, the angles Φ between the sides and the ratios of the sides given by $|\tau|$ have quantized values.

The quantization rules for the moduli of the higher genera is of exactly same form

$$\Omega_{ij} = K\left[\sum q_{ij} + i \times n_{ij} \times \frac{log(p)}{\pi}\right] \ , \tag{4.13}$$

If the quantization rules hold true also for the maxima of Kähler function in the real context or more precisely- in the intersection of real and p-adic variants of the "world of classical worlds" identified as partonic 2-surfaces at the boundaries of causal diamond plus the data about their 4-D tangent space, there are good hopes that the p-adicized expression for Δh is obtained by a simple algebraic continuation of the real formula. Thus p-adic length scale would characterize partonic surface X^2 rather than the light like causal determinant X_l^3 containing X^2. Therefore the idea that various p-adic primes label various X_l^3 connecting fixed partonic surfaces X_i^2 would not be correct.

Quite generally, the quantization of moduli means that the allowed 2-dimensional shapes form a lattice and are thus additive. It also means that the maxima of Kähler function would obey a linear superposition in an extreme abstract sense. The proposed number theoretical quantization is expected to apply for any complex space allowing some preferred complex coordinates. In particular, configuration space of 2-surfaces could allow this kind of quantization in the complex coordinates naturally associated with isometries and this could allow to define configuration space integration, at least the counterpart of integration in zero mode degrees of freedom, as a summation.

Number theoretic vision leads to the notion of multi-p-p-adicity in the sense that the same partonic 2-surface can correspond to several p-adic primes and that infinite primes code for these primes [5, 16]. At the level of the moduli space this corresponds to the replacement of p with an integer in the formulas so that one can interpret the formulas both in real sense and p-adic sense for the primes p dividing the integer. Also the exponent of given prime in the integer matters. The construction of generalized eigen modes of Chern-Simons Dirac operator leads to the proposal that the collection of infinite primes characterizing infinite prime characterizes the geometry of the orbit of partonic 2-surface [5]. It would not be too surprising if this connection would reduce to the proposed discretization of the modular parameters of the partonic 2-surface.

4.4 The calculation of the modular contribution $\langle\Delta h\rangle$ to the conformal weight

The quantization of the moduli implies that the integral over moduli can be defined as a sum over moduli. The theta function $\Theta[a,b](\Omega)_p(\tau_p)$ is proportional to $p^{a\cdot aIm(\Omega_{ij,p})} = p^{Kn_{ij}m(a)/4}$ for $a\cdot a = m(a)/4$, where $K=1$ *resp.* $K=4$ corresponds to the existence existence of elementary particle vacuum functionals *resp.* theta functions in R_p. These powers of p can be extracted from the thetas defining the vacuum functional. The numerator of the vacuum functional gives $(p^n)^{2K\sum_{a,b}m(a)}$. The numerator gives $(p^n)^{2K\sum_{a,b}m(a_0)}$, where a_0 corresponds to the minimum value of $m(a)$. $a_0=(0,0,..,0)$ is allowed and gives $m(a_0)=0$ so that the p-adic norm of the denominator equals to one. Hence one has

$$|\Omega_{vac}(\Omega_p)|_p = p^{-2nK\sum_{a,b}m(a)} \quad (4.14)$$

The sum converges extremely rapidly for large values of p as function of n so that in practice only few moduli contribute.

The definition of $log(\Omega_{vac})$ poses however problems since in $log(p)$ does not exist as a p-adic number in any p-adic number field. The argument of the logarithm should have a unit p-adic norm. The simplest manner to circumvent the difficulty is to use the fact that the p-adic norm $|\Omega_p|_p$ is also a modular invariant, and assume that the contribution to conformal weight depends on moduli as

$$\Delta h_p(\Omega_p) \propto log(\frac{\Omega_{vac}}{|\Omega_{vac}|_p}) . \quad (4.15)$$

The sum defining $\langle\Delta h_p\rangle$ converges extremely rapidly and gives a result of order $O(p)$ p-adically as required.

The p-adic expression for $\langle\Delta h_p\rangle$ should result from the corresponding real expression by an algebraic continuation. This encourages the conjecture that the allowed moduli are quantized for the maxima of Kähler function, so that the integral over the moduli space is replaced with a sum also in the real case, and that Δh given by the double thermodynamics as a function of moduli can be defined as in the p-adic case. The positive power of p multiplying the numerator could be interpreted as a degeneracy factor. In fact, the moduli are not primary dynamical variables in the case of the induced metric, and there must be a modular invariant weight factor telling how many 2-surfaces correspond to given values of moduli. The power of p could correspond to this factor.

5 General mass formulas for non-Higgsy contributions

In the sequel various contributions to the mass squared are discussed.

5.1 General mass squared formula

The thermal independence of Super Virasoro and modular degrees of freedom implies that mass squared for elementary particle is the sum of Super Virasoro, modular and Higgsy contributions:

$$M^2 = M^2(color) + M^2(SV) + M^2(mod) + M^2(Higgsy) \ . \tag{5.1}$$

Also small renormalization correction contributions might be possible.

5.2 Color contribution to the mass squared

The mass squared contains a non-thermal color contribution to the ground state conformal weight coming from the mass squared of CP_2 spinor harmonic. The color contribution is an integer multiple of $m_0^2/3$, where $m_0^2 = 2\Lambda$ denotes the 'cosmological constant' of CP_2 (CP_2 satisfies Einstein equations $G^{\alpha\beta} = \Lambda g^{\alpha\beta}$).

The color contribution to the p-adic mass squared is integer valued only if $m_0^2/3$ is taken as a fundamental p-adic unit of mass squared. This choice has an obvious relevance for p-adic mass calculations since the simplest form of the canonical identification does not commute with a division by integer. More precisely, the image of number xp in canonical identification has a value of order 1 when x is a non-trivial rational number whereas for $x = np$ the value is n/p and extremely is small for physically interesting primes.

The choice of the p-adic mass squared unit are no effects on zeroth order contribution which must vanish for light states: this requirement eliminates quark and lepton states for which the CP_2 contribution to the mass squared is not integer valued using m_0^2 as a unit. There can be a dramatic effect on the first order contribution. The mass squared $m^2 = p/3$ using $m_0^2/3$ means that the particle is light. The mass squared becomes $m^2 = p/3$ when m_0^2 is used as a unit and the particle has mass of order 10^{-4} Planck masses. In the case of W and Z^0 bosons this problem is actually encountered. For light states using $m_0^2/3$ as a unit only the second order contribution to the mass squared is affected by this choice.

5.3 Modular contribution to the mass of elementary particle

The general form of the modular contribution is derivable from p-adic partition function for conformally invariant degrees of freedom associated with the boundary components. The general form of the vacuum functionals as modular invariant functions of Teichmuller parameters was derived in [2] and the square of the elementary particle vacuum functional can be identified as a partition function. Even theta functions serve as basic building blocks and the functionals are proportional to the product of all even theta functions and their complex conjugates. The number of theta functions for genus $g > 0$ is given by

$$N(g) = 2^{g-1}(2^g + 1) \ . \tag{5.2}$$

One has $N(1) = 3$ for muon and $N(2) = 10$ for τ.

1. Single theta function is analogous to a partition function. This implies that the modular contribution to the mass squared must be proportional to $2N(g)$. The factor two follows from the presence of both theta functions and their conjugates in the partition function.

2. The factorization properties of the vacuum functionals imply that handles behave effectively as particles. For example, at the limit, when the surface splits into two pieces with g_1 and $g - g_1$ handles, the partition function reduces to a product of g_1 and $g - g_1$ partition functions. This implies that the contribution to the mass squared is proportional to the genus of the surface. Altogether one has

$$
\begin{aligned}
M^2(mod,g) &= 2k(mod)N(g)g\frac{m_0^2}{p} \ , \\
k(mod) &= 1 \ .
\end{aligned}
\tag{5.3}
$$

Here $k(mod)$ is some integer valued constant (in order to avoid ultra heavy mass) to be determined. $k(mod) = 1$ turns out to be the correct choice for this parameter.

Summarizing, the real counterpart of the modular contribution to the mass of a particle belonging to $g + 1$:th generation reads as

$$
\begin{aligned}
M^2(mod) &= 0 \ for \ e, \nu_e, u, d \ , \\
M^2(mod) &= 9\frac{m_0^2}{p(X))} \ for \ X = \mu, \nu_\mu, c, s \ , \\
M^2(mod) &= 60\frac{m_0^2}{p(X)} \ for \ X = \tau, \nu_\tau, t, b \ .
\end{aligned}
\tag{5.4}
$$

The requirement that hadronic mass spectrum and CKM matrix are sensible however forces the modular contribution to be the same for quarks, leptons and bosons. The higher order modular contributions to the mass squared are completely negligible if the degeneracy of massless state is $D(0, mod, g) = 1$ in the modular degrees of freedom as is in fact required by $k(mod) = 1$.

5.4 Thermal contribution to the mass squared

One can deduce the value of the thermal mass squared in order $O(p^2)$ (an excellent approximation) using the general mass formula given by p-adic thermodynamics. Assuming maximal p-adic temperature $T_p = 1$ one has

$$
\begin{aligned}
M^2 &= k(sp + Xp^2 + O(p^3)) \ , \\
s_\Delta &= \frac{D(\Delta+1)}{D(\Delta)} \ , \\
X_\Delta &= 2\frac{D(\Delta+2)}{D(\Delta)} - \frac{D^2(\Delta+1)}{D^2(\Delta)} \ , \\
k &= 1 \ .
\end{aligned}
\tag{5.5}
$$

Δ is the conformal weight of the operator creating massless state from the ground state.

The ratios $r_n = D(n+1)/D(n)$ allowing to deduce the values of s and X have been deduced from p-adic thermodynamics in [8]. Light state is obtained only provided $r(\Delta)$ is an integer. The remarkable result is that for lowest lying states this is the case. For instance, for Ramond representations the values of r_n are given by

$$
(r_0, r_1, r_2, r_3) = (8, 5, 4, \frac{55}{16}) \ . \tag{5.6}
$$

The values of s and X are

$$
\begin{aligned}
(s_0, s_1, s_2) &= (8, 5, 4) \ , \\
(X_0, X_1, X_2) &= (16, 15, 11 + 1/2)) \ .
\end{aligned}
\tag{5.7}
$$

The result means that second order contribution is extremely small for quarks and charged leptons having $\Delta < 2$. For neutrinos having $\Delta = 2$ the second order contribution is non-vanishing.

5.5 The contribution from the deviation of ground state conformal weight from negative integer

The interpretation inspired by p-adic mass calculations is that the squares λ_i^2 of the eigenvalues of the modified Dirac operator correspond to the conformal weights of ground states. Another natural physical interpretation of λ is as an analog of the Higgs vacuum expectation. The instability of the Higgs=0 phase would corresponds to the fact that $\lambda = 0$ mode is not localized to any region in which ew magnetic field or induced Kähler field is non-vanishing. A good guess is that induced Kähler magnetic field B_K dictates the magnitude of the eigenvalues which is thus of order $h_0 = \sqrt{B_K R}$, R CP_2 radius. The first guess is that eigenvalues in the first approximation come as $(n + 1/2)h_0$. Each region where induced Kähler field is non-vanishing would correspond to different scale mass scale h_0.

1. The vacuum expectation value of Higgs is only proportional to an eigenvalue λ, not equal to it. Indeed, Higgs and gauge bosons as elementary particles correspond to wormhole contacts carrying fermion and antifermion at the two wormhole throats and must be distinguished from the space-time correlate of its vacuum expectation as something proportional to λ. In the fermionic case the vacuum expectation value of Higgs does not seem to be even possible since fermions do not correspond to wormhole contacts between two space-time sheets but possess only single wormhole throat (p-adic mass calculations are consistent with this).

2. Physical considerations suggest that the vacuum expectation of Higgs field corresponds to a particular eigenvalue λ_i of modified Dirac operator so that the eigenvalues λ_i would define TGD counterparts for the minima of Higgs potential. Since the vacuum expectation of Higgs corresponds to a condensate of wormhole contacts giving rise to a coherent state, the vacuum expectation cannot be present for topologically condensed CP_2 type vacuum extremals representing fermions since only single wormhole throat is involved. This raises a hen-egg question about whether Higgs contributes to the mass or whether Higgs is only a correlate for massivation having description using more profound concepts. From TGD point of view the most elegant option is that Higgs does not give rise to mass but Higgs vacuum expectation value accompanies bosonic states and is naturally proportional to λ_i. With this interpretation λ_i could give a contribution to both fermionic and bosonic masses.

3. If the coset construction for super-symplectic and super Kac-Moody algebra implying Equivalence Principle is accepted, one encounters what looks like a problem. p-Adic mass calculations require negative ground state conformal weight compensated by Super Virasoro generators in order to obtain massless states. The tachyonicity of the ground states would mean a close analogy with both string models and Higgs mechanism. λ_i^2 is very natural candidate for the ground state conformal weights identified but would have wrong sign if the effective metric of X_l^3 defined by the inner products $T_K^{k\alpha} T_K^{l\beta} h_{kl}$ of the Kähler energy momentum tensor $T^{k\alpha} = h^{kl}\partial L_K/\partial h_\alpha^l$ and appearing in the modified Dirac operator D_K has Minkowskian signature.

   The situation changes if the effective metric has Euclidian signature. This seems to be the case for the light-like surfaces assignable to the known extremals such as MEs and cosmic strings. In this kind of situation light-like coordinate possesses Euclidian signature and real eigenvalue spectrum is replaced with a purely imaginary one. Since Dirac operator is in question both signs for eigenvalues are possible and one obtains both exponentially increasing and decreasing solutions. This is essential for having solutions extending from the past end of X_l^3 to its future end. Non-unitary time evolution is possible because X_l^3 does not strictly speaking represent the time evolution of 2-D dynamical object but actual dynamical objects (by light-likeness both interpretation as dynamical evolution and dynamical object are present). The Euclidian signature of the effective metric would be a direct analog for the tachyonicity of the Higgs in unstable minimum and the generation of Higgs vacuum expectation would correspond to the compensation of ground state conformal weight by conformal weights of Super Virasoro generators.

4. In accordance with this λ_i^2 would give constant contribution to the ground state conformal weight. What contributes to the thermal mass squared is the deviation of the ground state conformal weight from half-odd integer since the negative integer part of the total conformal weight can be

compensated by applying Virasoro generators to the ground state. The first guess motivated by cyclotron energy analogy is that the lowest conformal weights are of form $h_c = \lambda_i^2 = -1/2 - n + \Delta h_c$ so that lowest ground state conformal weight would be $h_c = -1/2$ in the first approximation. The negative integer part of the net conformal weight can be canceled using Super Virasoro generators but Δh_c would give to mass squared a contribution analogous to Higgs contribution. The mapping of the real ground state conformal weight to a p-adic number by canonical identification involves some delicacies.

5. p-Adic mass calculations are consistent with the assumption that Higgs type contribution is vanishing (that is small) for fermions and dominates for gauge bosons. This requires that the deviation of λ_i^2 with smallest magnitude from half-odd integer value in the case of fermions is considerably smaller than in the case of gauge bosons in the scale defined by p-adic mass scale $1/L(k)$ in question. Somehow this difference could relate to the fact that bosons correspond to pairs of wormhole throats.

5.6 General mass formula for Ramond representations

By taking the modular contribution from the boundaries into account the general p-adic mass formulas for the Ramond type states read for states for which the color contribution to the conformal weight is integer valued as

$$\begin{aligned}
\frac{m^2(\Delta = 0)}{m_0^2} &= (8 + n(g))p + Yp^2 \ , \\
\frac{m^2(\Delta = 1)}{m_0^2} &= (5 + n(g)p + Yp^2 \ , \\
\frac{m^2(\Delta = 2)}{m_0^2} &= (4 + n(g))p + (Y + \frac{23}{2})p^2 \ , \\
n(g) &= 3g \cdot 2^{g-1}(2^g + 1) \ .
\end{aligned} \tag{5.8}$$

Here Δ denotes the conformal weight of the operators creating massless states from the ground state and g denotes the genus of the boundary component. The values of $n(g)$ for the three lowest generations are $n(0) = 0$, $n(1) = 9$ and $n(2) = 60$. The value of second order thermal contribution is nontrivial for neutrinos only. The value of the rational number Y can, which corresponds to the renormalization correction to the mass, can be determined using experimental inputs.

Using m_0^2 as a unit, the expression for the mass of a Ramond type state reads in terms of the electron mass as

$$\begin{aligned}
M(\Delta, g, p)_R &= K(\Delta, g, p)\sqrt{\frac{M_{127}}{p}} m_e \\
K(0, g, p) &= \sqrt{\frac{n(g) + 8 + Y_R}{X}} \\
K(1, g, p) &= \sqrt{\frac{n(g) + 5 + Y_R}{X}} \\
K(2, g, p) &= \sqrt{\frac{n(g) + 4 + Y_R}{X}} \ , \\
X &= \sqrt{5 + Y(e)_R} \ .
\end{aligned} \tag{5.9}$$

Y can be assumed to depend on the electromagnetic charge and color representation of the state and is therefore same for all fermion families. Mathematica provides modules for calculating the real counterpart of the second order contribution and for finding realistic values of Y.

5.7 General mass formulas for NS representations

Using $m_0^2/3$ as a unit, the expression for the mass of a light NS type state for $T_p = 1$ ad $k_B = 1$ reads in terms of the electron mass as

$$\begin{aligned} M(\Delta, g, p, N)_R &= K(\Delta, g, p, N)\sqrt{\frac{M_{127}}{p}} m_e \\ K(0, g, p, 1) &= \sqrt{\frac{n(g) + Y_R}{X}} , \\ K(0, g, p, 2) &= \sqrt{\frac{n(g) + 1 + Y_R}{X}} , \\ K(1, g, p, 3) &= \sqrt{\frac{n(g) + 3 + Y_R}{X}} , \\ K(2, g, p, 4) &= \sqrt{\frac{n(g) + 5 + Y_R}{X}} , \\ K(2, g, p, 5) &= \sqrt{\frac{n(g) + 10 + Y_R}{X}} , \\ X &= \sqrt{5 + Y(e)_R} . \end{aligned} \tag{5.10}$$

Here N is the number of the 'active' NS sectors (sectors for which the conformal weight of the massless state is non-vanishing). Y denotes the renormalization correction to the boson mass and in general depends on the electro-weak and color quantum numbers of the boson.

The thermal contribution to the mass of W boson is too large by roughly a factor $\sqrt{3}$ for $T_p = 1$. Hence $T_p = 1/2$ must hold true for gauge bosons and their masses must have a non-thermal origin perhaps analogous to Higgs mechanism. Alternatively, the non-covariant constancy of charge matrices could induce the boson mass [8].

It is interesting to notice that the minimum mass squared for gauge boson corresponds to the p-adic mass unit $M^2 = m_0^2 p/3$ and this just what is needed in the case of W boson. This forces to ask whether $m_0^2/3$ is the correct choice for the mass squared unit so that non-thermally induced W mass would be the minimal $m_W^2 = p$ in the lowest order. This choice would mean the replacement

$$Y_R \rightarrow \frac{(3Y)_R}{3}$$

in the preceding formulas and would affect only neutrino mass in the fermionic sector. $m_0^2/3$ option is excluded by charged lepton mass calculation. This point will be discussed later.

5.8 Primary condensation levels from p-adic length scale hypothesis

p-Adic length scale hypothesis states that the primary condensation levels correspond to primes near prime powers of two $p \simeq 2^k$, k integer with prime values preferred. Black hole-elementary particle analogy [13] suggests a generalization of this hypothesis by allowing k to be a power of prime. The general number theoretical vision discussed in [17] provides a first principle justification for p-adic length scale hypothesis in its most general form. The best fit for the neutrino mass squared differences is obtained for $k = 13^2 = 169$ so that the generalization of the hypothesis might be necessary.

A particle primarily condensed on the level k can suffer secondary condensation on a level with the same value of k: for instance, electron ($k = 127$) suffers secondary condensation on $k = 127$ level. u, d, s quarks ($k = 107$) suffer secondary condensation on nuclear space-time sheet having $k = 113$). All quarks feed their color gauge fluxes at $k = 107$ space-time sheet. There is no deep reason forbidding the condensation of p on p. Primary and secondary condensation levels could also correspond to different but nearly identical values of p with the same value of k.

6 Fermion masses

In the earlier model the coefficient of $M^2 = kL_0$ had to be assumed to be different for various particle states. $k = 1$ was assumed for bosons and leptons and $k = 2/3$ for quarks. The fact that $k = 1$ holds true for all particles in the model including also super-symplectic invariance forces to modify the earlier construction of quark states. This turns out to be possible without affecting the earlier p-adic mass calculations whose outcome depend in an essential manner on the ground state conformal weights h_{gr} of the fermions (h_{gr} can be negative). The structure of lepton and quark states in color degrees of freedom was discussed in [8].

6.1 Charged lepton mass ratios

The overall mass scale for lepton and quark masses is determined by the condensation level given by prime $p \simeq 2^k$, k prime by length scale hypothesis. For charged leptons k must correspond to $k = 127$ for electron, $k = 113$ for muon and $k = 107$ for τ. For muon $p = 2^{113} - 1 - 4 * 378$ is assumed (smallest prime below 2^{113} allowing $\sqrt{2}$ but not $\sqrt{3}$). So called Gaussian primes are to complex integers what primes are for the ordinary integers and the Gaussian counterparts of the Mersenne primes are Gaussian primes of form $(1 \pm i)^k - 1$. Rather interestingly, $k = 113$ corresponds to a Gaussian Mersenne so that all charged leptons correspond to generalized Mersenne primes.

For $k = 1$ the leptonic mass squared is integer valued in units of m_0^2 only for the states satisfying

$$p \; mod \; 3 \neq 2 \ .$$

Only these representations can give rise to massless states. Neutrinos correspond to (p,p) representations with $p \geq 1$ whereas charged leptons correspond to $(p, p+3)$ representations. The earlier mass calculations demonstrate that leptonic masses can be understood if the ground state conformal weight is $h_{gr} = -1$ for charged leptons and $h_{gr} = -2$ for neutrinos.

The contribution of color partial wave to conformal weight is $h_c = (p^2+2p)/3$, $p \geq 1$, for neutrinos and $p = 1$ gives $h_c = 1$ (octet). For charged leptons $h_c = (p^2 + 5p + 6)/3$ gives $h_c = 2$ for $p = 0$ (decuplet). In both cases super-symplectic operator O must have a net conformal weight $h_{sc} = -3$ to produce a correct conformal weight for the ground state. p-adic considerations suggests the use of operators O with super-symplectic conformal weight $z = -1/2 - i \sum n_k y_k$, where $s_k = 1/2 + iy_k$ corresponds to zero of Riemann ζ. If the operators in question are color Hamiltonians in octet representation net super-symplectic conformal weight $h_{sc} = -3$ results. The tensor product of two octets with conjugate super-symplectic conformal weights contains both octet and decuplet so that singlets are obtained. What strengthens the hopes that the construction is not adhoc is that the same operator appears in the construction of quark states too.

Using CP_2 mass scale m_0^2 [8] as a p-adic unit, the mass formulas for the charged leptons read as

$$\begin{aligned}
M^2(L) &= A(\nu)\frac{m_0^2}{p(L)} \ , \\
A(e) &= 5 + X(p(e)) \ , \\
A(\mu) &= 14 + X(p(\mu)) \ , \\
A(\tau) &= 65 + X(p(\tau)) \ .
\end{aligned} \tag{6.1}$$

$X(\cdot)$ corresponds to the yet unknown second order corrections to the mass squared.

The following table gives the basic parameters as determined from the mass of electron for some values of Y_e. The mass of top quark favors as maximal value of CP_2 mass which corresponds to $Y_e = 0$.

Y_e	0	.5	.7798
$(m_0/m_{Pl}) \times 10^3$	.2437	.2323	.2266
$K \times 10^{-7}$	2.5262	2.7788	2.9202
$(L_R/\sqrt{G}) \times 10^{-4}$	3.1580	3.3122	3.3954

Table 1. Table gives the values of CP_2 mass m_0 using Planck mass $m_{Pl} = 1/\sqrt{G}$ as unit, the ratio $K = R^2/G$ and CP_2 geodesic length $L = 2\pi R$ for $Y_e \in \{0, 0.5, 0.7798\}$.

The following table lists the lower and upper bounds for the charged lepton mass ratios obtained by taking second order contribution to zero or allowing it to have maximum possible value. The values of lepton masses are $m_e = .510999$ MeV, $m_\mu = 105.76583$ MeV, $m_\tau = 1775$ MeV.

$$\begin{aligned}
\frac{m(\mu)_+}{m(\mu)} &= \sqrt{\frac{15}{5}2^7\frac{m_e}{(\mu)}} \simeq 1.0722 \ , \\
\frac{m(\mu)_-}{m(\mu)} &= \sqrt{\frac{14}{6}2^7\frac{m_e}{m(\mu)}} \simeq 0.9456 \ , \\
\frac{m(\tau)_+}{m(\tau)} &= \sqrt{\frac{66}{5}2^{10}\frac{m_e}{m(\tau)}} \simeq 1.0710 \ , \\
\frac{m(\tau)_-}{m(\tau)} &= \sqrt{\frac{65}{6}2^{10}\frac{m_e}{m(\tau)}} \simeq .9703 \ .
\end{aligned} \tag{6.2}$$

For the maximal value of CP_2 mass the predictions for the mass ratio are systematically too large by a few per cent. From the formulas above it is clear that the second order corrections to mass squared can be such that correct masses result.

τ mass is least sensitive to $X(p(e)) \equiv Y_e$ and the maximum value of $Y_e \equiv Y_{e,max}$ consistent with τ mass corresponds to $Y_{e,max} = .7357$ and $Y_\tau = 1$. This means that the CP_2 mass is at least a fraction .9337 of its maximal value. If Y_L is same for all charged leptons and has the maximal value $Y_{e,max} = .7357$, the predictions for the mass ratios are

$$\begin{aligned}
\frac{m(\mu)_{pr}}{m(\mu)} &= \sqrt{\frac{14+Y_{e,max}}{5+Y_{e,max}}} \times 2^7\frac{m_e}{m(\mu)} \simeq .9922 \ , \\
\frac{m(\tau)_{pr}}{m(\tau)} &= \sqrt{\frac{65+Y_{e,max}}{5+Y_e(max}} \times 2^{10}\frac{m_e}{m(\tau)} \simeq .9980 \ .
\end{aligned} \tag{6.3}$$

The error is .8 per cent *resp.*.2 per cent for muon *resp.* τ.

The argument leading to estimate for the modular contribution to the mass squared [8] leaves two options for the coefficient of the modular contribution for $g = 2$ fermions: the value of coefficient is either $X = g$ for $g \leq 1$, $X = 3g - 3$ for $g \geq 2$ or $X = g$ always. For $g = 2$ the predictions are $X = 2$ and $X = 3$ in the two cases. The option $X = 3$ allows slightly larger maximal value of Y_e equal to $Y_{e,max}^{1)} = Y_{e,max} + (5 + Y_{e,max})/66$.

6.2 Neutrino masses

The estimation of neutrino masses is difficult at this stage since the prediction of the primary condensation level is not yet possible and neutrino mixing cannot yet be predicted from the basic principles. The cosmological bounds for neutrino masses however help to put upper bounds on the masses. If one takes seriously the LSND data on neutrino mass measurement of [54, 46] and the explanation of the atmospheric ν-deficit in terms of $\nu_\mu - \nu_\tau$ mixing [50, 48] one can deduce that the most plausible condensation level of μ and τ neutrinos is $k = 167$ or $k = 13^2 = 169$ allowed by the more general form of the p-adic length scale hypothesis suggested by the blackhole-elementary particle analogy. One can also deduce information about the mixing matrix associated with the neutrinos so that mass predictions become rather precise. In particular, the mass splitting of μ and τ neutrinos is predicted correctly if one assumes that the mixing matrix is a rational unitary matrix.

6.2.1 Super Virasoro contribution

Using $m_0^2/3$ as a p-adic unit, the expression for the Super Virasoro contribution to the mass squared of neutrinos is given by the formula

$$\begin{aligned} M^2(SV) &= (s + (3Yp)_R/3)\frac{m_0^2}{p} , \\ s &= 4 \text{ or } 5 , \\ Y &= \frac{23}{2} + Y_1 , \end{aligned} \tag{6.4}$$

where m_0^2 is universal mass scale. One can consider two possible identifications of neutrinos corresponding to $s(\nu) = 4$ with $\Delta = 2$ and $s(\nu) = 5$ with $\Delta = 1$. The requirement that CKM matrix is sensible forces the asymmetric scenario in which quarks and, by symmetry, also leptons correspond to lowest possible excitation so that one must have $s(\nu) = 4$. Y_1 represents second order contribution to the neutrino mass coming from renormalization effects coming from self energy diagrams involving intermediate gauge bosons. Physical intuition suggest that this contribution is very small so that the precise measurement of the neutrino masses should give an excellent test for the theory.

With the above described assumptions and for $s = 4$, one has the following mass formula for neutrinos

$$\begin{aligned} M^2(\nu) &= A(\nu)\frac{m_0^2}{p(\nu))} , \\ A(\nu_e) &= 4 + \frac{(3Y(p(\nu_e)))_R}{3} , \\ A(\nu_\mu) &= 13 + \frac{(3Y(p(\nu_\mu)))_R}{3} , \\ A(\nu_\tau) &= 64 + \frac{(3Y(p(\nu_\tau)))_R}{3} , \\ 3Y &\simeq \frac{1}{2} . \end{aligned} \tag{6.5}$$

The predictions must be consistent with the recent upper bounds [43] of order 10 eV, 270 keV and 0.3 MeV for ν_e, ν_μ and ν_τ respectively. The recently reported results of LSND measurement [46] for $\nu_e - > \nu_\mu$ mixing gives string limits for $\Delta m^2(\nu_e, \nu_\mu)$ and the parameter $sin^2(2\theta)$ characterizing the mixing: the limits are given in the figure 30 of [46]. The results suggests that the masses of both electron and muon neutrinos are below 5 eV and that mass squared difference $\Delta m^2 = m^2(\nu_\mu) - m^2(\nu_e)$ is between $.25 - 25\ eV^2$. The simplest possibility is that ν_μ and ν_e have common condensation level (in analogy with d and s quarks). There are three candidates for the primary condensation level: namely $k = 163, 167$ and $k = 169$. The p-adic prime associated with the primary condensation level is assumed to be the nearest prime below 2^k allowing p-adic $\sqrt{2}$ but not $\sqrt{3}$ and satisfying $p\ mod\ 4 = 3$. The following table gives the values of various parameters and unmixed neutrino masses in various cases of interest.

k	p	$(3Y)_R/3$	$m(\nu_e)/eV$	$m(\nu_\mu)/eV$	$m(\nu_\tau)/eV$
163	$2^{163} - 4*144 - 1$	1.36	1.78	3.16	6.98
167	$2^{167} - 4*144 - 1$	.34	.45	.79	1.75
169	$2^{169} - 4*210 - 1$	.17	.22	.40	.87

6.2.2 Could neutrino topologically condense also in other p-adic length scales than $k = 169$?

One must keep mind open for the possibility that there are several p-adic length scales at which neutrinos can condense topologically. Biological length scales are especially interesting in this respect. In fact, all intermediate p-adic length scales $k = 151, 157, 163, 167$ could correspond to metastable neutrino states. The point is that these p-adic lengths scales are number theoretically completely exceptional in the sense that there exist Gaussian Mersenne $2^k \pm i$ (prime in the ring of complex integers) for all these values of k. Since charged leptons, atomic nuclei ($k = 113$) , hadrons and intermediate gauge bosons correspond to ordinary or Gaussian Mersennes, it would not be surprising if the biologically important Gaussian Mersennes would correspond to length scales giving rise to metastable neutrino states. Of course, one can keep mind open for the possibility that $k = 167$ rather than $k = 13^2 = 169$ is the length scale defining the stable neutrino physics.

6.2.3 Neutrino mixing

Consider next the neutrino mixing. A quite general form of the neutrino mixing matrix D given by the table below will be considered.

	ν_e	ν_μ	ν_τ
ν_e	c_1	s_1c_3	s_1s_3
ν_μ	$-s_1c_2$	$c_1c_2c_3 - s_2s_3exp(i\delta)$	$c_1c_2s_3 + s_2c_3exp(i\delta)$
ν_τ	$-s_1s_2$	$c_1s_2c_3 + c_2s_3exp(i\delta)$	$c_1s_2s_3 - c_2c_3exp(i\delta)$

Physical intuition suggests that the angle δ related to CP breaking is small and will be assumed to be vanishing. Topological mixing is active only in modular degrees of freedom and one obtains for the first order terms of mixed masses the expressions

$$\begin{aligned} s(\nu_e) &= 4+9|U_{12}|^2+60|U_{13}|^2 = 4+n_1 \ , \\ s(\nu_\mu) &= 4+9|U_{22}|^2+60|U_{23}|^2 = 4+n_2 \ , \\ s(\nu_\tau) &= 4+9|U_{32}|^2+60|U_{33}|^2 = 4+n_3 \ . \end{aligned} \tag{6.6}$$

The requirement that resulting masses are not ultraheavy implies that $s(\nu)$ must be small integers. The condition $n_1+n_2+n_3 = 69$ follows from unitarity. The simplest possibility is that the mixing matrix is a rational unitary matrix. The same ansatz was used successfully to deduce information about the mixing matrices of quarks. If neutrinos are condensed on the same condensation level, rationality implies that $\nu_\mu - \nu_\tau$ mass squared difference must come from the first order contribution to the mass squared and is therefore quantized and bounded from below.

The first piece of information is the atmospheric ν_μ/ν_e ratio, which is roughly by a factor 2 smaller than predicted by standard model [50]. A possible explanation is the CKM mixing of muon neutrino with τ-neutrino, whereas the mixing with electron neutrino is excluded as an explanation. The latest results from Kamiokande [50] are in accordance with the mixing $m^2(\nu_\tau) - m^2(\nu_\mu) \simeq 1.6 \cdot 10^{-2}\ eV^2$ and mixing angle $sin^2(2\theta) = 1.0$: also the zenith angle dependence of the ratio is in accordance with the mixing interpretation. If mixing matrix is assumed to be rational then only $k = 169$ condensation level is allowed for ν_μ and ν_τ. For this level $\nu_\mu - \nu_\tau$ mass squared difference turns out to be $\Delta m^2 \simeq 10^{-2}\ eV^2$ for $\Delta s \equiv s(\nu_\tau) - s(\nu_\mu) = 1$, which is the only acceptable possibility and predicts $\nu_\mu - \nu_\tau$ mass squared difference correctly within experimental uncertainties! The fact that the predictions for mass squared differences are practically exact, provides a precision test for the rationality assumption.

What is measured in LSND experiment is the probability $P(t,E)$ that ν_μ transforms to ν_e in time t after its production in muon decay as a function of energy E of ν_μ. In the limit that ν_τ and ν_μ masses are identical, the expression of $P(t,E)$ is given by

$$\begin{aligned} P(t,E) &= sin^2(2\theta)sin^2(\frac{\Delta Et}{2}) \ , \\ sin^2(2\theta) &= 4c_1^2s_1^2c_2^2 \ , \end{aligned} \tag{6.7}$$

where ΔE is energy difference of ν_μ and ν_e neutrinos and t denotes time. LSND experiment gives stringent conditions on the value of $\sin^2(2\theta)$ as the figure 30 of [46] shows. In particular, it seems that $sin^2(2\theta)$ must be considerably below 10^{-1} and this implies that s_1^2 must be small enough.

The study of the mass formulas shows that the only possibility to satisfy the constraints for the mass squared and $sin^2(2\theta)$ given by LSND experiment is to assume that the mixing of the electron neutrino with the tau neutrino is much larger than its mixing with the muon neutrino. This means that s_3 is quite near to unity. At the limit $s_3 = 1$ one obtains the following (nonrational) solution of the mass squared conditions for $n_3 = n_2 + 1$ (forced by the atmospheric neutrino data)

$$\begin{aligned} s_1^2 &= \frac{69 - 2n_2 - 1}{60} \ , \\ c_2^2 &= \frac{n_2 - 9}{2n_2 - 17} \ , \\ sin^2(2\theta) &= \frac{4(n_2 - 9)}{51}\frac{(34 - n_2)(n_2 - 4)}{30^2} \ , \\ s(\nu_\mu) - s(\nu_e) &= 3n_2 - 68 \ . \end{aligned} \tag{6.8}$$

The study of the LSND data shows that there is only one acceptable solution to the conditions obtained by assuming maximal mass squared difference for ν_e and ν_μ

$$\begin{aligned} n_1 &= 2 \ n_2 = 33 \ n_3 = 34 \ , \\ s_1^2 &= \frac{1}{30} \ c_2^2 = \frac{24}{49} \ , \\ sin^2(2\theta) &= \frac{24}{49}\frac{2}{15}\frac{29}{30} \simeq .0631 \ , \\ s(\nu_\mu) - s(\nu_e)) &= 31 \leftrightarrow .32 \ eV^2 \ . \end{aligned} \tag{6.9}$$

That c_2^2 is near $1/2$ is not surprise taking into account the almost mass degeneracy of ν_{mu} and ν_τ. From the figure 30 of [46] it is clear that this solution belongs to 90 per cent likelihood region of LSND experiment but $sin^2(2\theta)$ is about two times larger than the value allowed by Bugey reactor experiment. The study of various constraints given in [46] shows that the solution is consistent with bounds from all other experiments. If one assumes that $k > 169$ for ν_e $\nu_\mu - \nu_e$ mass difference increases, implying slightly poorer consistency with LSND data.

There are reasons to hope that the actual rational solution can be regarded as a small deformation of this solution obtained by assuming that c_3 is non-vanishing. $s_1^2 = \frac{69-2n_2-1}{60-51c_3^2}$ increases in the deformation by $O(c_3^2)$ term but if c_3 is positive the value of $c_2^2 \simeq \frac{24-102c_1^0c_2^0s_2^0c_3}{49} \sim \frac{24-61c_3}{49}$ decreases by $O(c_3)$ term so that it should be possible to reduce the value of $sin^2(2\theta)$. Consistency with Bugey reactor experiment requires $.030 \leq sin^2(2\theta) < .033$. $sin^2(2\theta) = .032$ is achieved for $s_1^2 \simeq .035, s_2^2 \simeq .51$ and $c_3^2 \simeq .068$. The construction of U and D matrices for quarks shows that very stringent number theoretic conditions are obtained and as in case of quarks it might be necessary to allow complex CP breaking phase in the mixing matrix. One might even hope that the solution to the conditions is unique.

For the minimal rational mixing one has $s(\nu_e) = 5$, $s(\nu_\mu) = 36$ and $s(\nu_\tau) = 37$ if unmixed ν_e corresponds to $s = 4$. For $s = 5$ first order contributions are shifted by one unit. The masses ($s = 4$ case) and mass squared differences are given by the following table.

k	$m(\nu_e)$	$m(\nu_\mu)$	$m(\nu_\tau)$	$\Delta m^2(\nu_\mu - \nu_e)$	$\Delta m^2(\nu_\tau - \nu_\mu)$
169	.27 eV	.66 eV	.67 eV	.32 eV^2	.01 eV^2

Predictions for neutrino masses and mass squared splittings for $k = 169$ case.

6.2.4 Evidence for the dynamical mass scale of neutrinos

In recent years (I am writing this towards the end of year 2004 and much later than previous lines) a great progress has been made in the understanding of neutrino masses and neutrino mixing. The pleasant news from TGD perspective is that there is a strong evidence that neutrino masses depend on environment [41]. In TGD framework this translates to the statement that neutrinos can suffer topological condensation in several p-adic length scales. Not only in the p-adic length scales suggested by the number theoretical considerations but also in longer length scales, as will be found.

The experiments giving information about mass squared differences can be divided into three categories [41].

1. There along baseline experiments, which include solar neutrino experiments [49, 52, 53] and [56] as well as earlier studies of solar neutrinos. These experiments see evidence for the neutrino mixing and involve significant propagation through dense matter. For the solar neutrinos and KamLAND the mass splittings are estimated to be of order $O(8 \times 10^{-5})$ eV^2 or more cautiously 8×10^{-5} eV^2 $< \delta m^2 < 2 \times 10^{-3}$ eV^2. For K2K and atmospheric neutrinos the mass splittings are of order $O(2 \times 10^{-3})eV^2$ or more cautiously $\delta m^2 > 10^{-3}$eV^2. Thus the scale of mass splitting seems to be smaller for neutrinos in matter than in air, which would suggest that neutrinos able to propagate through a dense matter travel at space-time sheets corresponding to a larger p-adic length scale than in air.

2. There are null short baseline experiments including CHOOZ, Bugey, and Palo Verde reactor experiments, and the higher energy CDHS, JARME, CHORUS, and NOMAD experiments, which involve muonic neutrinos (for references see [41]. No evidence for neutrino oscillations have been seen in these experiments.

3. The results of LSND experiment [46] are consistent with oscillations with a mass splitting greater than $3 \times 10^{-2}eV^2$. LSND has been generally been interpreted as necessitating a mixing with sterile neutrino. If neutrino mass scale is dynamical, situation however changes.

If one assumes that the p-adic length scale for the space-time sheets at which neutrinos can propagate is different for matter and air, the situation changes. According to [41] a mass 3×10^{-2} eV in air could explain the atmospheric results whereas mass of of order .1 eV and $.07eV^2 < \delta m^2 < .26eV^2$ would explain the LSND result. These limits are of the same order as the order of magnitude predicted by $k = 169$ topological condensation.

Assuming that the scale of the mass splitting is proportional to the p-adic mass scale squared, one can consider candidates for the topological condensation levels involved.

1. Suppose that $k = 169 = 13^2$ is indeed the condensation level for LSND neutrinos. $k = 173$ would predict $m_{\nu_e} \sim 7 \times 10^{-2}$ eV and $\delta m^2 \sim .02$ eV^2. This could correspond to the masses of neutrinos propagating through air. For $k = 179$ one has $m_{\nu_e} \sim .8 \times 10^{-2}$ eV and $\delta m^2 \sim 3 \times 10^{-4}$ eV^2 which could be associated with solar neutrinos and KamLAND neutrinos.

2. The primes $k = 157, 163, 167$ associated with Gaussian Mersennes would give $\delta m^2(157) = 2^6 \delta m^2(163) = 2^{10} \delta m^2(167) = 2^{12} \delta m^2(169)$ and mass scales $m(157) \sim 22.8$ eV, $m(163) \sim 3.6$ eV, $m(167) \sim .54$ eV. These mass scales are unrealistic or propagating neutrinos. The interpretation consistent with TGD inspired model of condensed matter in which neutrinos screen the classical Z^0 force generated by nucleons would be that condensed matter neutrinos are confined inside these space-time sheets whereas the neutrinos able to propagate through condensed matter travel along $k > 167$ space-time sheets.

6.2.5 The results of MiniBooNE group as a support for the energy dependence of p-adic mass scale of neutrino

The basic prediction of TGD is that neutrino mass scale can depend on neutrino energy and the experimental determinations of neutrino mixing parameters support this prediction. The newest results (11 April 2007) about neutrino oscillations come from MiniBooNE group which has published its first findings [42] concerning neutrino oscillations in the mass range studied in LSND experiments [40].

1. The motivation for MiniBooNE

Neutrino oscillations are not well-understood. Three experiments LSND, atmospheric neutrinos, and solar neutrinos show oscillations but in widely different mass regions (1 eV^2 , 3×10^{-3} eV^2, and 8×10^{-5} eV^2).

In TGD framework the explanation would be that neutrinos can appear in several p-adically scaled up variants with different mass scales and therefore different scales for the differences Δm^2 for neutrino masses so that one should not try to try to explain the results of these experiments using single neutrino mass scale. In single-sheeted space-time it is very difficult to imagine that neutrino mass scale would depend on neutrino energy since neutrinos interact so extremely weakly with matter. The best known attempt to assign single mass to all neutrinos has been based on the use of so called sterile neutrinos which do not have electro-weak couplings. This approach is an ad hoc trick and rather ugly mathematically and excluded by the results of MiniBooNE experiments.

2. The result of MiniBooNE experiment

The purpose of the MiniBooNE experiment was to check whether LSND result $\Delta m^2 = 1eV^2$ is genuine. The group used muon neutrino beam and looked whether the transformations of muonic neutrinos to electron neutrinos occur in the mass squared region $\Delta m^2 \simeq 1$ eV^2. No such transitions were found but there was evidence for transformations at low neutrino energies.

What looks first as an over-diplomatic formulation of the result was *MiniBooNE researchers showed conclusively that the LSND results could not be due to simple neutrino oscillation, a phenomenon in which one type of neutrino transforms into another type and back again.* rather than direct refutation of LSND results.

3. LSND and MiniBooNE are consistent in TGD Universe

The habitant of the many-sheeted space-time would not regard the previous statement as a mere diplomatic use of language. It is quite possible that neutrinos studied in MiniBooNE have suffered topological condensation at different space-time sheet than those in LSND if they are in different energy range (the preferred rest system fixed by the space-time sheet of the laboratory or Earth). To see whether this is the case let us look more carefully the experimental arrangements.

1. In LSND experiment 800 MeV proton beam entering in water target and the muon neutrinos resulted in the decay of produced pions. Muonic neutrinos had energies in 60-200 MeV range [40].

2. In MiniBooNE experiment [42] 8 GeV muon beam entered Beryllium target and muon neutrinos resulted in the decay of resulting pions and kaons. The resulting muonic neutrinos had energies the range 300-1500 GeV to be compared with 60-200 MeV.

Let us try to make this more explicit.

1. Neutrino energy ranges are quite different so that the experiments need not be directly comparable. The mixing obeys the analog of Schrödinger equation for free particle with energy replaced with $\Delta m^2/E$, where E is neutrino energy. The mixing probability as a function of distance L from the source of muon neutrinos is in 2-component model given by

$$P = sin^2(\theta)sin^2(1.27\Delta m^2 L/E) \ .$$

The characteristic length scale for mixing is $L = E/\Delta m^2$. If L is sufficiently small, the mixing is fifty-fifty already before the muon neutrinos enter the system, where the measurement is carried out and no mixing is detected. If L is considerably longer than the size of the measuring system, no mixing is observed either. Therefore the result can be understood if Δm^2 is much larger or much smaller than E/L, where L is the size of the measuring system and E is the typical neutrino energy.

2. MiniBooNE experiment found evidence for the appearance of electron neutrinos at low neutrino energies (below 500 MeV) which means direct support for the LSND findings and for the dependence of neutron mass scale on its energy relative to the rest system defined by the space-time sheet of laboratory.

3. Uncertainty Principle inspires the guess $L_p \propto 1/E$ implying $m_p \propto E$. Here E is the energy of the neutrino with respect to the rest system defined by the space-time sheet of the laboratory. Solar neutrinos indeed have the lowest energy (below 20 MeV) and the lowest value of Δm^2. However,

atmospheric neutrinos have energies starting from few hundreds of MeV and $\Delta;m^2$ is by a factor of order 10 higher. This suggests that the the growth of Δm^2 with E^2 is slower than linear. It is perhaps not the energy alone which matters but the space-time sheet at which neutrinos topologically condense. For instance, MiniBooNE neutrinos above 500 MeV would topologically condense at space-time sheets for which the p-adic mass scale is higher than in LSND experiments and one would have $\Delta m^2 >> 1$ eV^2 implying maximal mixing in length scale much shorter than the size of experimental apparatus.

4. One could also argue that topological condensation occurs in condensed matter and that no topological condensation occurs for high enough neutrino energies so that neutrinos remain massless. One can even consider the possibility that the p-adic length scale L_p is proportional to E/m_0^2, where m_0 is proportional to the mass scale associated with non-relativistic neutrinos. The p-adic mass scale would obey $m_p \propto m_0^2/E$ so that the characteristic mixing length would be by a factor of order 100 longer in MiniBooNE experiment than in LSND.

6.2.6 Comments

Some comments on the proposed scenario are in order: some of the are written much later than the previous text.

1. Mass predictions are consistent with the bound $\Delta m(\nu_\mu, \nu_e) < 2\ eV^2$ coming from the requirement that neutrino mixing does not spoil the so called r-process producing heavy elements in Super Novae [55].

2. TGD neutrinos cannot solve the dark matter problem: the total neutrino mass required by the cold+hot dark matter models would be about 5 eV. In [4] a model of galaxies based on string like objects of galaxy size and providing a more exotic source of dark matter, is discussed.

3. One could also consider the explanation of LSND data in terms of the interaction of ν_μ and nucleon via the exchange of $g = 1$ W boson. The fraction of the reactions $\bar{\nu}_\mu + p \rightarrow e^+ + n$ is at low neutrino energies $P \sim \frac{m_W^4(g=0)}{m_W^4(g=1)} sin^2(\theta_c)$, where θ_c denotes Cabibbo angle. Even if the condensation level of $W(g=1)$ is $k = 89$, the ratio is by a factor of order .05 too small to explain the average $\nu_\mu \rightarrow \nu_e$ transformation probability $P \simeq .003$ extracted from LSND data.

4. The predicted masses exclude MSW and vacuum oscillation solutions to the solar neutrino problem unless one assumes that several condensation levels and thus mass scales are possible for neutrinos. This is indeed suggested by the previous considerations.

6.3 Quark masses

The prediction or quark masses is more difficult due the facts that the deduction of even the p-adic length scale determining the masses of these quarks is a non-trivial task, and the original identification was indeed wrong. Second difficulty is related to the topological mixing of quarks. The new scenario leads to a unique identification of masses with top quark mass as an empirical input and the thermodynamical model of topological mixing as a new theoretical input. Also CKM matrix is predicted highly uniquely.

6.3.1 Basic mass formulas

By the earlier mass calculations and construction of CKM matrix the ground state conformal weights of U and D type quarks must be $h_{gr}(U) = -1$ and $h_{gr}(D) = 0$. The formulas for the eigenvalues of CP_2 spinor Laplacian imply that if m_0^2 is used as a unit, color conformal weight $h_c \equiv m_{CP_2}^2$ is integer for $p\ mod\ = \pm 1$ for U type quark belonging to $(p+1, p)$ type representation and obeying $h_c(U) = (p^2 + 3p + 2)/3$ and for $p\ mod\ 3 = 1$ for D type quark belonging $(p, p+2)$ type representation and obeying $h_c(D) = (p^2 + 4p + 4)/3$. Only these states can be massless since color Hamiltonians have integer valued conformal weights.

In the recent case the minimal $p = 1$ states correspond to $h_c(U) = 2$ and $h_c(D) = 3$. $h_{gr}(U) = -1$ and $h_{gr}(D) = 0$ reproduce the previous results for quark masses required by the construction of CKM matrix.

This requires super-symplectic operators O with a net conformal weight $h_{sc} = -3$ just as in the leptonic case. The facts that the values of p are minimal for spinor harmonics and the super-symplectic operator is same for both quarks and leptons suggest that the construction is not had hoc. The real justification would come from the demonstration that $h_{sc} = -3$ defines null state for SCV: this would also explain why h_{sc} would be same for all fermions.

Consider now the mass squared values for quarks. For $h(D) = 0$ and $h(U) = -1$ and using $m_0^2/3$ as a unit the expression for the thermal contribution to the mass squared of quark is given by the formula

$$\begin{aligned} M^2 &= (s+X)\frac{m_0^2}{p} \ , \\ s(U) &= 5 \ , \ s(D) = 8 \ , \\ X &\equiv \frac{(3Yp)_R}{3} \ , \end{aligned} \tag{6.10}$$

where the second order contribution Y corresponds to renormalization effects coming and depending on the isospin of the quark. When m_0^2 is used as a unit X is replaced by $X = (Y_p)_R$.

With the above described assumptions one has the following mass formula for quarks

$$\begin{array}{l} M^2(q) = A(q)\frac{m_0^2}{p(q)} \ , \\ \\ A(u) = 5 + X_U(p(u) \ , \quad A(c) = 14 + X_U(p(c)) \ , \quad A(t) = 65 + X_U(p(t)) \ , \\ A(d) = 8 + X_D(p(d)) \ , \quad A(s) = 17 + X_D(p(s)) \ , \quad A(b) = 68 + X_D(p(b)) \ . \end{array} \tag{6.11}$$

p-Adic length scale hypothesis allows to identify the p-adic primes labelling quarks whereas topological mixing of U and D quarks allows to deduce topological mixing matrices U and D and CKM matrix V and precise values of the masses apart from effects like color magnetic spin orbit splitting, color Coulombic energy, etc..

Integers n_{q_i} satisfying $\sum_i n(U_i) = \sum_i n(D_i) = 69$ characterize the masses of the quarks and also the topological mixing to high degree. The reason that modular contributions remain integers is that in the p-adic context non-trivial rationals would give CP_2 mass scale for the real counterpart of the mass squared. In the absence of mixing the values of integers are $n_d = n_u = 0$, $n_s = n_c = 9$, $n_b = n_t = 60$.

The fact that CKM matrix V expressible as a product $V = U^\dagger D$ of topological mixing matrices is near to a direct sum of 2×2 unit matrix and 1×1 unit matrix motivates the approximation $n_b \simeq n_t$. The large masses of top quark and of $t\bar{t}$ meson encourage to consider a scenario in which $n_t = n_b = n \leq 60$ holds true.

The model for topological mixing matrices and CKM matrix predicts U and D matrices highly uniquely and allows to understand quark and hadron masses in surprisingly detailed level.

1. $n_d = n_u = 60$ is not allowed by number theoretical conditions for U and D matrices and by the basic facts about CKM matrix but $n_t = n_b = 59$ allows almost maximal masses for b and t. This is not yet a complete hit. The unitarity of the mixing matrices and the construction of CKM matrix to be discussed in the next section forces the assignments

$$(n_d, n_s, n_b) = (5, 5, 59) \ , \quad (n_u, n_c, n_t) = (5, 6, 58) \ . \tag{6.12}$$

   fixing completely the quark masses apart possible Higgs contribution [12]. Note that top quark mass is still rather near to its maximal value.

2. The constraint that valence quark contribution to pion mass does not exceed pion mass implies the constraint $n(d) \leq 6$ and $n(u) \leq 6$ in accordance with the predictions of the model of topological mixing. $u - d$ mass difference does not affect $\pi^+ - \pi^0$ mass difference and the quark contribution to $m(\pi)$ is predicted to be $\sqrt{(n_d + n_u + 13)/24} \times 136.9$ MeV for the maximal value of CP_2 mass (second order p-adic contribution to electron mass squared vanishes).

6.3.2 The p-adic length scales associated with quarks and quark masses

The identification of p-adic length scales associated with the quarks has turned to be a highly non-trivial problem. The reasons are that for light quarks it is difficult to deduce information about quark masses for hadron masses and that the unknown details of the topological mixing (unknown until the advent of the thermodynamical model [12]) made possible several p-adic length scales for quarks. It has also become clear that the p-adic length scale can be different form free quark and bound quark and that bound quark p-adic scale can depend on hadron.

Two natural constraints have however emerged from the recent work.

1. Quark contribution to the hadron mass cannot be larger than color contribution and for quarks having $k_q \neq 107$ quark contribution to mass is added to color contribution to the mass. For quarks with same value of k conformal weight rather than mass is additive whereas for quarks with different value of k masses are additive. An important implication is that for diagonal mesons $M = q\overline{q}$ having $k(q) \neq 107$ the condition $m(M) \geq \sqrt{2}m_q$ must hold true. This gives strong constraints on quark masses.

2. The realization that scaled up variants of quarks explain elegantly the masses of light hadrons allows to understand large mass splittings of light hadrons without the introduction of strong isospin-isospin interaction.

The new model for quark masses is based on the following identifications of the p-adic length scales.

1. The nuclear p-adic length scale $L(k)$, $k = 113$, corresponds to the p-adic length scale determining the masses of u, d, and s quarks. Note that $k = 113$ corresponds to a so called Gaussian Mersenne. The interpretation is that quark massivation occurs at nuclear space-time sheet at which quarks feed their em fluxes. At $k = 107$ space-time sheet, where quarks feed their color gauge fluxes, the quark masses are vanishing in the first p-adic order. This could be due to the fact that the p-adic temperature is $T_p = 1/2$ at this space-time sheet so that the thermal contribution to the mass squared is negligible. This would reflect the fact that color interactions do not involve any counterpart of Higgs mechanism.

   p-Adic mass calculations turn out to work remarkably well for massive quarks. The reason could be that M_{107} hadron physics means that *all* quarks feed their color gauge fluxes to $k = 107$ space-time sheets so that color contribution to the masses becomes negligible for heavy quarks as compared to Super-Kac Moody and modular contributions corresponding to em gauge flux feeded to $k > 107$ space-time sheets in case of heavy quarks. Note that Z^0 gauge flux is feeded to space-time sheets at which neutrinos reside and screen the flux and their size corresponds to the neutrino mass scale. This picture might throw some light to the question of whether and how it might be possible to demonstrate the existence of M_{89} hadron physics.

   One might argue that $k = 107$ is not allowed as a condensation level in accordance with the idea that color and electro-weak gauge fluxes cannot be feeded at the space-time space time sheet since the classical color and electro-weak fields are functionally independent. The identification of η' meson as a bound state of scaled up $k = 107$ quarks is not however consistent with this idea unless one assumes that $k = 107$ space-time sheets in question are separate.

2. The requirement that the masses of diagonal pseudoscalar mesons of type $M = q\overline{q}$ are larger but as near as possible to the quark contribution $\sqrt{2}m_q$ to the valence quark mass, fixes the p-adic primes $p \simeq 2^k$ associated with c, b quarks but not t since toponium does not exist. These values of k are "nominal" since k seems to be dynamical. c quark corresponds to the p-adic length scale $k(c) = 104 = 2^3 \times 13$. b quark corresponds to $k(b) = 103$ for $n(b) = 5$. Direct determination of p-adic scale from top quark mass gives $k(t) = 94 = 2 \times 47$ so that secondary p-adic length scale is in question.

   Top quark mass tends to be slightly too low as compared to the most recent experimental value of $m(t) = 169.1$ GeV with the allowed range being [164.7, 175.5] GeV [44]. The optimal situation corresponds to $Y_e = 0$ and $Y_t = 1$ and happens to give top mass exactly equal to the most probable experimental value. It must be emphasized that top quark is experimentally in a unique position since toponium does not exist and top quark mass is that of free top.

In the case of light quarks there are good reasons to believe that the p-adic mass scale of quark is different for free quark and bound state quark and that in case of bound quark it can also depend on hadron. This would explain the notions of valence (constituent) quark and current quark mass as masses of bound state quark and free quark and leads also to a TGD counterpart of Gell-Mann-Okubo mass formula [12].

1. Constituent quark masses

Constituent quark masses correspond to masses derived assuming that they are bound to hadrons. If the value of k is assumed to depend on hadron one obtains nice mass formula for light hadrons as will be found later. The table below summarizes constituent quark masses as predicted by this model.

2. Current quark masses

Current quark masses would correspond to masses of free quarks which tend to be lower than valence quark masses. Hence k could be larger in the case of light quarks. The table of quark masses in Wikipedia [35] gives the value ranges for current quark masses depicted in the table below together with TGD predictions for the spectrum of current quark masses.

q	d	u	s
$m(q)_{exp}/MeV$	4-8	1.5-4	80-130
$k(q)$	(122,121,120)	(125,124,123,122)	(114,113,112)
$m(q)/MeV$	(4.5,6.6,9.3)	(1.4,2.0,2.9,4.1)	(74,105,149)
q	c	b	t
$m(q)_{exp}/MeV$	1150-1350	4100-4400	1691
$k(q)$	(106,105)	(105,104)	92
$m(q)/MeV$	(1045,1477)	(3823,5407)	167.8×10^3

Table 3. The experimental value ranges for current quark masses [35] and TGD predictions for their values assuming $(n_d, n_s, n_b) = (5, 5, 59)$, $(n_u, n_c, n_t) = (5, 6, 58)$, and $Y_e = 0$. For top quark $Y_t = 0$ is assumed. $Y_t = 1$ would give 169.2 GeV.

Some comments are in order.

1. The long p-adic length associated with light quarks seem to be in conflict with the idea that quarks have sizes smaller than hadron size. The paradox disappears when one realized that $k(q)$ characterizes the electromagnetic "field body" of quark having much larger size than hadron.

2. u and d current quarks correspond to a mass scale not much higher than that of electron and the ranges for mass estimates suggest that u could correspond to scales $k(u) \in (125, 124, 123, 122) = (5^3, 4\times31, 3\times41, 2\times61)$, whereas d would correspond to $k(d) \in (122, 121, 120) = (2\times61, 11^2, 3\times5\times8)$.

3. The TGD based model for nuclei based on the notion of nuclear string leads to the conclusion that exotic copies of $k = 113$ quarks having $k = 127$ are present in nuclei and are responsible for the color binding of nuclei [15, 20] ,[20].

4. The predicted values for c and b masses are slightly too low for $(k(c), k(b)) = (106, 105) = (2 \times 53, 3 \times 5 \times 7)$. Second order Higgs contribution could increase the c mass into the range given in [35] but not that of b.

5. The mass of top quark has been slightly below the experimental estimate for long time. The experimental value has been coming down slowly and the most recent value obtained by CDF [45] is $m_t = 165.1 \pm 3.3 \pm 3.1$ GeV and consistent with the TGD prediction for $Y_e = Y_t = 0$.

One can talk about constituent and current quark masses simultaneously only if they correspond to dual descriptions. $M^8 - H$ duality [8] has been indeed suggested to relate the old fashioned low energy description of hadrons in terms of $SO(4)$ symmetry (Skyrme model) and higher energy description of hadrons based on QCD. In QCD description the mass of say baryon would be dominated by the mass associated with super-symplectic quanta carrying color. In $SO(4)$ description constituent quarks would carry most of the hadron mass.

6.3.3 Can Higgs field develop a vacuum expectation in fermionic sector at all?

An important conclusion following from the calculation of lepton and quark masses is that if Higgs contribution is present, it can be of second order p-adically and even negligible, perhaps even vanishing. There is indeed an argument forcing to consider this possibility seriously. The recent view about elementary particles is following.

1. Fermions correspond to CP_2 type vacuum extremals topologically condensed at positive/negative energy space-time sheets carrying quantum numbers at light-like wormhole throat. Higgs and gauge bosons correspond to wormhole contacts connecting positive and negative energy space-time sheets and carrying fermion and anti-fermion quantum numbers at the two light-like wormhole throats.

2. If the values of p-adic temperature are $T_p = 1$ and $T_p = 1/n$, $n >$ 1f or fermions and bosons the thermodynamical contribution to the gauge boson mass is negligible.

3. Different p-adic temperatures and Kähler coupling strengths for fermions and bosons make sense if bosonic and fermionic partonic 3-surfaces meet only along their ends at the vertices of generalized Feynman diagrams but have no other common points [3]. This forces to consider the possibility that fermions cannot develop Higgs vacuum expectation value although they can couple to Higgs. This is not in contradiction with the modification of sigma model of hadrons based on the assumption that vacuum expectation of σ field gives a small contribution to hadron mass [9] since this field can be assigned to some bosonic space-time sheet pair associated with hadron.

4. Perhaps the most elegant interpretation is that ground state conformal is equal to the square of the eigenvalue of the modified Dirac operator. The ground state conformal weight is negative and its deviation from half odd integer value gives contribution to both fermion and boson masses. The Higgs expectation associated with coherent state of Higgs like wormhole contacts is naturally proportional to this parameter since no other parameter with dimensions of mass is present. Higgs vacuum expectation determines gauge boson masses only apparently if this interpretation is correct. The contribution of the ground state conformal weight to fermion mass square is near to zero. This means that λ is very near to negative half odd integer and therefore no significant difference between fermions and gauge bosons is implied.

q	d	u	s	c	b	t
n_q	4	5	6	6	59	58
s_q	12	10	14	11	67	63
$k(q)$	113	113	113	104	103	94
$m(q)/GeV$	.105	.092	.105	2.191	7.647	167.8

Table 2. Constituent quark masses predicted for diagonal mesons assuming $(n_d, n_s, n_b) = (5, 5, 59)$ and $(n_u, n_c, n_t) = (5, 6, 58)$, maximal CP_2 mass scale($Y_e = 0$), and vanishing of second order contributions.

7 Higgsy aspects of particle massivation

7.1 Can p-adic thermodynamics explain the masses of intermediate gauge bosons?

The requirement that the electron-intermediate gauge boson mass ratios are sensible, serves as a stringent test for the hypothesis that intermediate gauge boson masses result from the p-adic thermodynamics. It seems that the only possible option is that the parameter k has same value for both bosons, leptons, and quarks:

$$k_B = k_L = k_q = 1 \ .$$

In this case all gauge bosons have $D(0) = 1$ and there are good changes to obtain boson masses correctly. $k = 1$ together with $T_p = 1$ implies that the thermal masses of very many boson states are extremely heavy so that the spectrum of the boson exotics is reduced drastically. For $T_p = 1/2$ the thermal contribution to the mass squared is completely negligible.

Contrary to the original optimistic beliefs based on calculational error, it turned out impossible to predict W/e and Z/e mass ratios correctly in the original p-adic thermodynamics scenario. Although the errors are of order 20-30 percent, they seemed to exclude the explanation for the massivation of gauge bosons using p-adic thermodynamics.

1. The thermal mass squared for a boson state with N active sectors (non-vanishing vacuum weight) is determined by the partition function for the tensor product of N NS type Super Virasoro algebras. The degeneracies of the excited states as a function of N and the weight Δ of the operator creating the massless state are given in the table below.

2. Both W and Z must correspond to $N = 2$ active Super Virasoro sectors for which $D(1) = 1$ and $D(2) = 3$ so that (using the formulas of p-adic thermodynamics the thermal mass squared is $m^2 = k_B(p + 5p^2)$ for $T_p = 1$. The second order contribution to the thermal mass squared is extremely small so that Weinberg angle vanishes in the thermal approximation. $k_B = 1$ gives Z/e mass-ratio which is about 22 per cent too high. For $T_p = 1/2$ thermal masses are completely negligible.

3. The thermal prediction for W-boson mass is the same as for Z^0 mass and thus even worse since the two masses are related $M_W^2 = M_Z^2 cos^2(\theta_W)$.

7.2 Comparison of TGD Higgs and with MSSM Higgs

The notion of Higgs in TGD framework differs from that of standard model and super-symmetric extension in several respects. Very concisely, the two complex $SU(2)_V$ doublets are replaced with scalar and pseudoscalar triplet and singlet so that the number of field components is same. The Higgs possibly developing vacuum expectation is now uniquely the scalar singlet unless one allows parity breaking. The number of remaining Higgs field components is 5 as in the minimal supersymmetric extension of the standard model.

7.2.1 TGD based particle concept very briefly

Before attempt to clarify the differences between TGD and standard model Higgs it is good to list the basic ideas behind TGD based notion of particle.

1. Bosonic emergence means that gauge bosons and Higgs and their super partners can be in the first approximation regarded as wormhole contacts with the throats carrying quantum numbers of fermion and antifermion. A given throat carrying fermionic quantum numbers. Also many fermion states are possible and have interpretation in terms of a supersymmetry extending the ordinary space-time supersymmetry in which super-generators are simply the fermionic oscillator operators assignable to the partonic 2-surface. These generators can be used to construct various

super-conformal algebras. Right-handed neutrinos define the analog of ordinary space-time super-supersymmetry as it is encountered in MSSM. In topological condensation also fermions become wormhole contacts with second throat carrying purely bosonic quantum numbers.

2. The weak form of electric magnetic duality forces the conclusion that wormhole throats carry Kähler magnetic charges which much be neutralized by opposite Kähler magnetic charge. The natural idea is that monopole confinement is also behind color confinement and electroweak screening. In the case of color confinement the valence quarks would form wormhole throats connected by color magnetic flux tubes having total Kähler magnetic charge. Weak screening would mean that the throat compensating he Kähler magnetic charge of fermionic throat contains a neutrino-antineutrino pairs screening the weak isospin. This leaves to Z^0 coupling $I_L^3 + sin^2(\theta_W)Q_{em}$ and if classical Z^0 field is present this leads to an interaction distinguishing between TGD and standard model.

3. Particle massivation is described by p-adic thermodynamics. p-Adic thermodynamics cannot explain gauge boson masses and if it contributes the contribution is small and corresponds to low p-adic temperatures $T_p = 1/n$. It is not yet completely clear whether the generation of vacuum contribution to ground state conformal weight implying deviation from half-integer value is responsible for weak gauge boson masses. It might be sensible to speak about coherent state of Higgs bosons in zero energy ontology and also in the case of fermions if interacting fermions have suffered topological condensation. If this is the case Higgs vacuum expectation value defining the coherent state can contribute to the particle mass but only in the case of weak gauge bosons give a dominating contribution. It is not clear whether the generation of non-half-integer vacuum conformal weight and Higgs mechanisms could be seen descriptions of one and same thing.

7.2.2 Scalar and pseudo-scalar triplet and singlet instead of two doublets

TGD based notion of Higgs differs from its standard model and MSSM counterpart because the notion of spinor is different. If one believes on the following arguments, the basic implication is that two Higgs doublest of MSSM are replaced with scalar and pseudo-scalar triplet and singlet.

1. In TGD framework space-time spinors are induced spinors and therefore spinors of 8-D space $M^4 \times CP_2$. The mixing of M^4 chiralities in the modified Dirac equation in the space-time interior serves as a tell-tale signature for the massivation and does not imply mixing of the imbedding space chiralities identified in terms of leptons and quarks.

2. Group theoretically gauge bosons and Higgs itself corresponds to a tensor product of two $M^4 \times CP_2$ spinors giving rise to a spin singlet. In electroweak degrees of freedom one has a tensor product of right and left handed doublets decomposing to triplet and singlet under $SU(2)_V$. The first guess would be that one obtains just triplet 3 and singlet 1 whereas in standard model one has a complex $SU(2)_V$ doublet. In MSSM the cancellation of anomalies requires two doublets. As noticed, TGD allows supersymmetry generalizing the usual space-time supersymmetry and also no anomaly cancellation argument allows to expect a pairs of triplets and singlets.

3. One can assign fermion with "upper" throat and antifermion with the "lower" throat or vice versa and one can have both the sum or difference or these two states. This does not however imply additional degeneracy. Fermionic statistics requires the antisymmetry of the state with respect to the exchange of all quantum numbers. Spin and isospin triplets (singlets) are symmetric (antisymmetric) under the exchange of spin quantum numbers and singlets antisymmetric. In the case of Higgs triplet (singlet) the sum (difference) of these states must be assumed and there is no additional degeneracy.

4. One can construct gauge bosons and Higgs type particles from both quarks and leptons. The requirement that the gauge bosons couple to both quarks and leptons implies that they correspond to sums of these Higgses and behave like H-vectors for one has $\Gamma_9 = 1$. One can however ask whether also H-axial vector gauge bosons and Higgs with $\Gamma_9 = -1$ should be allowed. They are not suggested by the study of the modified Dirac equation and it seems that this leads to physically non-sensical results. First of all, the exchanges of vectorial and axial Higgses between leptons and quarks would be of opposite sign and at high energies the sum over these exchanges would approach

zero so that quark and lepton sectors would separate into non-interacting worlds. It is also difficult to imagine how one could avoid H-axial massless photon. p-Adic thermodynamics would allow the H-axial photon to become massive but it is not possible to understand how the H-axial scalar Higgs could transform to a longitudinal degree of freedom of the resulting H-axial photon.

5. One can construct the most general candidate for a Higgs particle using as charge matrix contracted between spinors associated with the opposite wormhole throats the product of a vector in the tangent space of CP_2 represented as sum of constant gamma matrices γ_A and electroweak charge matrix. One can express the products of the CP_2 gamma matrices and charge matrices in terms of CP_2 gamma matrices γ_A and and $\gamma_5(CP_2)\gamma_A$. The action of CP_2 gamma matrix γ_5 however reduces to that of $\epsilon\gamma_5(M^4)$, where the sign factor $\epsilon = \pm 1$ depends on H-chirality. Therefore one would have scalar Higgs and pseudo-scalar Higgs and the couplings of pseudo-scalar Higgs are of opposite sign for quarks and leptons. In unitary gauge one would have neutral scalar Higgs and 4 pseudo-scalar Higgses with the same charge spectrum as in MSSM. One can indeed construct Higgs particles as fermion-antifermion pairs by using products of charge matrices and CP_2 tangent space vector and transform them to scalar and pseudoscalar multiplets.

6. In Higgs mechanism the key idea is that one can represent the directional degrees of freedom of Higgs field in terms of coset space G/H, now $SU(2)_L \times U(1)/U(1)_{em}$. Therefore Higgs field can be written as in the form $exp(\sum_{T_a \in t} T^a \xi_a/v)(\rho + v)$, $t = g - h$, where v is the expectation value of the Higgs field fixing a preferred direction. The gauge transformation $g = exp(-\sum_{T_a \in t} T^a \xi_a/v)$ transforms Higgs to $\rho + v$ so that the degrees of freedom corresponding to the direction of Higgs are "eaten" by charge gauge potentials. In the resulting gauge the action contains only the YM part and Higgs term restricted to the fluctuations of Higgs around vacuum in the direction of v.

   In the recent case the coset space would be the coset space of the holonomy group of CP_2 divided by the subgroup defined by electromagnetic charge commuting with the vacuum expectation value which is therefore linear combination of γ_0 and γ_3 in the most general case. The condition that entire $SU(2)_V$ leaves invariant the preferred direction fixes this direction to γ_0 which corresponds to the radial coordinate of CP_2 in the standard vielbein basis. In the recent case CP_2 holonomy group naturally defines a preferred direction of Higgs field and it seems that vacuum expectation value is not necessary for the elimination of the charged Higgs. Neutral Higgs would essentially correspond to the magnitude of the Higgs field.

7. If the TGD based description of radiative corrections relying on the notion of generalized Feynman diagram is approximately equivalent with QFT based description and if should not differ too dramatically from those of MSSM in the approximation of $N = 1$ supersymmetry meaning that only the super partners obtained using right handed neutrinos and antineutrinos are taken into account. At high energies the the action of γ_5 gives only a minus sign telling the M^4 chirality of approximately massless particle and one has right to expect that the effects of pseudo-scalar exchange in loops do not differ dramatically from those of a scalar exchange.

7.2.3 Can one identify a classical correlate for the Higgs?

The natural question is whether one can identify classical correlates for the Higgs field and massivation. Kähler action does not allow to identify any obvious correlates whereas Kähler Dirac action does.

1. Kähler Dirac action in the interior of space-time surface should contain the counterpart of Higgs term whose signature is that it mixes M^4 chiralities. The interaction term analogous to that appearing in the ordinary Dirac action coupled to gauge fields is

$$\begin{aligned}
L_{int} &= \overline{\Psi}\hat{\Gamma}^{\alpha}A_{\alpha}\Psi \ , \\
\hat{\Gamma}^{\alpha} &= T^{\alpha k}\gamma_k = T^{\alpha k}\gamma_k(M^4) + T^{\alpha k}\gamma_k(CP_2) \ , \\
T^{\alpha k} &= \frac{\partial L_K}{\partial(\partial_{\alpha}h^k)} \ .
\end{aligned} \tag{7.1}$$

Here A_α are the components of the induced spinor connection. $T^{\alpha k}$ denotes canonical momentum densities and conserved momentum and color currents are closely related to them. They are required by internal consistency (in particular, by the consistency with the vacuum degeneracy of Kähler action) and super-symmetry. If action were defined by the volume of space-time surface in the induced metric, modified gamma matrices would reduce to induced gamma matrices coding information about classical gravitational fields. Also now information about gravitation is coded besides the dynamics of Kähler action associated with zero modes. Kähler field can indeed be said to characterize zero modes locally whereas quantum fluctuating degrees contributing to the WCW metric and therefore identifiable as gravitational degrees of freedom in generalized sense of the word [1].

The modified gamma matrices decompose to two parts corresponding to M^4 and CP_2 gamma matrices and the presence of CP_2 gamma matrices implies the mixing of M^4 chiralities so that massivation is unavoidable once one has a space-time surface which does not correspond to the canonically imbedded M^4. Also the kinetic part of $\hat{\Gamma}^\alpha\partial_\alpha$ contains a term mixing M^4 chiralities having no obvious counterpart in the ordinary Dirac equation. The important conclusion is that whatever the dynamical details of massivation are it must take place.

2. The interaction term $T^{\alpha k}\gamma_k(CP_2)A_\alpha$ of the modified Dirac action defined by the contraction of canonically conjugate momenta with gauge potentials mixes M^4 chiralities so that it is in this sense analogous to Higgs coupling. In gauge transformations the gauge potentials however transform inhomogenously. Does this mean that the term in question can be interpreted only as a signature for the presence of particle massivation or is also the identification as the classical counterpart of Higgs field sensical?

   Optimist could argue that there is a natural preferred gauge associated with the classical spinor connection. For instance, the Coulomb interaction term for Kähler action vanishes in preferred gauge for the general solution ansatz implying the reduction of Kähler function to Chern-Simons term for extremal in presence of a constraint expressing the weak form of electric-magnetic duality. For Kähler gauge potential this gauge is highly unique. Also, if one imagines adding to the induced gauge potential quantum fluctuating part representing the quantum field, one could say that the classical Higgs field transforms homogenously and that quantum part is gauge transformed inhomogenously. The situation remains unsettled.

3. The classical correlate for the Higgs field in TGD is not a genuine scalar field but defines a vector in the 4-D tangent space of CP_2. This allows to speak about CP_2 polarization. If the notion of unitary gauge meaning that an electro-weak gauge rotation takes Higgs to a standard direction invariant under $SU(2)_V$ rotations - in particular those induced by the vectorial isospin I_V^3 appearing in electromagnetic charge- then one can say that CP_2 polarization is always in the same direction for the scalar Higgs. In the case of pseudo-scalar Higgs all four CP_2 polarizations a possible.

7.3 How TGD based description of particle massivation relates to Higgs mechanism

In TGD framework p-adic thermodynamics gives the dominating contribution to fermion masses, which is something completely new. In the case of gauge bosons thermodynamic contribution is small since the inverse integer valued p-adic temperature is $T = 1/2$ for bosons or even lower: for fermions one has $T = 1$.

Whether Higgs can contribute to the masses is not completely clear. In TGD framework Mexican hat potential however looks like trick. One must however keep in mind that any other mechanism must explain the ratio of W and Z^0 masses and how these bosons receive their longitudinal polarizations. One must also consider seriously the possibility that all components for the TGD counterpart of Higgs boson are transformed to the longitudinal polarizations of the gauge bosons. Twistorial approach to TGD indeed strongly suggests that also the gauge bosons regarded usually as massless have a small mass guaranteeing cancellation of IR singularities. As I started write to write this piece of text I believed that photon does not eat Higgs but had to challenge my beliefs. Maybe there is no Higgs to be found at LHC! Only pseudo-scalar partner of Higgs would remain to be discovered.

The weak form of electric magnetic duality implying that each wormhole throat carrying fermionic quantum numbers is accompanied by a second wormhole throat carrying opposite magnetic charge and neutrino pair screening weak isospin and making gauge bosons massive. Concerning the implications the following view looks the most plausible one at this moment.

1. Neutral Higgs-if not eaten by photon- could develop a coherent state meaning vacuum expectation value and this is naturally proportional to the inverse of the p-adic length scale as are boson masses. This contribution can be assigned to the magnetic flux tube mentioned above since it screens weak force - or equivalently - makes them massive. Higgs expectation would not cause boson massivation. Rather, massivation and Higgs vacuum expectation would be caused by the presence of the magnetic flux tubes. Standard model would suffer from a causal illusion. Even a worse illusion is possible if the photon eats the neutral Higgs.

2. The "stringy" magnetic flux tube connecting fermion wormhole throat and the wormhole throat containing neutrino pair would give to the vacuum conformal weight a small contribution and therefore to the mass squared of both fermions and gauge bosons (dominating one for the latter). This contribution would be small in the p-adic sense (proportional $1/p^2$ rather than $1/p$). I cannot calculate this "stringy" contribution but stringy formula in weak scale is very suggestive.

3. In the case of light fermions and massless gauge bosons the stringy contribution must vanish and therefore must correspond to $n = 0$ string excitation (string does not vibrate at all) : otherwise the mass of fermion would be of order weak boson mass. For weak bosons $n = 1$ would look like a natural identification but also $n = 0$ makes sense since $h \pm 1$ states corresponds opposite three-momenta for massless fermion and antifermion so that the state is massive. The mechanism bringing in the $h = 0$ helicity of gauge boson would be the TGD counterpart for the transformation of Higgs component to a longitudinal polarization. $n \geq 0$ excited states of fermions and $n \geq 1$ excitations of bosons having masses above weak boson masses are predicted and would mean new physics becoming possibly visible at LHC.

7.4 The identification of Higgs

Consider now the identification of Higgs in TGD framework.

1. In TGD framework Higgs particles do not correspond to complex $SU(2)$ doublets but to triplet and singlet having same quantum numbers as gauge bosons. Therefore the idea that photon eats neutral Higgs is suggestive. Also a pseudo-scalar variant of Higgs is predicted. Let us see how these states emerge from weak strings.

2. The two kinds of massive states corresponding to $n = 0$ and $n = 1$ give rise to massive spin 1 and spin 2 particles. First of all, the helicity doublet $(1, -1)$ is necessarily massive since the 3-momenta for massless fermion and anti-fermion are opposite. For $n = L = 0$ this gives two states but helicity zero component is lacking. For $n = L = 1$ one has tensor product of doublet $(1, -1)$ and angular momentum triplet formed by $L = 1$ rotational state of the weak string. This gives 2×3 states corresponding to $J = 0$ and $J = 2$ multiplets. Note however than in spin degrees of freedom the Higgs candidate is not a genuine elementary scalar particle.

3. Fermion and antifermion can have parallel three momenta summing up to a massless 4-momentum. Spin vanishes so that one has Higgs like particle also now. This particle is however pseudo-scalar being group theoretically analogous to meson formed as a pair of quark and antiquark. p-Adic thermodynamics gives a contribution to the mass squared. By taking a tensor product with rotational states of strings one would obtain Regge trajectory containing pseudoscalar Higgs as the lowest state.

7.5 Do all gauge bosons possess small mass?

Consider now the problem how the gauge bosons can eat the Higgs boson to get their longitudinal component.

1. ($J = 0, n = 1$) Higgs state can be combined with $n = 0$ $h = \pm 1$ doublet to give spin 1 massive triplet provided the masses of the two states are same. This will be discussed below.

2. Also gauge bosons usually regarded as massless can eat the scalar Higgs so that Higgs like particle could disappear completely. There would be no Higgs to be discovered at LHC! But is this a real prediction? Could it be that it is not possible to have exactly massless photons and gluons? The mixing of M^4 chiralities for Chern-Simons Dirac equation implies that also collinear massless fermion and antifermion can have helicity ± 1. The problem is that the mixing of the chiralities is a signature of massivation!

   Could it really be that even the gauge bosons regarded as massless have a small mass characterized by the length scale of the causal diamond defining the physical IR cutoff and that the remaining Higgs component would correspond to the longitudinal component of photon? This would mean the number of particles in the final states for a particle reaction with a fixed initial state is always bounded from above. This is important for the twistorial aesthetics of generalized Feynman diagrammatics implied by zero energy ontology. Also the vanishing of IR divergences is guaranteed by a small physical mass [19]. Maybe internal consistency allows only pseudo-scalar Higgs.

7.6 Weak Regge trajectories

The weak form of electric-magnetic duality suggests strongly the existence of weak Regge trajectories.

1. The most general mass squared formula with spin-orbit interaction term $M_{L-S}^2 L \cdot S$ reads as

$$M^2 = nM_1^2 + M_0^2 + M_{L-S}^2 L \cdot S \ , \ n = 0, 2, 4 \text{ or } n = 1, 3, 5, ..., \ . \tag{7.2}$$

   M_1^2 corresponds to string tension and M_0^2 corresponds to the thermodynamical mass squared and possible other contributions. For a given trajectory even (odd) values of n have same parity and can correspond to excitations of same ground state. From ancient books written about hadronic string model one vaguely recalls that one can have several trajectories (satellites) and if one has something called exchange degeneracy, the even and odd trajectories define single line in $M^2 - J$ plane. As already noticed TGD variant of Higgs mechanism combines together $n = 0$ states and $n = 1$ states to form massive gauge bosons so that the trajectories are not independent.

2. For fermions, possible Higgs, and pseudo-scalar Higgs and their super partners also p-adic thermodynamical contributions are present. M_0^2 must be non-vanishing also for gauge bosons and be equal to the mass squared for the $n = L = 1$ spin singlet. By applying the formula to $h = \pm 1$ states one obtains

$$M_0^2 = M^2(boson) \ . \tag{7.3}$$

   The mass squared for transversal polarizations with $(h, n, L) = (\pm 1, n = L = 0, S = 1)$ should be same as for the longitudinal polarization with $(h = 0, n = L = 1, S = 1, J = 0)$ state. This gives

$$M_1^2 + M_0^2 + M_{L-S}^2 L \cdot S = M_0^2 \ . \tag{7.4}$$

   From $L \cdot S = [J(J+1) - L(L+1) - S(S+1)]/2 = -2$ for $J = 0, L = S = 1$ one has

$$M_{L-S}^2 = -\frac{M_1^2}{2} \ . \tag{7.5}$$

   Only the value of weak string tension M_1^2 remains open.

3. If one applies this formula to arbitrary $n = L$ one obtains total spins $J = L+1$ and $L-1$ from the tensor product. For $J = L-1$ one obtains

$$M^2 = (2n+1)M_1^2 + M_0^2 \ .$$

For $J = L+1$ only M_0^2 contribution remains so that one would have infinite degeneracy of the lightest states. Therefore stringy mass formula must contain a non-linear term making Regge trajectory curved. The simplest possible generalization which does not affect n=0 and n=1 states is of from

$$M^2 = n(n-1)M_2^2 + (n - \frac{L \cdot S}{2})M_1^2 + M_0^2 \ . \tag{7.6}$$

The challenge is to understand the ratio of W and Z^0 masses, which is purely group theoretic and provides a strong support for the massivation by Higgs mechanism.

1. The above formula and empirical facts require

$$\frac{M_0^2(W)}{M_0^2(Z)} = \frac{M^2(W)}{M^2(Z)} = cos^2(\theta_W) \ . \tag{7.7}$$

in excellent approximation. Since this parameter measures the interaction energy of the fermion and antifermion decomposing the gauge boson depending on the net quantum numbers of the pair, it would look very natural that one would have

$$M_0^2(W) = g_W^2 M_{SU(2)}^2 \ , \ M_0^2(Z) = g_Z^2 M_{SU(2)}^2 \ . \tag{7.8}$$

Here $M_{SU(2)}^2$ would be the fundamental mass squared parameter for $SU(2)$ gauge bosons. p-Adic thermodynamics of course gives additional contribution which is vanishing or very small for gauge bosons.

2. The required mass ratio would result in an excellent approximation if one assumes that the mass scales associated with $SU(2)$ and $U(1)$ factors suffer a mixing completely analogous to the mixing of $U(1)$ gauge boson and neutral $SU(2)$ gauge boson W_3 leading to γ and Z^0. Also Higgs, which consists of $SU(2)$ triplet and singlet in TGD Universe, would very naturally suffer similar mixing. Hence $M_0(B)$ for gauge boson B would be analogous to the vacuum expectation of corresponding mixed Higgs component. More precisely, one would have

$$\begin{aligned} M_0(W) &= M_{SU(2)} \ , \\ M_0(Z) &= cos(\theta_W)M_{SU(2)} + sin(\theta_W)M_{U(1)} \ , \\ M_0(\gamma) &= -sin(\theta_W)M_{SU(2)} + cos(\theta_W)M_{U(1)} \ . \end{aligned} \tag{7.9}$$

The condition that photon mass is very small and corresponds to IR cutoff mass scale gives $M_0(\gamma) = \epsilon cos(\theta_W)M_{SU(2)}$, where ϵ is very small number, and implies

$$\begin{aligned} \frac{M_{U(1)}}{M(W)} &= tan(\theta_W) + \epsilon \ , \\ \frac{M(\gamma)}{M(W)} &= \epsilon \times cos(\theta_W) \ , \\ \frac{M(Z)}{M(W)} &= \frac{1 + \epsilon \times sin(\theta_W)cos(\theta_W)}{cos(\theta_W)} \ . \end{aligned} \tag{7.10}$$

There is a small deviation from the prediction of the standard model for W/Z mass ratio but by the smallness of photon mass the deviation is so small that there is no hope of measuring it. One can of course keep mind open for $\epsilon = 0$. The formulas allow also an interpretation in terms of Higgs vacuum expectations as it must. The vacuum expectation would most naturally correspond to interaction energy between the massless fermion and antifermion with opposite 3-momenta at the throats of the wormhole contact and the challenge is to show that the proposed formulas characterize this interaction energy. Since CP_2 geometry codes for standard model symmetries and their breaking, it woul not be surprising if this would happen. One cannot exclude the possibility that p-adic thermodynamics contributes to $M_0^2(boson)$. For instance, ϵ might characterize the p-adic thermal mass of photon.

If the mixing applies to the entire Regge trajectories, the above formulas would apply also to weak string tensions, and also photons would belong to Regge trajectories containing high spin excitations.

3. What one can one say about the value of the weak string tension M_1^2? The naive order of magnitude estimate is $M_1^2 \simeq m_W^2 \simeq 10^4$ GeV^2 is by a factor 1/25 smaller than the direct scaling up of the hadronic string tension about 1 GeV^2 scaled up by a factor 2^{18}. The above argument however allows also the identification as the scaled up variant of hadronic string tension in which case the higher states at weak Regge trajectories would not be easy to discover since the mass scale defined by string tension would be 512 GeV to be compared with the recent beam energy 7 TeV. Weak string tension need of course not be equal to the scaled up hadronic string tension. Weak string tension - unlike its hadronic counterpart- could also depend on the electromagnetic charge and other characteristics of the particle.

7.7 Is the earlier conjectured pseudoscalar Higgs there at all?

Spin 1 gauge bosons and Higgs differ only by different spin direction of fermions at opposite wormhole throats. For spin 1 gauge bosons the 3-momenta at two wormhole throats cannot be parallel if if one wants non-vanishing spin component in the direction of moment. 3-momenta are most naturally opposite for the massless states at throats. This forces massivation for all gauge bosons and even graviton and this in turn requires Higgs even in the case of gluons.

This inspires the question whether the parity properties of the couplings of gauge boson and corresponding Higgs transforming like 3+1 under SU(2) (this is due to the special character of imbedding space spinors) could be exactly the same? Higgs would couple like a mixture of scalar and pseudoscalar to fermions just as weak gauge bosons couple and the mixture would be just the same. If there are no axial variants of vector gauge bosons there should exist no pseudoscalar Higgs. The nonexistence of axial variants of vector gauge bosons is suggested by quantum classical correspondence: only gauge bosons having classical space-time correlates as induced gauge potentials should be allowed, nothing else. Note that color variant of Higgs would exist and would be eaten by gluons to get mass.

The similarity of the construction of gauge bosons and Higgsinos as pairs of wormhole throats containing fermion and antifermion encourages to think that Higgs mechanism is invariant under supersymmetries. If so, also Higgsinos would be eaten and one would have massive super-symmetric gauge theory with fermions with photon and other massless particle possessing a tiny mass. This looks very simple. The testable implication would be that only weak gauginos should contribute to muon g-2 anomaly.

7.8 Higgs issue after Europhysics 2011

The general feeling at the Eve of Europhysics 2011 conference was that this meeting might become one of the key events in the history of physics. This might turn out to be the case. CDF and D0 were the groups representing the data from p-pbar collisions at Tevatron whereas ATLAS and CMS represented the data about p-p collisions at LHC. The blog participation transformed the conference from a closed meeting of specialists to a media event inspiring intense blog discussions and viXra log became the most interesting discussion forum thanks to the excellent postings of Phil Gibbs giving focused summaries of various reports about SUSY and Higgs.

The hope was that two basic questions would receive a unique answer. Does Higgs exist and if so what is its mass? Is the standard view about SUSY correct: in other words do the super-partners exist with masses below TeV scale? It was clear that negative answer to even the Higgs issue would force a thorough reconsideration of the status of not only MSSM but also that of super string theory and M-theory because of the general role of Higgs mechanism in the description massivation and symmetry breaking for the QFT limits of these theories. The implications are far reaching also for the inflationary cosmology where scalar fields and Higgs mechanism are taken as granted. Actually the non-existence of Higgs forces to reconsider the entire quantum field theoretic description of particle massivation.

Already before the conference several anomalies had emerged and the question was whether LHC data gives a support for these anomalies.

- A 145 GeV bump with 4 sigma significance in the mass distribution of jet pairs jj in Wjj final states was reported by CDF [34] but not confirmed by D0 [39]. The interpretation as Higgs was excluded and some of the proposed identifications of 145 GeV bump was as decay products of leptophobic Z′ boson or of technicolor pion. There were also indications for 300 GeV bump in the mass distribution of Wjj states themselves suggesting cascade like decay.

- Both CDF and D0 had reported two bumps at almost same mass about 325 GeV [36, 37] having no obvious interpretation in standard model framework. Technicolor approach and also TGD suggests an interpretation as pionlike state.

- CDF and D0 had also reported anomalous forward-backward asymmetry in top-pair production in p-pbar collisions suggesting the existence of new kind of flavor changing colored neutral currents [38, 51]. TGD based explanation of family replication phenomenon combined with bosonic emergence predicts that gauge bosons should appear as flavor singlets and octets. Octets would indeed induce flavor changing currents and asymmetry. Also many other indications for new physics such as anomalously large CP breaking in BBbar system had been reported and one should not forget long list of forgotten anomalies from previous years, say the two and half year old CDF anomaly which D0 failed to observe. Recall also that proton has shown no signs of decaying.

What did we learn during these days? Already before the conference it was clear that standard SUSY had transformed from the healer of the standard model to a patient. The parameter space for MSSM (minimal supersymmetric extension of standard model predicting two Higgs multiplets) had been narrowed down by strong lower limits on squark and sgluon masses to the extent that the original basic motivation for MSSM (stability of Higgs mass against radiative corrections) had been lost as well as the explanation for the anomaly of g-2 of muon. During the conference the bounds on SUSY parameters were tightened further and the rough conclusion is that squark and gluinos masses must be above 1 TeV. Even Lubos Motl was forced to conclude that the probability that LCH discovers standard SUSY is 50 per cent instead of 90 per cent or more of 2008 blog posting. In TGD framework simple p-adic scaling arguments lead to the proposal that the only sfermions with mass below 1 TeV are selectron and sneutrinos with selectron having mass equal to 262 GeV. Low sneutrino masses allow in principle to understand g-2 of muon. Selectron could decay to electron plus neutralino for which mass must be larger than 46 GeV neutralino would eventually decay to photon or virtual Z plus neutrino.

The Higgs issue became the central theme of the conference and the three days from Thursday to Sunday were loaded with excitement. After many twists, the final conclusion was that there is 2.5 sigma evidence from ATLAS for a state in the mass range 140-150 GeV, which might be Higgs or something else. The press release of Fermi lab at Friday announced that they have confined Higgs to the interval 120-137 GeV. After the announcement of ATLAS both D0 and CDF discovered suddenly evidence for Higgs in 140-150 GeV mass range. The evidence for this mass range emerged from the decays of a might-be Higgs to WW pairs decaying in turn to lepton pairs. The proponent of technicolor would of course see this as evidence for an off mass shell state of a neutral pion like state explaining also the jj bump in Wjj system and at 145 GeV mass and not allowing an interpretation as Higgs. In TGD framework the experience with earlier anomalies such as two year old CDF anomaly encouraged the interpretation in terms p-adic mass octaves of the pion of p-adically scaled up variant of hadron physics with mass scale 512 times higher than that of the ordinary hadron physics. Somewhat frustratingly, the final conclusion about the Higgs issue was promised to emerge only towards the end of the next year but it is clear that already now standard model might well be inconsistent with all data irrespective of the mass of Higgs. MSSM would allow additional flexibility but is also in difficulties.

The surprise of the first conference day was additional evidence for the bump at 327 GeV reported already earlier by CDF. This state is a complete mystery in standard model framework and therefore extremely interesting. The proponents of technicolor would probably suggest interpretation as exotic ρ or ω meson. in TGD framework both 145 GeV pion and 325 GeV ρ and ω appear as mesons of M_{89} hadron physics if one assumes that the u and d quarks of M_{89} physics have masses corresponding to the p-adic length scale $k = 93$ (mass is 102 GeV and should be visible as a preferred quark jet mass). I would not be surprised if technicolor models would experience a brief renessaince but fail experimentally since a lot of new states and elementary particles is implied by the extension of the color gauge group. The mere p-adic scaling does not imply anything like this.

Also super string inspired predictions of various exotics such as microscopic black holes, strong gravity, large extra dimensions, Randall-Sundrum gravitons, split supersymmetry, and whatever were tested. No evidence was found. Neither there was evidence for lepto-quarks, heavier partners of intermediate gauge bosons, and various other exotics.

To my view, the results of the conference force to re-consider the basic assumptions of the approach followed during last more than three decades. Is it possible be find a more realistic physical interpretation of the mathematically extremely attractive supersymmetry? Unitarity requires new physics in TeV scale: is this new physics technicolor or its TGD analog without gauge group extenssion or something else? To me however the mother of all questions concerns the microscopic description of massivation. The description in terms of Higgs is after all a phenomenological description borrowed from condensed matter physics. It does not work for extended objects like strings but require quantum field theory limit. p-Adic thermodynamics for conformal weight (to which mass squared is proportional to) should be an essential element of the microscopic approach too since it is a description working for the fundamental objects and in presence of super-conformal invariance.

What actually happens in the massivation: could it be that all components of Higgs, of its super partners, and of its higher spin generalizations are eaten in a process in which massless multiplets with various spins combine to form only massive multiplets? Here twistor approach might provide the guideline since its applicability requires that massive particles should allow an interpretation as bound states of massless ones. Perhaps the simple observation that spin one bound states of massless fermion and anti-fermion are automatically massive might help to get to the deeper waters.

What next? Standard model Higgs is more or less excluded and the same fate is very probably waiting the SUSY Higgs. I would not be surprised if technicolor models would experience a brief renaissance but fail experimentally since very many new hadronlike states and new elementary particles are implied by the extension of the color gauge group. Sooner or later the simple p-adic scaling of the ordinary hadron physics probably turns out to be the only realistic option. If technicolor becomes in fashion, the hadrons of M_{89} hadron physics will be however found as a side product of this search.

Eventually this requires giving up the Planck length scale reductionism as the basic philosophy and replacing it with p-adic fractality as the basic guiding principle tying together physics at very short and at very long length scales making possible the long sough for ultraviolet completion of known physics. This led to the landscape catastrophe in M-theory since very many physics in long length scales had the same UV completion. Some general principle fixing the long range physics is obviously missing. p-Adic smoothness for which infinite in real sense is infinitesimal selects the unique long length scale physics among infinitely many alternatives. The real problems are really much much deeper than finding proper parameters for SUSY and it would be a high time for theoreticians to finally realize this.

8 Calculation of hadron masses and topological mixing of quarks

The calculation of quark masses is not enough since one must also understand CKM mixing of quarks in order to calculate hadron massess. A model for CKM matrix and hadron masses is constructed in [12] and here only a brief summary about basic ideas involved is given.

8.1 Topological mixing of quarks

In TGD framework CKM mixing is induced by topological mixing of quarks (that is 2-dimensional topologies characterized by genus). The strongest number theoretical constraint on mixing matrices would be that they are rational. Perhaps a more natural constraint is that they are expressible in terms of roots of unity for some finite dimensional algebraic extension of rationals and therefore also p-adic numbers.

Number theoretical constraints on topological mixing can be realized by assuming that topological mixing leads to a thermodynamical equilibrium subject to constraints from the integer valued modular contributions remaing integer valued in the mixing. This gives an upper bound of 1200 for the number of different U and D matrices and the input from top quark mass and $\pi^+ - \pi^0$ mass difference implies that physical U and D matrices can be constructed as small perturbations of matrices expressible as direct sum of essentially unique 2×2 and 1×1 matrices. The maximally entropic solutions can be found numerically by using the fact that only the probabilities p_{11} and p_{21} can be varied freely. The solutions are unique in the accuracy used, which suggests that the system allows only single thermodynamical phase.

The matrices U and D associated with the probability matrices can be deduced straightforwardly in the standard gauge. The U and D matrices derived from the probabilities determined by the entropy maximization turn out to be unitary for most values of integers n_1 and n_2 characterizing the lowest order contribution to quark mass. This is a highly non-trivial result and means that mass and probability constraints together with entropy maximization define a sub-manifold of $SU(3)$ regarded as a sub-manifold in 9-D complex space. The choice $(n(u), n(c)) = (4, n)$, $n < 9$, does not allow unitary U whereas $(n(u), n(c)) = (5, 6)$ does. This choice is still consistent with top quark mass and together with $n(d) = n(s) = 5$ it leads to a rather reasonable CKM matrix with a value of CP breaking invariant within experimental limits. The elements V_{i3} and V_{3i}, $i = 1, 2$ are however roughly twice larger than their experimental values deduced assuming standard model. V_{31} is too large by a factor 1.6. The possibility of scaled up variants of light quarks could lead to too small experimental estimates for these matrix elements. The whole parameter space has not been scanned so that better candidates for CKM matrices might well exist.

8.2 Higgsy contribution to fermion masses is negligible

There are good reasons to believe that Higgs expectation for the fermionic space-time sheets is vanishing although fermions couple to Higgs. Thus p-adic thermodynamics would explain fermion masses completely. This together with the fact that the prediction of the model for the top quark mass is consistent with the most recent limits on it, fixes the CP_2 mass scale with a high accuracy to the maximal one obtained if second order contribution to electron's p-adic mass squared vanishes. This is very strong constraint on the model.

8.3 The p-adic length scale of quark is dynamical

The assumption about the presence of scaled up variants of light quarks in light hadrons leads to a surprisingly successful model for pseudo scalar meson masses using only quark masses and the assumption mass squared is additive for quarks with same p-adic length scale and mass for quarks labelled by different primes p. This conforms with the idea that pseudo scalar mesons are Goldstone bosons in the sense that color Coulombic and magnetic contributions to the mass cancel each other. Also the mass differences between hadrons containing different numbers of strange and heavy quarks can be understood if s, b and c quarks appear as several scaled up versions.

This hypothesis yields surprisingly good fit for meson masses but for some mesons the predicted mass is slightly too high. The reduction of CP_2 mass scale to cure the situation is not possible since top quark mass would become too low. In case of diagonal mesons for which quarks correspond to same p-adic prime, quark contribution to mass squared can be reduced by ordinary color interactions and in the case of non-diagonal mesons one can require that quark contribution is not larger than meson mass.

It should be however made clear that the notion of quark mass is problematic. One can speak about current quark masses and constituent quark masses. For u and d quarks constituent quark masses have scale 10^2 GeV are much higher than current quark masses having scale 10 GeV. For current quarks the dominating contribution to hadron mass would come from super-symplectic bosons at quantum level and at more phenomenological level from hadronic string tension. The open question is which option to choose or whether one should regard the two descriptions as duals of each other based on $M^8 - H$ duality. M^8 description would be natural at low energies since $SO(4)$ takes the role of color group. One could also say that current quarks are created in deconfinement phase transition which involves change of the p-adic length scale characterizing the quark. Somewhat counter intuitively but in accordance with Uncertainty Principle this length scale would increase but one could assign it the color magnetic field body of the quark.

8.4 Super-symplectic bosons at hadronic space-time sheet can explain the constant contribution to baryonic masses

Current quarks explain only a small fraction of the baryon mass and that there is an additional contribution which in a good approximation does not depend on baryon. This contribution should correspond to the non-perturbative aspects of QCD which could be characterized in terms of constituent quark masses in M^8 picture and in terms of current quark masses and string tension or super-symplectic bosons in $M^4 \times CP_2$ picture.

Super-symplectic gluons provide an attractive description of this contribution. They need not exclude more phenomenological description in terms of string tension. Baryonic space-time sheet with $k = 107$ would contain a many-particle state of super-symplectic gluons with net conformal weight of 16 units. This leads to a model of baryons masses in which masses are predicted with an accuracy better than 1 per cent. Super-symplectic gluons also provide a possible solution to the spin puzzle of proton.

Hadronic string model provides a phenomenological description of non-perturbative aspects of QCD and a connection with the hadronic string model indeed emerges. Hadronic string tension is predicted correctly from the additivity of mass squared for $J = 2$ bound states of super-symplectic quanta. If the topological mixing for super-symplectic bosons is equal to that for U type quarks then a 3-particle state formed by 2 super-symplectic quanta from the first generation and 1 quantum from the second generation would define baryonic ground state with 16 units of conformal weight.

In the case of mesons pion could contain super-symplectic boson of first generation preventing the large negative contribution of the color magnetic spin-spin interaction to make pion a tachyon. For heavier bosons super-symplectic boson need not to be assumed. The preferred role of pion would relate to the fact that its mass scale is below QCD Λ.

8.5 Description of color magnetic spin-spin splitting in terms of conformal weight

What remains to be understood are the contributions of color Coulombic and magnetic interactions to the mass squared. There are contributions coming from both ordinary gluons and super-symplectic gluons and the latter is expected to dominate by the large value of color coupling strength.

Conformal weight replaces energy as the basic variable but group theoretical structure of color magnetic contribution to the conformal weight associated with hadronic space-time sheet ($k = 107$) is same as in case of energy. The predictions for the masses of mesons are not so good than for baryons, and one might criticize the application of the format of perturbative QCD in an essentially non-perturbative situation.

The comparison of the super-symplectic conformal weights associated with spin 0 and spin 1 states and spin 1/2 and spin 3/2 states shows that the different masses of these states could be understood in terms of the super-symplectic particle contents of the state correlating with the total quark spin. The resulting model allows excellent predictions also for the meson masses and implies that only pion and kaon can be regarded as Goldstone boson like states. The model based on spin-spin splittings is consistent with the model.

To sum up, the model provides an excellent understanding of baryon and meson masses. This success is highly non-trivial since the fit involves only the integers characterizing the p-adic length scales of quarks and the integers characterizing color magnetic spin-spin splitting plus p-adic thermodynamics and topological mixing for super-symplectic gluons. The next challenge would be to predict the correlation of hadron spin with super-symplectic particle content in case of long-lived hadrons.

References

Books related to TGD

[1] M. Pitkänen. Construction of Configuration Space Kähler Geometry from Symmetry Principles. In *Quantum Physics as Infinite-Dimensional Geometry.* Onlinebook. http://tgd.wippiespace.com/public_html/tgdgeom/tgdgeom.html#compl1, 2006.

[2] M. Pitkänen. Construction of elementary particle vacuum functionals. In *p-Adic length Scale Hypothesis and Dark Matter Hierarchy.* Onlinebook. http://tgd.wippiespace.com/public_html/paddark/paddark.html#elvafu, 2006.

[3] M. Pitkänen. Construction of Quantum Theory: M-matrix. In *Towards M-Matrix.* Onlinebook. http://tgd.wippiespace.com/public_html/tgdquant/tgdquant.html#towards, 2006.

[4] M. Pitkänen. Cosmic Strings. In *Physics in Many-Sheeted Space-Time.* Onlinebook. http://tgd.wippiespace.com/public_html/tgdclass/tgdclass.html#cstrings, 2006.

[5] M. Pitkänen. Does the Modified Dirac Equation Define the Fundamental Action Principle? In *Quantum Physics as Infinite-Dimensional Geometry.* Onlinebook. http://tgd.wippiespace.com/public_html/tgdgeom/tgdgeom.html#Dirac, 2006.

[6] M. Pitkänen. Does the QFT Limit of TGD Have Space-Time Super-Symmetry? In *Towards M-Matrix.* Onlinebook. http://tgd.wippiespace.com/public_html/tgdquant/tgdquant.html#susy, 2006.

[7] M. Pitkänen. Identification of the Configuration Space Kähler Function. In *Quantum Physics as Infinite-Dimensional Geometry.* Onlinebook. http://tgd.wippiespace.com/public_html/tgdgeom/tgdgeom.html#kahler, 2006.

[8] M. Pitkänen. Massless states and particle massivation. In *p-Adic Length Scale Hypothesis and Dark Matter Hierarchy.* Onlinebook. http://tgd.wippiespace.com/public_html/paddark/paddark.html#mless, 2006.

[9] M. Pitkänen. New Particle Physics Predicted by TGD: Part I. In *p-Adic Length Scale Hypothesis and Dark Matter Hierarchy.* Onlinebook. http://tgd.wippiespace.com/public_html/paddark/paddark.html#mass4, 2006.

[10] M. Pitkänen. Nuclear String Hypothesis. In *p-Adic Length Scale Hypothesis and Dark Matter Hierarchy.* Onlinebook. http://tgd.wippiespace.com/public_html/paddark/paddark.html#nuclstring, 2006.

[11] M. Pitkänen. *p-Adic length Scale Hypothesis and Dark Matter Hierarchy.* Onlinebook. http://tgd.wippiespace.com/public_html/paddark/paddark.html, 2006.

[12] M. Pitkänen. p-Adic Particle Massivation: Hadron Masses. In *p-Adic Length Scale Hypothesis and Dark Matter Hierarchy.* Onlinebook. http://tgd.wippiespace.com/public_html/paddark/paddark.html#mass3, 2006.

[13] M. Pitkänen. p-Adic Physics: Physical Ideas. In *TGD as a Generalized Number Theory.* Onlinebook. http://tgd.wippiespace.com/public_html/tgdnumber/tgdnumber.html#phblocks, 2006.

[14] M. Pitkänen. Riemann Hypothesis and Physics. In *TGD as a Generalized Number Theory.* Onlinebook. http://tgd.wippiespace.com/public_html/tgdnumber/tgdnumber.html#riema, 2006.

[15] M. Pitkänen. TGD and Nuclear Physics. In *p-Adic Length Scale Hypothesis and Dark Matter Hierarchy.* Onlinebook. http://tgd.wippiespace.com/public_html/paddark/paddark.html#padnucl, 2006.

[16] M. Pitkänen. TGD as a Generalized Number Theory: Infinite Primes. In *TGD as a Generalized Number Theory*. Onlinebook. http://tgd.wippiespace.com/public_html/tgdnumber/tgdnumber.html#visionc, 2006.

[17] M. Pitkänen. TGD as a Generalized Number Theory: p-Adicization Program. In *TGD as a Generalized Number Theory*. Onlinebook. http://tgd.wippiespace.com/public_html/tgdnumber/tgdnumber.html#visiona, 2006.

[18] M. Pitkänen. Was von Neumann Right After All. In *Towards M-Matrix*. Onlinebook. http://tgd.wippiespace.com/public_html/tgdquant/tgdquant.html#vNeumann, 2006.

[19] M. Pitkänen. Yangian Symmetry, Twistors, and TGD. In *Towards M-Matrix*. Onlinebook. http://tgd.wippiespace.com/public_html/tgdquant/tgdquant.html#Yangian, 2006.

Articles about TGD

[20] M. Pitkänen. Further Progress in Nuclear String Hypothesis. http://tgd.wippiespace.com/public_html/articles/nuclstring.pdf, 2007.

[21] M. Pitkänen. Physics as Generalized Number Theory II: Classical Number Fields. https://www.createspace.com/3569411, July 2010.

[22] M. Pitkänen. Physics as Infinite-dimensional Geometry I: Identification of the Configuration Space Kähler Function. https://www.createspace.com/3569411, July 2010.

[23] M. Pitkänen. Physics as Infinite-dimensional Geometry II: Configuration Space Kähler Geometry from Symmetry Principles. https://www.createspace.com/3569411, July 2010.

[24] M. Pitkänen. Physics as Generalized Number Theory I: p-Adic Physics and Number Theoretic Universality. https://www.createspace.com/3569411, July 2010.

[25] M. Pitkänen. Physics as Generalized Number Theory III: Infinite Primes. https://www.createspace.com/3569411, July 2010.

[26] M. Pitkänen. Physics as Infinite-dimensional Geometry III: Configuration Space Spinor Structure. https://www.createspace.com/3569411, July 2010.

[27] M. Pitkänen. Physics as Infinite-dimensional Geometry IV: Weak Form of Electric-Magnetic Duality and Its Implications. https://www.createspace.com/3569411, July 2010.

[28] M. Pitkänen. The Geometry of CP_2 and its Relationship to Standard Model. https://www.createspace.com/3569411, July 2010.

Mathematics

[29] R. Accola. *Riemann Surfaces, Theta functions and Abelian Automorphism Groups. Lecture Notes in Mathematics 483.* Springer Verlag, 1975.

[30] Moore G. Alvarez-Gaume L. and Vafa C. Theta functions, Modular invariance and Strings. *Comm. Math. Phys.*, 106, 1986.

[31] M. Farkas, H. and I. Kra. *Riemann Surfaces.* Springer Verlag, 1980.

[32] D. Mumford. *Tata Lectures on Theta I,II,III.* Birkhäuser, 1983.

[33] N. Pope, C. Eigenfunctions and $Spin^c$ Structures on CP_2, 1980.

Particle and Nuclear Physics

[34] Invariant Mass Distribution of Jet Pairs Produced in Association with a W boson in ppbar Collisions at $\sqrt{s} = 1.96$ TeV. http://arxiv.org/abs/1104.0699.

[35] Quark. http://en.wikipedia.org/wiki/Current_quark_mass.

[36] D0: 2.5-sigma evidence for a 325 GeV top prime quark. http://motls.blogspot.com/2011/04/d0-3-sigma-evidence-for-325-gev-top.html, 2011.

[37] If That Were A Higgs At 200 GeV... http://www.science20.com/quantum_diaries_survivor/if_were_higgs_200_gev, 2011.

[38] CDF collaboration. Evidence for a Mass Dependent Forward-Backward Asymmetry in Top Quark Pair Production. http://arxiv.org/abs/1101.0034, 2011.

[39] D0 collaboration. Study of the dijet invariant mass distribution in $p\overline{p} \rightarrow W(\rightarrow l\nu) + jj$ final states at $\sqrt{s} = 1.96$ TeV. http://www-d0.fnal.gov/Run2Physics/WWW/results/final/HIGGS/H11B, 2011.

[40] LSND Collaboration. Evidence for $\nu_{\mu} - \nu_e$ oscillations from LSND. http://arxiv.org/absnucl-ex/9709006, 1997.

[41] A. E. Nelson D. B. Kaplan and N. Weiner. Neutrino Oscillations as a Probe of Dark Energy. http://arxiv.org/abs/hep-ph/0401099, 2004.

[42] R. Van de Water. Updated Anti-neutrino Oscillation Results from MiniBooNE. http://indico.cern.ch/getFile.py/access?contribId=208&sessionId=3&resId=0&materialId=slides&confId=73981, 2010.

[43] A. D. Dolgov and I. Z. Rothstein. *Phys. Rev. Lett.*, 71(4), 1993.

[44] T. Dorigo. Rumsfeld hadrons. http://dorigo.wordpress.com/2007/06/20/rumsfeld-hadrons/, 2007.

[45] T. Dorigo. Top quark mass measured with neutrino phi weighting. http://dorigo.wordpress.com/2008/12/08/top-quark-mass-measured-with-neutrino-phi-weighting/, 2008.

[46] C. Athanassopoulos et al. Evidence for Neutrino Oscillations from Muon Decay at Rest. http://arxiv.org/abs/nucl-ex/9605001, 1996.

[47] Decamp et al. The number of neutrino species. Aleph Collaboration,CERN-EP/89-141, 1989.

[48] L. Borodovsky et al. *Phys. Rev. Lett.*, 68:274, 1992.

[49] SNO: Q. R Ahmad et al. *Phys. Rev. Lett.*, 89:11301, 2002.

[50] Y. Fukuda et al. *Phys. Lett. B*, 335:237, 1994.

[51] Y.Takeuchi et al. Measurement of the Forward Backward Asymmetry in Top Pair Production in the Dilepton Decay Channel using 5.1 fb^{-1}. http://www-cdf.fnal.gov/physics/new/top/2011/DilAfb/, 2011.

[52] K. Eguchi et al (KamLAND). *Phys. Rev. Lett.*, 90:21802, 2003.

[53] R. J. Wilkes (K2K). http://arxiv.org/abs/hep-ex/0212035, 2002.

[54] W. C. Louis. In *Proceedings of the XVI Conference on Neutrino Physics and Astrophysics. Eilat, Israel*, 1994.

[55] Q. Z. Qian and G. M. Fuller. *Phys. Rev. D*, 51:1479, 1995.

[56] M. B. Smy (Super-Kamiokande). *Nucl. Phys. Proc. Suppl.*, 118:25, 2003.

Neuroscience and Consciousness

[57] M. Pitkänen et al H. Abdelmeik. Changes in electrical properties of frog sciatic nerves at low temperature: superconductor-like behaviour. http://tgd.wippiespace.com/public_html/articles/hafpaper.pdf, 2003.

Made in the USA
Columbia, SC
05 April 2019